Origins of Giant Planets, Volume 1

Disks, dust, and planetesimals

AMERICAN ASTRONOMICAL SOCIETY | IOP ebooks™

About the program:

AAS-IOP Astronomy ebooks is the official book program of the American Astronomical Society (AAS) and aims to share in depth the most fascinating areas of astronomy, astrophysics, solar physics, and planetary science. The program includes publications in the following topics:

- GALAXIES AND COSMOLOGY
- INTERSTELLAR MATTER AND THE LOCAL UNIVERSE
- STARS AND STELLAR PHYSICS
- EDUCATION, OUTREACH, AND HERITAGE
- HIGH-ENERGY PHENOMENA AND FUNDAMENTAL PHYSICS
- THE SUN AND THE HELIOSPHERE
- THE SOLAR SYSTEM, EXOPLANETS, AND ASTROBIOLOGY
- LABORATORY ASTROPHYSICS, INSTRUMENTATION, SOFTWARE, AND DATA

Books in the program range in level from short introductory texts on fast-moving areas, graduate and upper-level undergraduate textbooks, research monographs, and practical handbooks.

For a complete list of published and forthcoming titles, please visit iopscience.org/books/aas.

About the American Astronomical Society

The American Astronomical Society (aas.org), established 1899, is the major organization of professional astronomers in North America. The membership (~7,000) also includes physicists, mathematicians, geologists, engineers, and others whose research interests lie within the broad spectrum of subjects now comprising the contemporary astronomical sciences. The mission of the Society is to enhance and share humanity's scientific understanding of the universe.

Editorial Advisory Board

Origins of Giant Planets, Volume 1

Disks, dust, and planetesimals

Sarah Dodson-Robinson
Department of Physics and Astronomy, University of Delaware, Newark, DE, USA

IOP Publishing, Bristol, UK

ISBN 978-0-7503-2136-5 (ebook)
ISBN 978-0-7503-2134-1 (print)
ISBN 978-0-7503-2137-2 (myPrint)
ISBN 978-0-7503-2135-8 (mobi)

DOI 10.1088/2514-3433/ac1db7

Version: 20220101

AAS–IOP Astronomy
ISSN 2514-3433 (online)
ISSN 2515-141X (print)

British Library Cataloguing-in-Publication Data: A catalogue record for this book is available from the British Library.

Published by IOP Publishing, wholly owned by The Institute of Physics, London

IOP Publishing, Temple Circus, Temple Way, Bristol, BS1 6HG, UK

US Office: IOP Publishing, Inc., 190 North Independence Mall West, Suite 601, Philadelphia, PA 19106, USA

Cover image: Dark and Stormy Jupiter. Image credit: NASA/JPL-Caltech/SwRI/MSSS/Kevin M. Gill.

This book is dedicated to Peter Bodenheimer, who refused to write it in 2008 because "We don't know enough about planet formation."

Contents

Acknowledgments

I am grateful to my husband, Eric Dodson-Robinson, for his support during the writing of this manuscript.

Author biography

Sarah Dodson-Robinson

Sarah Dodson-Robinson received her PhD in Astronomy & Astrophysics from the University of California at Santa Cruz in 2008, then took a Spitzer Postdoctoral Fellowship at the NASA Exoplanet Science Institute. She is now an Associate Professor of Physics and Astronomy at the University of Delaware. In 2013, she won the American Astronomical Society's Annie Jump Cannon award for her contributions to the study of planet formation. Dr. Dodson-Robinson conducts numerical simulations of the chemical and dynamical evolution of planet-forming disks and participates in observational studies of debris disks. She also develops and tests methods for distinguishing exoplanet discoveries from stellar noise.

Chapter 1

The Scientific Legacy of Giant-planet Research

But who shall dwell in these Worlds if they be inhabited? Are We, or They, Lords of the World? And how are all things made for Man?

—Johannes Kepler, quoted by H. G. Wells in *The War of the Worlds*, 1898 (Dick 1996)

In a 2010 feature in the *Smithsonian* magazine, science writer Laura Helmuth presents the intrepid reader with a list of the 10 most disturbing scientific discoveries in history.[1] Her top-ranked creepy discovery, beating out microbes whose evolution outpaces that of our pharmacopeia and a climate already altered by carbon dioxide emissions, is the simple fact that Earth is not the center of the universe. "We've had more than 400 years to get used to the idea, but it's still a little unsettling," writes Helmuth. Unsettling indeed—adopting the Sun-centered cosmology proposed by Nicolaus Copernicus in the 1543 text *De revolutionibus orbium coelestium* required admitting that there were, literally, other worlds besides Earth.[2] "We revolve around the Sun like any other planet," Copernicus concluded in *Commentary on the Theories of the Motions of Heavenly Objects from Their Arrangements*, a 1514 precursor to *De revolutionibus* that he shared only with friends (Hawking 2002). *De revolutionibus* was published as Copernicus was dying; fortunately, Pope Paul III approved of Copernicus' work during his lifetime, even going so far as to consult him on ecclesiastical calendar reform. But Giordano Bruno, who supported Copernicanism and asserted in *De l'Infinito, Universo e Mondi* (1584) that the universe contained an infinite number of intelligent civilizations, was not so lucky: Bruno was convicted of heresy by the Roman

[1] https://www.smithsonianmag.com/science-nature/the-ten-most-disturbing-scientific-discoveries-214212/
[2] In an early draft of *De revolutionibus*, Copernicus named the astronomer Aristarchus of Samos (310–230 BCE) and the philosopher Philolaus (470–385 BCE) as fellow nongeocentrists (Gingerich 1985).

doi:10.1088/2514-3433/ac1db7ch1

Inquisition and burned at the stake in 1600. Here we explore the scientific and philosophical revolutions made possible by observations of the giant planets and present a modern motivation for studying the formation of these supersized worlds.

By the 17th century CE, Europe was roiling with the violent conflict between Protestant and Roman Catholic states. The precarious position of the Roman Catholic Church formed the backdrop for a scientific and religious drama in which Jupiter and Saturn took center stage. On 1610 January 7, Galileo Galilei began a two-month-long investigation that put the nail in the coffin of Earth-centric cosmology. Turning his new "spyglass" to Jupiter, he observed

> …planets
> flying around the star of Jupiter at unequal intervals
> and periods with wonderful swiftness;
> which, unknown by anyone until this day,
> the first author detected recently …

By the 11th observing night, Galileo had "arrived at the conclusion, entirely beyond doubt, that in the heavens there are three stars wandering around Jupiter like Venus and Mercury around the Sun."[3] In the concluding paragraph of his book *Sidereus Nuncius* (Starry Messenger, Galileo 1610), he wrote

> We have moreover an excellent and splendid argument for taking away the scruples of those who, while tolerating with equanimity the revolution of the planets around the Sun in the Copernican system, are so disturbed by the attendance of one Moon around the Earth while the two together complete the annual orb around the Sun that they conclude that this constitution of the Universe must be overthrown as impossible.

Galileo's conclusion that he had discovered a system of satellites orbiting Jupiter received immediate support from Johannes Kepler, who detailed corroborating telescopic observations in *Narratio de Observatis Quatuor Jovis Satellitibus* (*Narration about Four Satellites of Jupiter Observed*) in 1611. Figure 1.1 shows some of Galileo's rough sketches of Jupiter and its moons. His view of Earth as one among many planets was also bolstered by his discoveries of Venus' phases and the Moon's mountains and valleys.

> *I'm convinced that a controlled disrespect for authority is essential to a scientist.*
> —Luis Alvarez (1911–1988)
>
> Professor of Physics, University of California, Berkeley; winner of 1968 Nobel Prize for physics; co-author of the giant-impact hypothesis for the Cretaceous–Tertiary extinction

[3] Callisto, the outermost Galilean satellite, would first emerge on night 13.

Figure 1.1. The top of Galileo's scrap paper contains a rough draft of a letter from Galileo to Leonardo Donato, Doge of Venice, describing the telescope and its military uses. On the bottom, Galileo sketched the relative positions of Jupiter and its associated "stars" from one week of observations, then depicted the satellites' motion as it would appear from Jupiter in the lower right. Making these drawings may have led Galileo to realize that the new stars were actually Jupiter's moons. This single-sheet manuscript resides in the University of Michigan Special Collections Library, https://www.lib.umich.edu/special-collections-research-center/galileo-manuscript. The digital reproduction has not been altered and is distributed under a Creative Commons Attribution license (https://creativecommons.org/licenses/by/4.0/).

Galileo had, perhaps, underestimated the "scruples" of the church leaders who supported the pre-Copernican, Earth-centric model. He received his first judgment from the Roman Inquisition under Pope Paul V in 1616. Delivered by Cardinal Bellarmino, the injunction ordered him to "...abstain completely from teaching or defending this doctrine ...that the Sun stands still at the center of the world and the Earth moves ..." (Heilbron 2010). In 1632, believing that the scientific climate under Pope Urban VIII was friendlier to Copernicanism and that Grand Duke Ferdinando II's patronage was secure, Galileo published *Dialogue Concerning the Two Chief World Systems*, a fictionalized debate between the empiricist Salviati, who advocates for the Copernican worldview, and the traditionalist Simplicio, who uses religious arguments to defend the Earth-centric cosmology. *Dialogue* became a bestseller and is considered by some to be the first popular science book ever written.[4] But although Galileo had received tentative permission to publish *Dialogue* from the Secretary of the Vatican in 1630, he was ultimately summoned before the Roman Inquisition, forced to recant his discoveries under threat of physical torture, and sentenced to lifetime house arrest.

> *I knew of no sadder picture in the history of science than that of the old man, Galileo, worn by a long life of scientific research, weak and feeble, trembling before that tribunal whose frown was torture, and declaring that to be false which he knew to be true.*
>
> —Maria Mitchell (1818–1889)
> Professor of Astronomy and Director of Vassar College Observatory; American Academy of Arts and Sciences member; American Philosophical Society member.

Yet Copernicus, Galileo, and Kepler had given the revolutionary idea that Earth was just one among many worlds so much momentum that the church could no longer control the cosmological narrative—or even the scientists themselves. Despite being forbidden by the Inquisition to publish any more work, Galileo made one more distinguished contribution to scientific literature while under house arrest: *Discorsi e Dimostrazioni Matematiche Intorno a Due Nuove Scienze*, or *Discourses and Mathematical Demonstrations Relating to Two New Sciences* (Galileo 1638). The manuscript was smuggled out of Italy by Lodowijk Elzevir and published in protestant Holland. According to Stephen Hawking, *Discourses* "anticipated Isaac Newton's laws of motion" (Hawking 2002); one of its legacies to planetary science is the introduction of scaling relations of the type now used to calculate the outcomes of planet-forming collisions (e.g., Stewart & Leinhardt 2009). And Kepler took the idea of other worlds so seriously that he started writing the first-ever science fiction novel, *Somnium*, in 1593—almost two decades before Galileo first deployed his telescope. A draft of *Somnium*, which featured night-feeding giants living on the Moon, was published in 1609 and most unfortunately provided ammunition against Kepler's

[4] http://www.pbs.org/empires/medici/renaissance/galileo.html

mother, Katharina, in her trial for witchcraft (Brake & Hook 2007). Katharina Kepler was eventually released after more than a year of imprisonment in 1621.

> *...innumerable suns exist; innumerable earths revolve around these suns in a manner similar to the way the seven planets revolve around our Sun. Living beings inhabit these worlds ...*
>
> —Giordano Bruno (1548–1600) in *De L'infinito Universo E Mondi*, quoted by Dick (1996)

It was Galileo's work on Saturn that encouraged the young Dutch physicist Christiaan Huygens to begin studying optics in 1652. In *Systema Saturnium*, Huygens wrote

> Galileo had first seen [Saturn] shining, not as a single disk, but in what seemed to be a triple form, as two smaller stars in close proximity to, and on opposite sides of, a larger star, in line with its center ...And so I was also drawn by an urgent longing to behold these wonders of the heaven ...I, therefore, set myself to work with all the earnestness and seriousness I could command to learn the art by which glasses are fashioned for these uses ...

Galileo had, in fact, heightened the mystery surrounding Saturn's shape by circulating an anagram shortly after his first observations of the planet in July 1610:

> smaismrmilmepoetaleumibunenugttauiras.

The solution: *Altissimum planetam tergeminum observavi*, or "I have observed the highest planet triform" (Figure 1.2). His goal was to protect his work from being

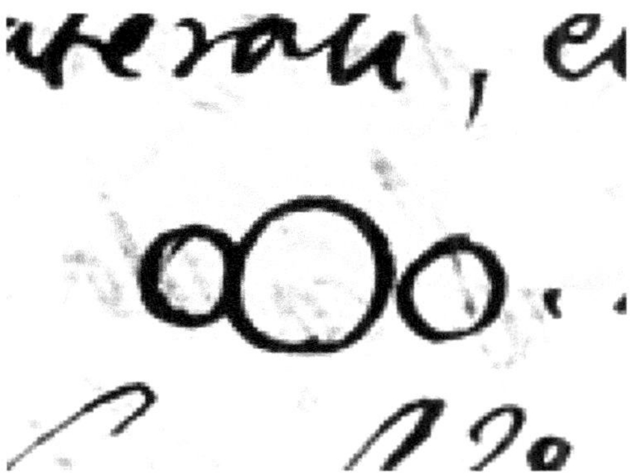

Figure 1.2. Galileo's rendering of Saturn's three-bodied shape in a July 1610 letter to Belisario Vinta, Secretary of State of the Grand Duchy of Tuscany. Image: Museo Galileo, https://brunelleschi.imss.fi.it/galileopalazzostrozzi/object/GalileoGalileiLetterToBelisarioVinta.html.

scooped: if another astronomer reported seeing Saturn's "triform" before Galileo was ready to publish, he could unscramble the anagram and prove he had observed Saturn's unique shape first.[5] Kepler would try and fail to decipher both the Saturn anagram and Galileo's December 1610 anagram announcing the phases of Venus, but his creative mistranslation led him to believe Galileo had discovered two moons orbiting Mars. (Phobos and Deimos would eventually be sighted in 1877.)

In 1612, Galileo was surprised to discover—after a two-month observing hiatus—that the triform had disappeared:

> I found [Saturn] solitary without the assistance of the supporting stars, and, in sum, perfectly round and clearly defined as Jupiter …Perhaps Saturn has devoured his own children? Or else it was an illusion and a fraud with which the glasses have for so long deceived me and so many others who have observed him with me many times? (Third letter on sunspots, Galileo 1613; Van Helden 1974)

Earth was crossing Saturn's ring-plane at the time; Galileo was, of course, viewing the rings edge on. According to the Ph.D. thesis of Johann Locher (University of Ingolstadt, 1614),

> "hitherto Saturn deceives or really mocks the astronomers out of hatred or malice."

Thus, when Christiaan Huygens began observing Saturn in 1655, he was tackling one of the highest-profile problems in science (Van Helden 1968), perhaps on par with today's goal of finding biosignatures in the spectra of exoplanets.

> *No other planet looks as unworldly or serene as Saturn. When you see it floating in your telescope, you feel as if you've uncovered mystery in the cosmos.*[6]
>
> —Carolyn Porco
> Imaging science team lead for the Cassini mission; CEO and President of Diamond Sky Productions, LLC

On 1655 February 3, Huygens and his brother Constantijn finished manufacturing the objective lens for a new 12 ft telescope that would have superior resolving power to Galileo's instrument. The Huygens brothers used a diamond to inscribe on the lens a quotation from the Roman poet Ovid: *Admovere oculis distantia sidera nostris*, or "they brought the distant stars closer to our eyes" (Louwman 2004). Christiaan Huygens discovered Saturn's moon Titan on 1655 March 25. In true

[5] http://galileo.rice.edu/sci/observations/saturn.html
[6] https://www.theatlantic.com/science/archive/2016/01/a-major-correction/422514/

Galilean fashion, Huygens disclosed his discovery in a letter to Oxford professor J. Wallis using an anagram that *included* the Ovid line:

Admovere oculis distantia sidera nostris vvvvvvv ccc rr h n b q x

He left Prof. Wallis to ponder the puzzle for nine months, finally sending the solution on 1656 March 15: *Saturno luna sua circunducitur diebus sexdecim horis quatuor*, or "A moon revolves around Saturn in 16 days and four hours." In 1656, possibly with the aid of a new 23 ft telescope (Louwman 2004), Huygens finally viewed Saturn with a high enough resolving power to discern the ring structure. His formal report of the discovery of Titan, *De Saturni luna observatio nova* (Huygens 1656), included (what else?) an anagram describing Saturn's shape:

aaaaaaa ccccc d eeeee h iiiiiii llll mm nnnnnnnnn oooo pp q rr s ttttt uuuuu,

with the solution "Annulo cingitur, tenui, plano, nusquam cohaerente, ad eclipticam inclinato," or "It is surrounded by a thin flat ring, nowhere touching, and inclined to the ecliptic." Huygens allowed Jean Chapelain to present the ring theory to the Montmortian Academy in 1658 before publishing *Systema Saturnium* (Huygens 1659), which explained not only the origin of Saturn's "ansae" (handles) but their disappearances in 1612 and 1642 (Van Helden 1974). (Pierre Gassendi had rekindled interest in Saturn's shape after observing it ringless in 1642, writing in his notebook "I have immediately advised my friends ... of this matter, to observe for themselves this noteworthy thing ....") Figure 1.3 shows Huygens' reconstruction of Saturn's appearance at different orbital phases. Huygens even followed Bruno and Kepler into astrobiology, suggesting in *Cosmotheoros* that extraterrestrial life might exist and that liquid water would be an essential ingredient for the origin of life (Huygens 1695). *Cosmotheoros* also included a scale drawing showing the sizes of the planets compared with the Sun (Figure 1.4), further demoting Earth to a minor role in the solar system (Van Helden 2009).

> *This [new scale of the solar system] shows us how vast those Orbs must be, and how inconsiderable this Earth is, the Theatre upon which all our mighty Designs, all our Navigations, and all our Wars are transacted when compared to them. A very fit Consideration, and matter of Reflection, for those Kings and Princes who sacrifice the Lives of so many People, only to flatter their Ambition in being Masters of some pitiful corner of this small Spot.*
>
> —Christiaan Huygens (1629–1695) in *Cosmotheoros*, quoted by Owen & Bolton (2010)

The list of major discoveries spurred by studies of the giant planets continues with the realization that light has a finite speed. In 1671, Giovanni Domenico Cassini—who would later discover four Saturnian moons and whose namesake NASA mission would begin exploring the Saturn system in 2004—presided over the opening of the Paris Observatory, which was funded by King Louis XIV of France. Testing a method of

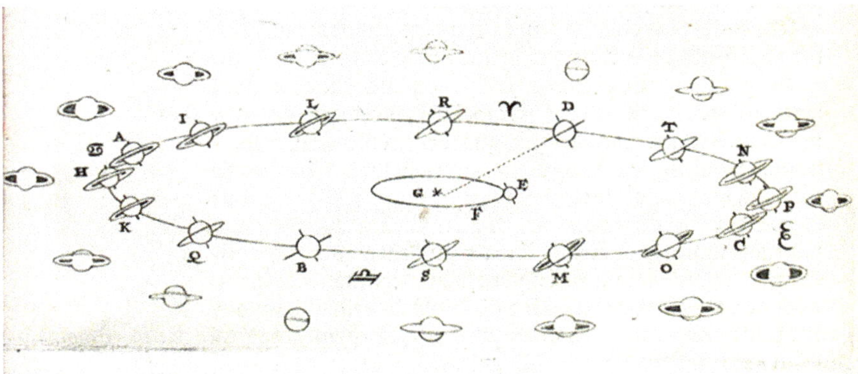

Figure 1.3. Christiaan Huygens' diagram of the appearance of Saturn's rings as a function of orbital phase (*Oeuvres complètes de Christiaan Huygens*, 1658–1666).

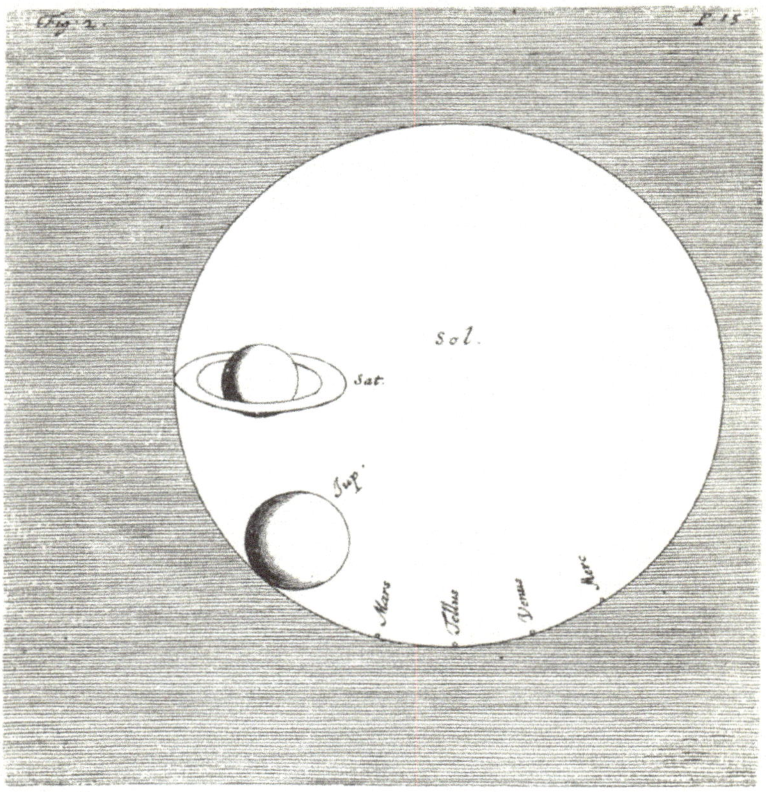

Figure 1.4. Scale drawing showing the relative sizes of the planets and the Sun (Huygens 1695, Cosmotheoros).

measuring longitude proposed by Galileo, Cassini used the newly constructed observatory to record the precise times of Jupiter's eclipses of the Galilean satellites, while Jean Picard conducted simultaneous observations from Tycho Brahe's old observatory near Copenhagen. Picard's assistant, Ole Rømer, would later join Cassini at the Paris Observatory. Continuing his observations of Io, the innermost of Jupiter's moons, Rømer noticed anomalies in the eclipse timings: the intervals between eclipses were not always identical, but instead varied with the distance between Earth and Jupiter. Rømer realized his measurements could be explained if light from the Jovian system took a finite time to reach Earth and reported his discovery to the *Académie Royale des Sciences* in 1676 (Bobis & Lequeux 2008). The $O - C$ (observed minus calculated) timing technique he employed is still an important astronomical tool today, leading to the Nobel prize–winning discovery of a pulsar in a binary system (Hulse & Taylor 1975) and to precise exoplanet mass measurements by transit timing variations (e.g., Steffen et al. 2012).

In 1687, Isaac Newton linked the motions of the planets across the sky with the trajectories of Earthbound falling objects in *Philosophiæ Naturalis Principia Mathematica*, the first-ever theoretical physics textbook and the most important publication in the history of science, according to Stephen Hawking (2002). The motivation to publish *Principia* came from astronomer, geophysicist, and inventor Edmond Halley, who—along with architect and mathematician Christopher Wren and engineer, microscopist, and astronomer Robert Hooke—was vexed by a need to find the physical underpinning for the elliptical orbits that Kepler had so meticulously mapped. In 1684, the three scientists met at a coffee house to discuss the merits of a Sun-centered "emanating force" that decreased with the square of distance. Hooke, who had been attacking Newton since the latter's very first presentation to England's Royal Society in 1672 (Koyré 1952), then infuriated Halley by saying that he had derived the inverse-square form of gravity from Kepler's laws but would not share the proof with his colleagues (Hawking 2002). Halley's response was to consult with Hooke's foremost scientific rival posthaste. The mathematician Abraham de Moivre described the conversation between Halley and Newton in a 1727 memorandum:

> …the Dr [Halley] asked him [Newton] what he thought the curve would be that would be described by the Planets supposing the force of attraction towards the Sun to be reciprocal to the square of their distance from it. Sr Isaac replied immediately that it would be an Ellipsis, the Doctor struck with joy & amazement asked him how he knew it, why saith he I have calculated it, whereupon Dr Halley asked him for his calculation without any farther delay, Sr Isaac looked among his papers but could not find it …

Halley gently encouraged Newton to redo his calculations and eventually financed the publication of *Principia*, though the book bore the imprimatur of the Royal Society. The impecunious society "repaid" Halley in copies of its previous literary flop, *The History of Fishes* (Bryson 2003).

After *Principia* Book 1 was presented to the Royal Society in 1686, Halley, who had already traveled to Poland to mediate a dispute between Hooke and Johannes Hevelius about the size of the solar system, found himself in the middle of another one of Hooke's quarrels. Hooke enraged Newton by announcing that the theory of gravity was partially a product of ideas from the pair's exchange of letters in 1679. (Newton would go on to refer to Hooke as an "ignoramus" in a letter to his other bitter adversary, Gottfried Wilhelm Leibniz, whom he would viciously attack in the calculus priority dispute; Blank 2009.) Although Newton famously wrote to Hooke, "If I have seen further it is by standing on the shoulders of Giants," the two were never really on good terms; Hawking (2002) argues that Newton suffered multiple nervous breakdowns as a result of their constant feuding. Halley stepped into the breach after Newton threatened to cancel the publication of Book 3, somehow convincing the "prickly, neurotic, and unyielding" theorist to complete his work (Blank 2009). Book 3 offered descriptions of the orbits of Jupiter's satellites, Saturn's satellites, the Moon, and finally all planets in terms of the inverse-square law, stating unequivocally that all planets orbit the Sun. Other momentous contributions to planetary science were the theory of tides and the explanation of rotational flattening. Although Newton was largely motivated by the desire to explain planetary motion, the scope of *Principia* went far beyond astronomy: Newton's Laws form the backbone of classical mechanics and are still given pride of place in the university physics curriculum over 300 years after being published. Halley would go on to use Newton's law of universal gravitation to predict the 1758 return of his now-eponymous comet.

> *Nature and Nature's Laws lay hid in Night:*
> *God said, "Let Newton be!" and all was light.*
>
> —Alexander Pope (1688–1744), Epitaph on Sir Isaac Newton

By the mid-1700s, a new planetary problem was attracting the attention of Europe's astronomers and mathematicians. While Kepler, Jeremiah Horrocks, Halley, and John Flamsteed (England's Astronomer Royal at the time *Principia* was published) had all noted "anomalies" in the mean motions of Jupiter and Saturn, the increasing accuracy of positional astronomy after James Bradley's discoveries of the aberration of starlight (1729) and the nutation of Earth's axis (1748) made the giant planets' deviations from perfect Keplerian ellipses much more obvious (Figure 1.5). Given Newton's statement in *Principia* Book 3 that "the action of Jupiter on Saturn is not altogether to be neglected," it seemed reasonable that the planets' mutual gravitational attraction caused the anomalies, an idea advanced by Joseph-Louis Lagrange in 1765 (Wilson 1985). But the changes in the planets' mean motions did not follow a 59-year cycle, which is the length of time it takes the two planets to return to a given configuration with respect to the Sun—a fact that led astronomer Jérôme Lalande to publicly reject the mutual gravitation hypothesis. Instead, many astronomers concluded that an interplanetary ether was sapping potential energy from both Jupiter and the Moon, which has a secular acceleration first noted by Halley in 1693. Luminaries such as Leonhard Euler and Jean d'Alembert

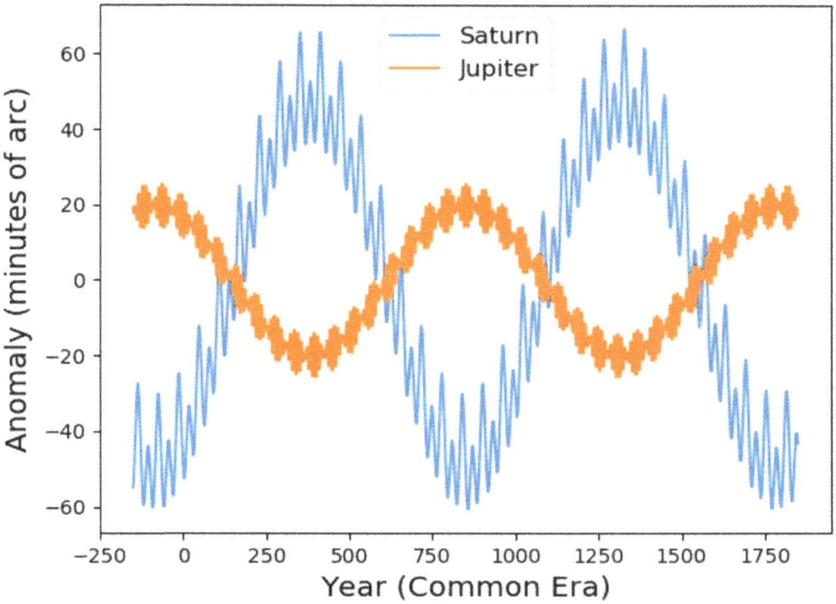

Figure 1.5. Anomalies in the orbits of Jupiter and Saturn: deviations from Keplerian orbits as a function of year in the common era. Adapted from Figure 2 of Wilson (1985).

considered the æther hypothesis plausible (Wilson 1985), but it could do nothing to explain Saturn's secular *deceleration*, which requires an increase in potential energy. Furthermore, the inevitable consequence of an energy-draining ether was that the planets would eventually spiral into the Sun—an outcome Euler decided would be consistent with the end of the world in Christian theology. Indeed, Newton himself thought divine intervention might be required to keep the solar system stable.

> *That shameless Saturn, as being more remote from the Sun or source of motion, seems to wish to be released from the absolute regularity that I hope to find in the other [planets].*[7]
>
> —Jeremiah Horrocks (1618–1641)
> Predictor of the 1639 transit of Venus, first person to prove that the Moon's orbit is elliptical

An explanation of the "Great Inequality" of Jupiter and Saturn would come from Pierre Simon Laplace, who left his theological studies at the University of Caen after publishing his first journal article on mathematics at age 19 or 20. Laplace, who would

[7] Letter to William Crabtree, 14 September 1639, quoted by Wilson (1985).

soon self-identify as an atheist or agnostic, believed that observed phenomena should have purely physical explanations (Hahn 2005). Although Laplace's investigations of probability theory would lead him to write papers on gambling, lotteries, and voting patterns, the science he would rename "celestial mechanics" was his most abiding interest (Hahn 2005). He had some false starts on the Great Inequality: in 1774, at age 25, he showed that Lagrange's 1765 results were incomplete, but then hypothesized a finite speed of gravitational force transmission to explain the Jovian and lunar anomalies. Once again, the theory could not explain Saturn's motion. After a several-year productive detour into studies of comet orbits, physics of heat, and attraction of spheroids (Gillispie 1997), Laplace's interest in the Great Inequality was rekindled when Lagrange pointed out the near 2:5 ratio of the planets' periods (Wilson 1985). Laplace identified an 877-year cycle in Jupiter and Saturn's eccentricities and apsidal positions that brought the predictions of Newtonian gravity into agreement with observations. With the Great Inequality revealed, it was clear that the planets could maintain stable orbits without help from a higher power. Meanwhile, the desire to understand the giant-planet orbits had led Euler, d'Alembert, and Lagrange to develop widely applicable methods of "*analyse*," including trigonometric series solutions to differential equations. Hahn (2005) considers the aforementioned three to be the real innovators in celestial mechanics and improbably calls Laplace a second-rank mathematician (p. 44), a charge that can easily be refuted by a quick search for the text "Laplace" in a differential equations textbook. And Laplace's "nebular hypothesis" of planet formation, in which the planets condensed out rings shed by a contracting disk surrounding the young Sun, is a forerunner to the theories presented in this book![8]

Heretofore we have considered only discoveries spurred by studies of Jupiter and Saturn, the two giant planets visible to the naked eye. While Laplace and other mathematicians were occupied with orbital mechanics, William Herschel was learning to build the telescope that would allow him to spot a new planet and launch one of the most productive careers in the history of astronomy. A native of Germany, Friedrich Wilhelm Herschel had joined his father Isaac and brother Jacob as an oboe player in the Hanover Guards band at age 14; when the French invaded Hanover, Isaac urged his sons to escape the violence by fleeing to England (Hoskin 2011). Friedrich Wilhelm took the name William and worked his way up through the music profession, becoming the organist at the Octagon Chapel in Bath in 1766. With the twin privileges of a secure paycheck and enough leisure time to study, Herschel was able to explore his numerous other interests: philosophy, languages, science, and mathematics. Having read Cambridge professor Robert Smith's music theory treatise *Harmonics* shortly after arriving in England, Herschel "was drawn from one branch of mathematics to another" (Lemonick 2009); in Smith's (a.k.a. Old Focus) *A Compleat System of Opticks*, he found the inspiration for his second career. He also formulated a plan to liberate his sister from dreary domestic drudgery.

Caroline Herschel's face was scarred from a childhood battle with smallpox. Her mother Anna, who was strongly opposed to education for women, assumed no one

[8] Immanuel Kant and Emmanuel Swetenborg had previously published similar theories, but it is unclear whether Laplace was aware of their work.

would want to marry Caroline and assigned her the role of family servant. Anna, herself illiterate, guarded against the possibility that Caroline might leave home by simply not allowing her to learn French and music, skills that would qualify her to be a governess. In the end, William and Alexander Herschel ransomed their sister by paying their mother to hire another maid, and Caroline, age 22, left Hanover for Bath on 1772 August 16 (Hoskin 2011). Under William's tutelage, she became a celebrated singer of oratorio. But William was becoming more and more captivated by astronomy, casting and grinding his own telescope mirrors starting in 1773 (using molds made from horse dung) and keeping a journal of observations from 1774 onward (Bennett 1976). In 1778 he polished the mirror of the 7 ft telescope that would reveal the planet Uranus. Caroline, who had hoped to build an independent music career for herself, was instead pressed into service as his assistant astronomer (not to mention music scribe, choral section leader, housekeeper, and nurse; Brock 2007). She would later write of William,

> ...by way of keeping him alife I was even obliged to feed him by putting the Vitals by bitts into his mouth—this was once the case when at the finishing of a 7 feet mirror he had not left his hands from it for 16 hours together ...

The new planet came on 1781 March 13. According to a paper William presented to the Bath Literary and Philosophical Society later that month,

> ...between ten and eleven in the evening, while I was examining the small stars in the neighborhood of H. Geminorum, I perceived one that appeared visibly larger than the rest: being struck with its uncommon [size], I compared it to H. Geminorum and the small star in the quartile between Auriga and Gemini, and finding it so much larger than either of them, suspected it to be a comet.

He checked back a few nights later and the "comet" had already moved; it was clearly a solar system object. But it was in the ecliptic and didn't have a fuzzy appearance, as was typical of comets. Despite the unprecedented clarity of his optics, William was still a novice astronomer. It was an expert user of what William would later call "common telescopes" with "great aberration"—the Astronomer Royal, Dr. Nevil Maskelyne—who would gradually realize that Uranus was a planet (Schaffer 1981). For his discovery, William was awarded the Royal Society's Copley Medal and a position as King's Astronomer to George III. Accepting the job meant a 50% pay cut and a move to Datchet, near Windsor, much to Caroline's dismay (Brock 2007). But she grudgingly grew to enjoy her new vocation, becoming an expert cataloguer—now a full-time career for many archive astronomers. Moreover, Caroline finally achieved her ambition to earn her own living and became the first professional British female scientist when George III offered her a pension of £50 per year for her work as a comet discoverer. William, now free to devote himself entirely to astronomy, went on to revolutionize telescope construction (Figure 1.6), pioneer stellar spectroscopy, and discover infrared radiation while conducting experiments on the temperatures of his colored filters when exposed to light.

Figure 1.6. The Herschels' 20 ft telescope. According to the Wellcome Library, the etching was copied in reverse from a black-and-white engraving published by William Herschel in February 1794 (Bennett 1976). The unknown colorist apparently misidentified the scope as the 40 ft model. This image is distributed by the Wellcome Collection under a Creative Commons License, https://creativecommons.org/, and has not been modified.

Since stars appear to be suns, and suns, according to the common opinion, are bodies that serve to enlighten, warm, and sustain a system of planets, we may have an idea of the numberless globes that serve for the habitation of living creatures.

—William Herschel, *On the Nature and Construction of the Sun and fixed Stars*, 1794 December 18.

…[Caroline] it was who planned the labour of each succeeding night; she it was who reduced every observation, made every calculation; she it was who arranged everything in systematic order; and she it was who helped [William] to obtain his unperishable name. …But her claims to our gratitude end not here; as an original observer she demands, and I am sure she has, our unfeigned thanks.

—J. South, Esq., upon presenting the Royal Astronomical Society's Honorary Medal to Caroline Herschel, 1828 February 8 (Brock 2006)

Our account of the scientific legacy of giant-planet astronomy would not be complete without the story of the Neptune Affair, an international priority dispute that would pit France's Académie des Sciences against Britain's Royal Greenwich Observatory. By the 1840s, Uranus had completed over two-thirds of an orbit since William Herschel first spotted it, with new position data being accumulated regularly. Furthermore, by combing old star charts and catalogs, astronomers such as Johann Bode and Alexis Bouvard identified 22 sightings between 1690 and 1771 in which Uranus was mistakenly marked as a star (Bourtembourg 2013). The abundance of data should have led to precise orbital mapping, but no single orbit fit both the prediscovery and postdiscovery data. Was there, perhaps, an undiscovered planet tugging on Uranus? Cambridge undergraduate John Adams thought so. His 1841 July 3 note to self reads (Fernie 1995):

> Formed a design in the beginning of this week, of investigating, as soon as possible after taking my degree, the irregularities in the motion of Uranus …in order to find whether they might be attributed to the action of an undiscovered planet beyond it …

In 1843, the newly graduated mathematician began his mathematical hunt for Neptune, starting with the Titius–Bode law for planetary positions and its "fair amount of arbitrariness" (Nieto 1972) to assume an orbital radius of $0.4 + (0.3 \times 2^n) = 38$ au for $n = 7$. Early calculations were promising, and in 1845, Adams shared his predictions with George Airy, Britain's Astronomer Royal, and James Challis, Plumian Professor of Astronomy at Cambridge. Here the historical accounts differ: either Adams' predictions were correct to within 2° and Challis, superintendent of the perfect search telescope, lacked the diligence to find the planet (e.g., Fernie 1995), or Adams churned out several rounds of predictions that differed by as much as 20°, leaving Challis and Airy without enough information to conduct a proper search (Rawlins 1992; Kollerstrom 2006; Krajnović 2016).[9]

What is indisputable is that in 1845, when a British search for Neptune might have commenced if Challis and Airy had followed up on Adams' work, Paris Observatory director François Arago was encouraging École Polytechnique mathematician Urbain Le Verrier to investigate the Uranus problem. The more experienced Le Verrier followed a similar problem-solving strategy to Adams, also using 38 au as his initial estimate for the phantom planet's orbital radius. On 1846 June 1, Le Verrier delivered his second memoir on Uranus to the Paris Academy, a recap of which reached Airy at Greenwich Observatory on June 23. Airy, who had been ambivalent about Adams' work and possibly insulted when Adams did not answer one of his letters (Fernie 1995), now decided to act. On July 9, he asked Challis to start a sky search, for which Adams computed an ephemeris. In France, Le Verrier was doing the same thing. He presented his final predictions to the Paris Academy on August 31, and 19 days later wrote a letter to astronomer Johann Galle at Berlin Observatory asking him to start the planet hunt. The

[9] See also http://www.dioi.org/kn/Neptune/index.htm.

strangeness of Le Verrier's having to importune a German colleague to conduct the observations instead of Arago and the Paris Observatory staff jumping on the discovery opportunity lessens when one considers that government-funded observatories of the 19th century were intended for time-keeping and sky mapping, not scientific research (Krajnović 2016). Plus, Galle had previously contacted Le Verrier, leading the latter to believe his entreaty would be successful. On the British side, Chapman (1988) argues that Airy's particularly strong sense of public duty was a large factor in his delay in starting the hunt for Neptune.

The observers now take center stage in the Neptune Affair. Challis started his desultory sky search on July 29, a full 20 days after Airy's urging. Even though Adam's predicted ephemeris was 2° 30′ off from Neptune's actual position,[10] Challis swept over Neptune three times and actually noticed that its appearance was disk-like without ever stopping to confirm. Meanwhile, Le Verrier's letter serendipitously reached Berlin Observatory on director Johann Encke's birthday. Encke, who was not particularly convinced by Le Verrier's predictions, went home to celebrate and turned the telescope over to the enthusiastic Galle and his assistant Heinrich d'Arrest for the night (Krajnović 2016). Just after midnight on 1846 September 24, the two young astronomers found Neptune: "It is not on the map!" d'Arrest exclaimed, referring to the fact that Neptune's changing sky position meant it was not charted along with the fixed stars. (However, according to Fernie 1995, one famous observer did count Neptune as a star—Galileo, who was studying Jupiter's moons in 1612 when the two planets were at conjunction!) Le Verrier's ephemeris had been less than 1° away from Neptune's actual position.

> *Almost all the greatest discoveries in astronomy have resulted from what we have elsewhere termed Residual Phenomena, of a qualitative or numerical kind, of such portions of the numerical or quantitative results of observation as remain outstanding and unaccounted for …*
>
> —John Herschel (1811–1877), *Outlines of Astronomy* (p. 548)

For the French, the discovery story was straightforward: Le Verrier accurately predicted Neptune's position and formally published his calculations (Le Verrier 1846a, 1846b, 1846c). Galle and d'Arrest, who found Neptune after less than an hour of searching, did not even count themselves as co-discoverers, and Arago boasted that Le Verrier "discovered the planet with the point of his pen" (Krajnović 2016). But nationalism was strong in the scientific community, and Airy and John Herschel, William's son, decided to claim British co-discovery. Herschel (1846) belatedly announced Adams' unpublished calculations in a Letter to *The Athenaeum*, while Adams and Challis proposed to name the new planet "Oceanus," an audacious move that led to one of the stormiest meetings of the Académie des Sciences on record and a condemnation of Britain's "odious national

[10] N. Kollerstrom, http://www.dioi.org/kn/neptune/witihin.htm.

jealousy" in the French press (Fernie 1995). In many ways the conflict paralleled the Calculus Priority Dispute: Adams, like Newton, began his work first but failed to publish, while Le Verrier and Leibniz's later calculations were written down and adopted by their peers. Fortunately, Adams lacked Newton's vicious temper and credited Le Verrier's discovery from the start; the two even became friends. The French came to acknowledge Adams' work, which, though not as precise as Le Verrier's, was still admirably prescient. Meanwhile, Airy (unintentionally) threw his own career under the bus by forwarding Adams' case. Despite their own astronomical achievements, he and Challis received bitter recriminations for failing to bag a planet for Britain.

Before we turn to modern-day planetary science, we must note that methods of observing giant planets that anticipated today's astronomical and mathematical techniques were not confined to Europe. From 69 BCE to 1342 CE, Chinese astronomers recorded 31 lunar occultations of Jupiter and 32 lunar occultations of Saturn (Stephenson & Baylis 2012). (Occultations of Venus, Mars, and, rarely, Mercury were also tallied.) They were able to spot occultations even during what we now call "bright time"—when the lunar phase is at least 0.95, rendering observations of low-surface-brightness objects impossible. Lunar occultations are now used to measure stellar diameters. Babylonian astronomers developed a method for predicting the ephemerides of Jupiter and Saturn that was similar to integrating a differential equation from the initial condition: starting with the position of the giant planet at a particular event (e.g., first visibility or first stationary point), the astronomer would repeatedly apply a zigzag function to advance the giant planet's position to the desired date (Steele 2010). Mayan priest-astronomers found prime factorizations of the synodic periods of all planets visible to the naked eye and used them to calculate their calendar "super-number," which is also a whole multiple of the Moon's synodic period. Their system of arithmetic could handle numbers of order 10^{14} (Chanier 2018).

What role do giant planets play in 21st century astronomy? The "hot Jupiters" (Mayor & Queloz 1995) were the vanguard of the now 4000-strong collection of known exoplanet systems orbiting main-sequence stars,[11] the vindications of pluralistic predictions by Bruno, Kepler, Huygens, and William Herschel. Dynamical studies of the solar system show that giant planets control the growth, composition, and orbital evolution of their smaller, potentially habitable neighbors. For example, a migrating Jupiter may have truncated the disk of planetesimals out of which Earth and Venus accreted, preventing an Earth-size planet from forming in Mars' position (Walsh et al. 2011). Neptune may have starved the inner solar system of inward-drifting pebbles, cutting off a supply of solids that could have formed a super-Earth (Raymond et al. 2018). As Jupiter grew, it directed volatile-rich asteroids into the inner solar system, providing a water source for the early Earth (Raymond & Izidoro 2017). Jupiter subsequently depleted the asteroid belt either by migrating or through orbital resonances (Walsh et al. 2011; O'Brien et al. 2007), reducing the

[11] As of 2019 April 15; exoplanet.eu/catalog.

eventual number of near-Earth objects and contributing to the long-term safety of the planet (Granvik et al. 2018). Although detections of biomarkers in exoplanet atmospheres are eagerly anticipated (e.g., López-Morales et al. 2019), it will be impossible to unravel the formation history of any habitable world without fully characterizing its giant-planet neighbors.

> *It's highly improbable in the limitless vastness of the Universe that we humans stand alone.*[12]
>
> —Charles Bolden, Astronaut, NASA Administrator (retired)

Furthermore, there is no reason we should exclude giant planets from the search for extraterrestrial life. Even if there are no living "gasbags" hunting, sinking, or floating their way through Jupiter's cloud tops (Sagan & Salpeter 1976), the subsurface oceans of giant-planet satellites may offer habitats similar to Earth's subglacial lakes—many of which are populated with chemotrophic bacteria (e.g., Karl et al. 1999; Skidmore et al. 2000; Gaidos et al. 2004; Skidmore 2011; Mikucki et al. 2016). The Europa Clipper mission, the hardware of which is currently under construction, will assess whether Europa could sustain life[13] (Pappalardo & Lopes 2014; Richey et al. 2018). Speculations about moon habitability no longer belong to the solar system alone: Kepler and Hubble Space Telescope observations suggest a possible exomoon orbiting the planet Kepler 1625b (Teachey & Kipping 2018).

This book, presented in two volumes, is a comprehensive overview of giant-planet formation. In Chapter 2, we discuss the essential qualities of planet nurseries, including disk lifetime, mass budget, temperature profile, viscous and wind-driven evolution, and turbulence. We also assess the viability of giant-planet formation via disk instability. In Chapter 3, we analyze the growth of solids from micron-size dust grains into macroscopic pebbles. Here we introduce the mathematics governing pebble drift, which may rob the disk of the raw ingredients required for building planets, and delve into the physics of collisional growth and fragmentation. Chapter 4 begins with a meteoritic timeline of planetesimal formation, then presents a detailed mathematical description of the formation of rubble-pile planetesimals via streaming instabilities (with alternative planetesimal formation mechanisms also considered). Planetesimal formation models are benchmarked against observational constraints from the asteroid belt and Kuiper Belt.

In Volume 1, the behavior of the disk's solids is entirely governed by gas; in Volume 2, the solids' self-gravity and mutual gravitational interactions become important. Chapter 5 will cover the growth of solid planet cores and consider the relative merits of pebble accretion versus collision-based models. Chapter 6 will

[12] https://www.theatlantic.com/politics/archive/2014/07/heres-how-nasa-thinks-well-find-aliens/442176/
[13] https://europa.nasa.gov/about-clipper/overview/

explain gaseous envelope formation and examine the problem of opacity-limited accretion efficiency. Finally, Volume 2 will conclude with an exploration of planet migration and dynamical instabilities and describe how both processes affect the exoplanet mass-semimajor axis distribution.

Though our main goal is to create a useful reference volume for scientists with a range of experience levels, we also hope to convey our enthusiasm for all things planet-related and to inspire readers to conduct their own research on giant-planet formation.

References

Bennett, J. A. 1976, JHA, 7, 75

Bobis, L., & Lequeux, J. 2008, JAHH, 11, 97

Bourtembourg, R. 2013, JHA, 44, 377

Blank, B. 2009, Notices of the American Mathematical Society, 56, 5

Brake, M., & Hook, N. 2007, PhyEd, 42, 245

Brock, C. 2006, History Workshop Journal, 61, 249

Brock, C. 2007, The Comet Sweeper: Caroline Herschel's Astronomical Ambition (London: Icon Books Ltd)

Bryson, B. 2003, A Short History of Nearly Everything (New York: Broadway)

Chanier, T. 2018, Math Intelligencer, 40, 18

Chapman, A. 1988, JHA, 19, 121

Dick, S. J. 1996, The Biological Universe (Cambridge: Cambridge Univ. Press)

Fernie, J. D. 1995, AmSci, 83, 116

Gaidos, E., Lanoil, B., Thorsteinsson, T., et al. 2004, AsBio, 4, 327

Galileo, G. 1610, Siderius Nuncius, Engl. transl. A. Van Helden 1989 (Chicago, IL: Chicago Univ. Press)

Galileo, G. 1613, Istoria e dimostrazioni intorno alle macchie solari e loro accidenti (Rome: Academy of the Lincei)

Galileo, G. 1638, Discorsi e Dimostrazioni Matematiche Intorno a Due Nuove Scienze (Leiden, South Holland: Appresso gli Elsevirii), Engl. transl. H. Crew & A. de Salvio 1954 (New York: Dover)

Gillispie, C. C. 1997, Pierre-Simon Laplace, 1749-1827: A Life in Exact Science (Princeton, NJ: Princeton Univ. Press)

Gingerich, O. 1985, JHA, 16, 68

Granvik, M., Morbidelli, A., Jedicke, R., et al. 2018, Icar, 312, 181

Hahn, R. 2005, Pierre Simon Laplace, 1749-1827: A Determined Scientist (Cambridge, MA: Harvard Univ. Press)

Hawking, S. 2002, On the Shoulders of Giants (Philadelphia, PA: Running Press)

Heilbron, J. L. 2010, Galileo (Oxford: Oxford Univ. Press)

Herschel, J. 1846, Letter to The Athenaeum, 988, 1019

Hoskin, M. 2011, Discoverers of the Universe: William and Caroline Herschel (Princeton, NJ: Princeton Univ. Press)

Hulse, R. A., & Taylor, J. H. 1975, ApJL, 195, L51

Huygens, C. 1656, De Saturni Luna Observatio Nova (The Hague)

Huygens, C. 1659, Systema Saturnium (The Hague), Engl. transl. H. Walden 1928

Huygens, C. 1925, Oeuvres Complètes de Christiaan Huygens, vol. 15 (The Hague: Martinus Nijhoff)
Huygens, C. 1695, Cosmotheoros (The Hague: Adriaan Moetjens)
Karl, D. M., Bird, D. F., Björkman, K., Houlihan, T., & Tupas, L. 1999, Sci, 286, 2144
Kollerstrom, N. 2006, HisSc, 44, 349
Koyré, A. 1952, Isis, 43, 312
Krajnović, D. 2016, A&G, 57, 5.28
Lee, C.-F., Li, Z.-Y., Hirano, N., et al. 2018, ApJ, 863, 94
Lemonick, M. D. 2009, The Georgian Star: How William and Caroline Herschel Revolutionized Our Understanding of the Cosmos (New York: W. W. Norton & Company, Inc.)
Le Verrier, U. J. 1846a, AN, 25, 53
Le Verrier, U. J. 1846b, AN, 25, 65
Le Verrier, U. J. 1846c, AN, 25, 85
López-Morales, M., Currie, T., Teske, J., et al. 2019, BAAS, 51, 162
Louwman, P. 2004, Titan—From Discovery to Encounter (ESA SP-1278) (Noordwijk: ESA), 103
Mayor, M., & Queloz, D. 1995, Natur, 378, 355
Mikucki, J. A., Lee, P. A., Ghosh, D., et al. 2016, RSPTA, 374, 20140290
Nieto, M. M. 1972, The Titius-Bode Law of Planetary Distances: Its History and Theory (Oxford: Pergamon Press Ltd)
O'Brien, D. P., Morbidelli, A., & Bottke, W. F. 2007, Icar, 191, 434
Owen, T., & Bolton, S. 2010, in IAU Symp. 269, Galileo's Medicean Moons: Their Impact on 400 Years of Discovery (Cambridge: Cambridge Univ. Press), 13
Pappalardo, R., & Lopes, R. 2014, 40th COSPAR Scientific Assembly B0, 40, 3–18–14
Postl, A. 1977, VA, 21, 325
Rawlins, D. 1992, DIO, 2, 98
Raymond, S. N., & Izidoro, A. 2017, Icar, 297, 134
Raymond, S. N., Izidoro, A., & Morbidelli, A. 2018, arXiv:1812.01033
Richey, C., Pappalardo, R. T., Senske, D., et al. 2018, AGU Fall Meeting, P53F-3018
Sagan, C., & Salpeter, E. E. 1976, ApJS, 32, 737
Schaffer, S. 1981, JHA, 12, 11
Skidmore, M. 2011, GMS, 192, 61
Skidmore, M. L., Foght, J. M., & Sharp, M. J. 2000, ApEnM, 66, 3214
Steele, J. M. 2010, JHA, 41, 261
Steffen, J. H., Fabrycky, D. C., Ford, E. B., et al. 2012, MNRAS, 421, 2342
Stephenson, R. F., & Baylis, J. T. 2012, JHA, 43, 455
Stewart, S. T., & Leinhardt, Z. M. 2009, ApJL, 691, L133
Teachey, A., & Kipping, D. M. 2018, SciA, 4, 1784
Van Helden, A. 1968, Notes and Records of the Royal Society of London, 23
Van Helden, A. 1974, JHA, 5, 175
Van Helden, A. 2009, ExA, 25, 3
Walsh, K. J., Morbidelli, A., Raymond, S. N., O'Brien, D. P., & Mandell, A. M. 2011, Natur, 475, 206
Wilson, C. 1985, AHES, 33, 15

Chapter 2

The Planet-forming Environment

2.1 Introduction

Guided by the low mutual inclinations of planets in the solar system (all planets except Mercury are within 3° of an invariable plane) and the principle of angular momentum conservation, philosopher Immanuel Kant realized that the solar system planets must be the remnants of high angular-momentum material that the young Sun could not absorb. Indeed, the orbits of the four outer planets contain 99.3% of the solar system's total angular momentum (Figure 2.1), while the Sun's rotation and the terrestrial planets' orbits combine to contribute only 0.7% (Pinto et al. 2011). If the Sun were to take up the angular momentum contained in Jupiter's and Saturn's orbits while keeping its current size, it would have to spin up to breakup speed.[1] Kant (1755) proposed that planets grow in flat disks, the natural products of rotating, collapsing gas clouds. More than two centuries later, Hubble Space Telescope (HST) observations of the Orion Nebula confirmed the existence of protoplanetary disks, the presence of which had already been inferred from spectral energy distributions (SEDs) of young stars (e.g., Kenyon & Hartmann 1987) but which were first actually imaged as silhouettes against background emission from gas ionized by the young, massive star θ^1 C Ori (O'dell et al. 1993; McCaughrean & O'dell 1996). Figure 2.2 shows HST images of Orion and ρ Ophiuchi protoplanetary disks in their native molecular clouds. Disks allow nascent stars to accrete mass by transporting angular momentum outward (Lynden-Bell & Pringle 1974), supply the gas for giant-planet atmospheres, and create a dynamically stable environment throughout the solid accretion process. Disks also provide the high densities required for frequent collisions between dust grains, which first forge pebbles, then planetesimals, and finally planet cores.

[1] At breakup speed, gas at the star's equator is in Keplerian orbit: $v_{\rm equator} = (GM_\odot/R_\odot)^{1/2}$. This calculation is approximate, as a star rotating at breakup speed would naturally assume an oblate spheroid shape.

doi:10.1088/2514-3433/ac1db7ch2

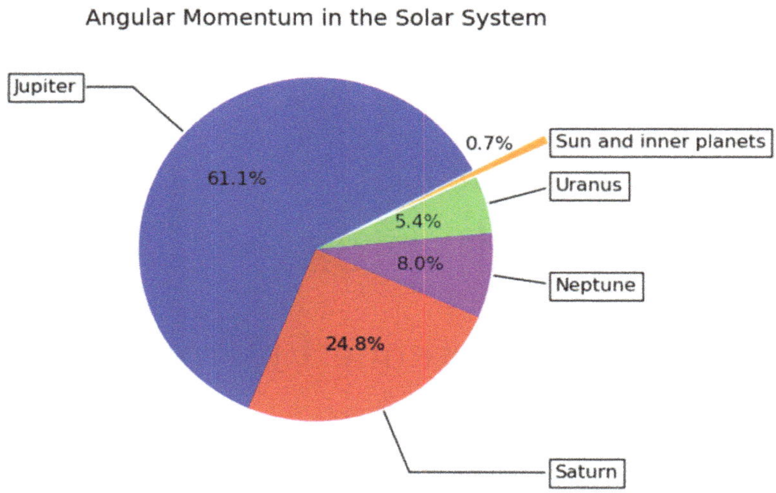

Figure 2.1. The solar system's angular momentum distribution.

In this chapter, we outline the characteristics of giant planet-forming disks. We first provide an overview of protostar and disk evolution with example SEDs of protostars in each evolutionary phase where disks are present (Section 2.2), alongside a discussion of how disks are first discovered and then followed up with more detailed observations. We then define a simple analytical model of an axisymmetric, mid-plane-symmetric protoplanetary disk that can be used for rough estimates of quantities that affect planet growth, such as temperature, density, and turbulent speed (Section 2.3). In Section 2.4 we extend our model to cover the observed complexity of astrophysical disks. We include radial pressure support (Section 2.4.1), gaps and inner holes (Section 2.4.2), warm surface layers (Section 2.4.4), accretion heating (Section 2.4.5), turbulence, vortices, and angular momentum transport (Section 2.4.6), self-gravity and spiral structure (Section 2.4.7), and finally disks in binary systems (Section 2.4.8). Throughout Section 2.4, we discuss which disk features may aid or inhibit planet formation.

Resources in this chapter include Table 2.1, which contains definitions and units of all disk-related variables, and explanatory insets on topics that may be new to students: signatures of disk accretion (inset box 2.1), the minimum-mass solar nebula (mmsn; which we advocate replacing with the maximum-mass solar nebula; inset box 2.2), hydrostatic equilibrium (inset box 2.3), the Atacama Large Millimeter Array (inset box 2.4), and opacity, optical depth, and intensity (inset box 2.5).

2.2 Disk Evolutionary Phases and Spectral Energy Distributions

A protoplanetary disk is a mixture of gas and dust in Keplerian rotation around a pre main-sequence star. The main mass reservoir in planet-forming disks is H_2 gas, which, along with helium, makes up about 99% of the total disk mass. "Abundant" molecules such as CO, which is the basis for most submillimeter kinematic and chemical studies of disks, and H_2O, which may have provided up to half of the solid

Figure 2.2. Top: protoplanetary disks in the Orion Nebula. Disks near θ^1 C Ori have Hα-emitting shells that face the young O star, while disks that are farther away from θ^1 C Ori are seen only as silhouettes. From the top down, the disks are 132-1832, 206-446, 180-331, 106-417, 231-838, and 181-825. Image credit: NASA, ESA, M. Robberto (Space Telescope Science Institute/ESA), the Hubble Space Telescope Orion Treasury Project Team and L. Ricci (ESO). Bottom: the edge-on "Flying Saucer" disk in the ρ Ophiuchi star-forming region. The dark shadow across the center of the disk is caused by the dust-rich midplane. Image credit: Digitized Sky Survey 2/NASA/ESA.

mass required to build Jupiter's core, are only $\sim 10^{-4}$ times as common as H_2 molecules (e.g., Dodson-Robinson et al. 2009). Fortunately for both planet formation and observability, protoplanetary disks also harbor microscopic dust grains that absorb visible and near-infrared starlight and reradiate it at longer wavelengths. Disks are first identified photometrically by searching for excess infrared emission over what the star photosphere produces (e.g., Beichman et al. 1986; Adams et al. 1987), then

Table 2.1. Definitions of Disk-related Variables

Variable	Units	Definition
Time and age		
t	s, yr, or Myr	time or disk age
$f(t)$		fraction of protostars with disks
τ_l	Myr	mean disk lifetime
Star and planet parameters		
M_*	$M_\odot$ or g	star mass
R_*	$R_\odot$, cm or km	star radius
L_*	$L_\odot$ or erg s^{-1}	star luminosity
$\dot{M}_*$	$M_\odot$ yr^{-1} or g s^{-1}	star accretion rate
L_{acc}	$L_\odot$ or erg s^{-1}	accretion luminosity
q		planet/star mass ratio
Embedded protostar parameters		
R_{core}	pc or au	radius of the star-forming cloud core
L_{bol}	$L_\odot$ or erg s^{-1}	bolometric luminosity of collapsing envelope+disk +protostar (Stage I)
T_{bol}	K	temperature of a blackbody with the same mean frequency as the protostar SED
ρ_{env}	g cm^{-3}	volume density of envelope gas
Disk coordinate system		
R	au or cm	distance from the star in cylindrical coordinates
r	au or cm	absolute distance from the star
z	au or cm	distance above the midplane
θ		angle between vectors $\vec{r}$ and $\vec{R}$ (see Figure 2.8)
ϕ		azimuthal coordinate[a]
Overall disk properties		
M_d	$M_\odot$ or g	disk mass
R_{out}	au or cm	outer radius of the disk
μ	g mol^{-1}	mean molar mass of gas[b]
ξ		surface density power-law index
q_T		temperature power-law index
γ		ratio of specific heats (assumed 7/5 for primarily H_2 gas)
c_V	erg g^{-1} K^{-1}	specific heat at constant volume
Disk properties that vary with R only		
Σ	g cm^{-2} or $M_\odot$ au^{-2}	surface density
Σ_{gap}	g cm^{-2} or $M_\odot$ au^{-2}	surface density in the gap opened by the planet
$\dot{\Sigma}_w$	g cm^{-2} s^{-1}	surface density rate of change due to photoevaporative wind
Ω_K	s^{-1}	Keplerian angular speed
Ω	s^{-1}	pressure-supported angular speed of midplane gas
v_R	cm s^{-1} or km s^{-1}	radial velocity of midplane gas
v_ϕ	cm s^{-1} or km s^{-1}	azimuthal velocity of midplane gas
j	cm^2 s^{-1} or km^2 s^{-1}	specific angular momentum of midplane gas
H	cm or au	density/pressure scale height

H_S	cm or au	height above midplane at which starlight is absorbed
φ		angle of incidence of starlight on the disk surface
τ	dyn cm	viscous torque on annulus of gas
$T_{\mathrm{acc,surf}}$	K	disk surface temperature if accretion were the only heat source
Disk properties that may vary in two or three dimensions		
g_z	g cm s^{-2}	vertical component of stellar gravity
$\vec{g}$	g cm s^{-2}	acceleration due to gravity
$\Omega(z)$	s^{-1}	angular speed as a function of height above the midplane
ρ	g cm^{-3}	gas density
P	dyn cm^{-2}	gas pressure
S	erg K^{-1}	entropy
T	K	temperature[c]
T_{bb}	K	temperature of a blackbody in thermal equilibrium with stellar radiation
$T_{\mathrm{eq,surf}}$	K	temperature in optically thin surface layers when in equilibrium with stellar radiation
T_{surf}	K	true temperature in optically thin surface layers
T_{acc}	K	temperature if accretion were the only heat source
c_s	cm s^{-1}	isothermal sound speed[c]
ν	cm^2 s^{-1}	turbulent viscosity
$\bar{\nu}$	cm^2 s^{-1}	mass-weighted vertical average turbulent viscosity
v_t	cm s^{-1}	characteristic turbulent speed
$\vec{v}$	cm s^{-1} or km s^{-1}	velocity vector
F_{visc}	dyn	viscous force
α		efficiency of energy extraction from velocity shear
ℓ	au or cm	eddy size/mixing length
κ_λ	cm^2 g^{-1}	opacity to photons of wavelength λ
κ_*	cm^2 g^{-1}	Planck mean opacity to starlight incident on the disk surface
κ_d	cm^2 g^{-1}	Planck mean opacity to thermally reprocessed starlight in optically thin surface layers
κ_R	cm^2 g^{-1}	Rosseland mean opacity to thermal radiation in an optically thick interior
λ_{mfp}	cm	photon mean free path
τ_λ		optical depth to photons of wavelength λ
$\tau_{1/2}$		midplane-to-surface optical depth to disk thermal radiation
τ_*		optical depth to incident starlight
S_λ	erg s^{-1} cm^{-3} ster^{-1}	source function for radiation at wavelength λ
J	erg s^{-1} cm^{-2} ster^{-1}	bolometric mean intensity
$q_{+,m}$	erg g^{-1} s^{-1}	viscous heating rate per unit mass
$q_{+,v}$	erg cm^{-3} s^{-1}	viscous heating rate per unit volume
F_{acc}	erg cm^{-2} s^{-1}	accretion flux

(*Continued*)

Table 2.1. (*Continued*)

Variable	Units	Definition
u	erg cm^{-3}	volume energy density carried by photons
$P_{\rm rad}$	erg cm^{-3}	radiation pressure
∇		thermodynamic gradient $d \ln T/d \ln P$
$\vec{F}$	dyn cm^{-3}	magnetic tension
$\vec{B}$, B	Gauss	magnetic flux density (vector and magnitude)
Λ_z		Elsässer number for the vertical direction
η	cm^2 s^{-1}	magnetic diffusivity
β		ratio of gas pressure to magnetic pressure
χ	cm^2 s^{-1}	thermal diffusivity
t_c	s or yr	radiative diffusion time (optically thick regions)
τ_c	s or yr	cooling time (optically thick or thin regions)
N	s^{-1}	Brunt–Väisälä frequency
Pe		Peclet number (ratio of advection time to thermal diffusion time)
L_J	cm or au	Jeans length
Q		Toomre Q
Φ	cm^2 s^{-2}	gravitational potential due to disk gas
Properties of perturbations		
ℓ_p	cm or au	length scale of perturbation
L_s	cm or au	minimum length of perturbation that can be destroyed by shear
τ_{ff}	s or yr	free-fall time
τ_s	s or yr	shear time
k	cm^{-1} or au^{-1}	radial wavenumber
ω	s^{-1} or yr^{-1}	perturbation growth rate
κ_e	s^{-1} or yr^{-1}	epicyclic frequency
Properties of observations		
λ	Å, μm, mm, or cm	observing wavelength
F_λ	erg s^{-1} cm^{-2} Å^{-1}	flux per unit wavelength
I_λ	erg s^{-1} cm^{-3} ster^{-1}	specific intensity at wavelength λ

Notes:
[a] □ is not used when disk is assumed axisymmetric.
[b] We assume $\mu = 2:33$ throughout the disk, but in reality changes slightly across ice lines.
[c] With isothermal assumption, T and c_s are functions of R only.

followed up with high-contrast visible or near-infrared imaging, infrared spectroscopy, and radio interferometry. High-contrast imaging maps starlight scattered off the disk surface and often exposes gaps, spiral structure, ripples in the disk surface, or shadows cast by the inner disk on the outer disk (e.g., Benisty et al. 2017, 2018; Currie et al. 2019). Infrared spectroscopy provides information about mineralogy, surface chemical composition, and accretion rate (box 2.1), and can reveal outflows (e.g., Bouwman et al. 2008; Pontoppidan et al. 2010; Pascucci et al. 2011; Salyk et al. 2013). Radio interferometry traces the spatial distribution of large dust particles located near the midplane and the rotational emission from cold gas in the outer disk, revealing the composition of giant planet-forming gas (e.g., Öberg et al. 2015; Bergin et al. 2016; Öberg et al. 2017; see review by Dutrey et al. 2014).

Yet high-resolution imaging and spectroscopy are only viable options for nearby, relatively bright sources. Because the nearest star-forming regions (Taurus, Lupus, and Ophiuchus, for example) and moving groups (TW Hya) come from low-mass clouds and have mainly low-mass protostars, disk imaging surveys are biased against massive young stars. The faint, often compact disks around young stars with $M_* \lesssim 0.5M_\odot$ are also poor candidates for high-resolution follow-up (spectral or spatial). Most disks are discovered and characterized by SEDs[2] assembled from unresolved photometry and by low-resolution (sub)millimeter maps where the emitting region size is a small multiple of the beam size. Here we briefly describe how to use SEDs to infer protostar and disk properties.

Figure 2.3 shows the SEDs of four protostars with disks. Panel (a) shows the Stage I[3] protostar IRAS 04016+2610, identified by the Infrared Astronomical Satellite (IRAS) as an infrared source in the L1489 dark cloud (Beichman et al. 1986; Benson & Myers 1989). Photometry and Owens Valley Radio Observatory maps from Eisner et al. (2005) are consistent with a disk embedded in an infalling envelope of gas and dust. Stage I is the earliest phase in low-mass star formation in which a disk is definitely present (e.g., Harsono et al. 2014; Lindberg et al. 2014; Aso et al. 2015). The combined mass of the disk and star exceeds the mass of the infalling envelope but the disk/star mass ratio may be as high as 50%. The outer edge of the disk is at the centrifugal radius, where the specific angular momentum of the innermost envelope gas matches the Keplerian value $j = \sqrt{GM_*R}$ (M_* is the star mass and R is the distance from the star). Stage I objects accrete, or gain mass, at a rate of $\dot{M}_* \gtrsim 10^{-6}M_\odot\ \mathrm{yr}^{-1}$ (Eisner et al. 2005; Robitaille et al. 2006), with accretion rates estimated either from the bolometric temperature T_{bol} and luminosity L_{bol} (where L_{bol} is found from integrating F_λ, the flux in a given wavelength bin, across all wavelengths, and T_{bol} is the temperature of a blackbody with the same mean frequency as the SED; Myers & Ladd 1993; Myers et al. 1998; Young & Evans 2005) or from a grid of SED models for accreting protostars (e.g., Whitney et al. 2003). The Stage I phase lasts between 0.09 Myr and 0.54 Myr (Kristensen & Dunham 2018; Evans et al. 2009). Although the star has a strong effect on the SED, heating the disk and envelope so that they are detectable in the near-infrared (for more on reprocessing of starlight by dust, see Natta et al. 2007), the envelope completely

[2] The SED is slightly different from the observed spectrum $F_\nu(\nu)$ (flux per frequency bin) or $F_\lambda(\lambda)$ (flux per wavelength bin). To accurately understand the energy output of our star/disk system across the spectrum, we correct for the fact that the same integrated flux F (units erg cm^{-2} s^{-1}) corresponds to different values of F_ν and F_λ, depending on the wavelength range. The SED is defined as $\nu F_\nu = \lambda F_\lambda$.

[3] The SED is identified as Class I by the slope of λF_λ between 2 and 20 μm; Class I SEDs usually, but not always, indicate sources in Stage I. Viewing angle strongly affects the SED so that (for example) a star/disk system with negligible remaining envelope, which would normally fall in Class II, can mimic a Class I SED if viewed at a high inclination (Whitney et al. 2003; Robitaille et al. 2006). Here we focus on physical characteristics of protostars (such as the presence or absence of disks and envelopes) rather than observational classifications; we therefore follow Robitaille et al. (2006) in referring to evolutionary stages rather than spectral classes.

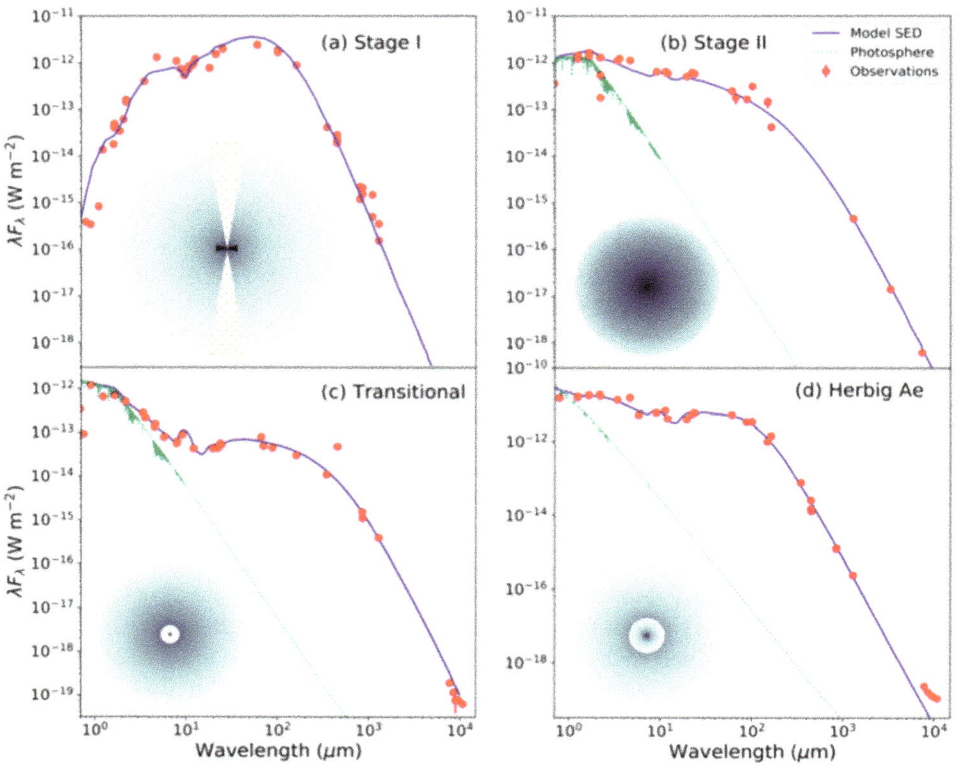

Figure 2.3. Spectral energy distributions of protostars at different evolutionary stages. (a) Stage I, where the nascent low-mass star and disk are embedded in an infalling envelope. (b) Stage II, where very little of the envelope remains and the low-mass star photosphere becomes visible. (c) Transitional, in which the optically thick disk has at least one gap in the radial dust distribution caused by photoevaporation, tidal forces from planets, or locally high magnetic flux. (d) Embedded Herbig Ae, similar to Stage I but with an intermediate-mass protostar ($2.3M_\odot$) instead of a low-mass protostar.

obscures the star so that the emission is not photospheric at any wavelength—hence the very red colors in the visible and near-IR. The inset in Figure 2.3, panel (a) depicts the structure of IRAS 04016+2610 modeled by Eisner et al. (2005): an envelope with a radius of 2000 au and a disk within a centrifugal radius of 100 au. The color scale shows $\log_{10}(\rho_{\rm env})$, where $\rho_{\rm env}$ is the envelope density; the disk is shown edge-on in black. The Eisner et al. (2005) model also includes a cavity carved by outflows perpendicular to the disk.

Before proceeding with Figure 2.3, we pause to consider the very youngest protostars in Stage 0 (SED not shown). Here envelope mass exceeds star mass and dust-induced reddening is so strong that the protostar is rarely detectable at wavelengths $\lambda < 10$ μm (André et al. 2000). Some Stage 0 sources appear to have disks (Segura-Cox et al. 2018; Maury et al. 2019), though it is often difficult to distinguish between a true disk in Keplerian motion and a rotationally flattened inner envelope. Most of the star's mass is assembled during Stage 0, when the rate of infall from the envelope onto the star/disk is at its peak. While the dynamical environment that creates planetary systems with coplanar orbits may not be in place until Stage I, the growth of solid bodies is already underway in Stage 0: in an analysis of archival interferometric data, Li et al. (2017) found that four of nine Stage 0 protostars in their sample had grains larger than could be found in the interstellar size distribution. The Stage 0 phase lasts between 0.05 Myr and 0.17 Myr (Kristensen & Dunham 2018; Enoch et al. 2009). We pinpoint "$t = 0$"—the start of a protostar's evolutionary clock, which we usually associate with the beginning of Stage 0—as the formation of a point mass in the middle of what was formerly a prestellar core with radius $R_{\rm core} \lesssim 10,000$ au (Ward-Thompson et al. 2007).

Moving on with Figure 2.3, we note that the infalling envelope is what distinguishes low-mass Stage I protostars from more evolved Stage II or classical T-Tauri stars. The classical T-Tauri star is still accreting from the disk at about $10^{-8} M_{\odot}\ {\rm yr}^{-1}$ (solar masses per year; Hartmann et al. 1998), but the star and disk are no longer accreting from an envelope, which has been dissipated by infall and/or outflows. (For more on signatures of accretion from the disk onto the star, see the inset box 2.1.) The star photosphere, which was dust-obscured in Stage I, now dominates the visible light SED in Stage II, while the optically thick disk appears at near-IR and longer wavelengths. Panel (b) of Figure 2.3 shows the SED of RU Lup, a $1.15 M_{\odot}$ star of age 1.2 Myr surrounded by a disk with a radius of 308 au (Woitke et al. 2019). The gently decreasing value of λF_{λ} between 2 μm and 20 μm is typical of Stage II. (Protostars with "flat" spectra (not shown), where λF_{λ} is approximately constant between 2 μm and 20 μm, are probably Stage I objects with rotationally flattened envelopes and/or disks seen at 30°–50° inclination; Calvet et al. 1994; Whitney et al. 2003). Dionatos et al. (2019) tabulated the photometry from previous literature sources (red points), and Woitke et al. (2019) used the `ProDiMo` code (*Pro*toplanetary *Di*sk *Mo*del; Woitke et al. 2009; Kamp et al. 2010) to fit a model to the RU Lup SED (blue line). The inset in Figure 2.3(b) is a face-on rendering of $\log_{10}[\Sigma(R)]$, the model disk surface density distribution (vertically integrated mass per cm^2 as a function of distance from the star R); the actual inclination of the disk is 19°. We will encounter RU Lup again in Section 2.4 and Figure 2.14. Although the Woitke et al. (2019) `ProDiMo` model is for a disk with monotonically decreasing $\Sigma(R)$, high-resolution Atacama Large Millimeter Array (ALMA; see the inset box 2.4) imaging shows concentric overdense rings in the millimeter-size dust grain distribution. Nevertheless, the young RU Lup disk has no optically thin gaps or holes in the dust surface density distribution—unlike the disk behind our next SED.

As disks evolve, their surface densities are sculpted by planets, magnetic winds, photoevaporation, and turbulence. The disk surrounding LkCa 15 (Figure 2.3(c)) is ~2 Myr old (Kraus & Ireland 2012; Drabek-Maunder et al. 2016) and contains at least one wide gap that leads to the observed flux deficit between 2 μm and 32 μm (Marsh & Mahoney 1992; Jensen & Mathieu 1997). The inset in Figure 2.3, panel (c) is a face-on rendering of the LkCa 15 surface density distribution, as modeled by Woitke et al. (2019). Pretransitional disks such as LkCa 15, which have an outer disk separated from a small interior disk by a wide gap (Espaillat et al. 2007), are a subclass of transitional disks, which are optically thick disks with inner holes (e.g., Calvet et al. 2005). Holes in transitional and pretransitional disks are not fully empty: the stars still manage to accrete 10^{-9}–$10^{-8} M_{\odot}$ yr^{-1} (e.g., Najita et. al. 2007; Andrews et al. 2011; see inset box 2.1), indicating that there must be a way to transport mass through regions that appear optically thin. One possibility is that wide holes are carved by tides from multiple planets, which shepherd nearby gas and dust into tidal tails that cross the planet orbits, allowing the optically thick outer disk to supply raw material for star accretion (Dodson-Robinson & Salyk 2011). The term "transitional" comes from the hypothesis that the inner hole is the precursor to complete disk dissipation (e.g., by photoevaporation, Hollenbach et al. 1994; Clarke et al. 2001; Alexander et al. 2006; Ercolano et al. 2008); the nascent planets in the disk would therefore be transitioning between the accretion phase and the contraction/differentiation phase (see Chapter 6). However, while transitional disks are on average older than Stage II disks (Muzerolle et al. 2010), it is by no means clear that all disks with gaps are immediately doomed: gaps can appear at any point in the disk evolution, including Stage I (Sheehan & Eisner 2017). Section 2.3.1 contains an in-depth discussion of disk lifetime, broken down by wavelength of infrared excess. Stage II disks should show excess at all wavelengths from 2 to 100 μm, while transitional disks may not always show detectable excess at 8–15 μm. Note that a transitional disk's host star is still designated a classical T-Tauri star, even though the disk does not usually carry the T-Tauri label.

SED models are degenerate, especially for the complex surface density distributions in transitional disks. For example, millimeter-wave interferometry by Isella et al. (2010) revealed a gap in the MWC 758 disk that was not apparent from the SED (Isella et al. 2008). In light of the fact that transitional disks may host young (multi)planetary systems, high-resolution follow-up is always desirable, but even then the results can be model dependent. Sparse aperture masking (SAM), in which high angular-resolution observations are obtained by fitting an optical telescope with an interferometric mask, suggested that planets might occupy the LkCa 15 gap (Kraus & Ireland 2012; Sallum et al. 2015), which was originally thought to extend from 10 to 40 au (Piétu et al. 2006; Espaillat et al. 2008). However, more recent high-contrast, unmasked near-IR images of starlight scattered from the disk surface, combined with Spitzer IRS spectra from 5 to 35 μm, suggest a three-ring disk rather than a wide gap occupied by planets. The new model consists of an inner ring spanning 0.12–3 au, surrounded by a gap and then a second ring spanning 20–40 au, then another gap, and finally the outer disk extending over 55–160 au (Thalmann et al. 2016; Currie et al. 2019). Brightness asymmetries in the middle (20–40 au) ring could masquerade as planets in SAM data when fit with point-source visibility models (Martinache 2010; Currie et al. 2019). Figure 2.4 shows adaptive optics –coronagraph observations of the LkCa 15 disk from the Subaru telescope (left) along with the best-fit model (right) by Currie et al. (2019). Dust grains in the disk are strongly forward scattering, leading the axisymmetric model annuli to appear as arcs. The varying interpretations of the SED, IRS spectra, and SAM visibilities provide a cautionary tale about disk observations: the same data may be consistent with a variety of physical models.

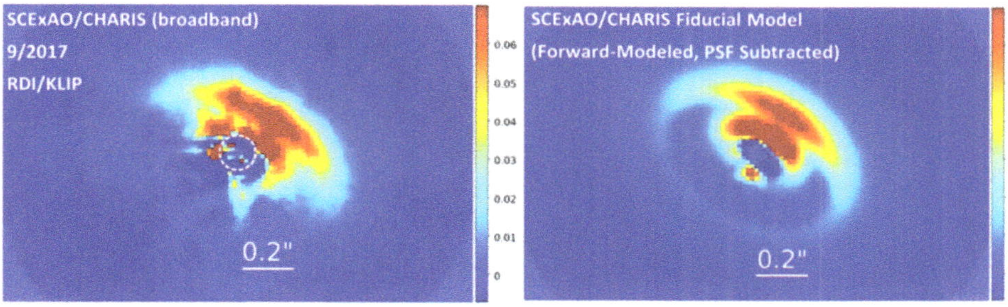

Figure 2.4. Left: broadband, near-infrared image of the LkCa 15 disk from the Subaru telescope. The dashed white circle shows the resolution limit of the coronagraph/adaptive optics system. Right: model of the disk surface as seen in scattered starlight. Although all annuli in the model disk have axisymmetric density distributions, they appear brightest on the side of the disk nearest Earth because dust grains primarily forward-scatter the incident starlight. Images from Currie et al. (2019) are reprinted with permission.

While the Stage II and transitional disk SEDs pictured in Figure 2.3(b) and (c) come from systems with low-mass protostars ($1.15M_\odot$ and $1.0M_\odot$, respectively), the final SED in Figure 2.3, panel (d) belongs to the intermediate-mass Herbig Ae star-disk system AB Aur ($M_* = 2.3M_\odot$; Andrews et al. 2013). Herbig Ae/Be protostars (Herbig 1960) are precursors to main-sequence A and B stars with $M_* \gtrsim 2M_\odot$. The ingredients of a Herbig Ae/Be system are similar to Stage I and II protostars: the young star is accreting from an optically thick disk, which may, in turn, be embedded in an envelope. However, Herbig Ae/Be disks evolve differently from T-Tauri disks, having high accretion rates and short lifetimes. In a study of 38 Herbig Ae/Be stars, Mendigutía et al. (2012) found a median accretion rate of $2.7 \times 10^{-7}M_\odot$ yr^{-1}, compared with $10^{-8}M_\odot$ yr^{-1} for T-Tauri stars (Hartmann et al. 1998). Using a larger sample of 106 Herbig Ae/Be stars, Arun et al. (2019) found a disk dissipation timescale of 1.9 ± 0.1 Myr, compared with at least 2.3 Myr and possibly as high as 5.5 Myr for T-Tauri disks (see Section 2.3.1). Hillenbrand et al. (1992) split Herbig Ae/Be objects into Group I, which is the high-mass counterpart to Stage II in that the SED is consistent with an isolated star/disk system, and Group II, where the SED indicates an additional cold cloud component. AB Aur is a Group II object; Woitke et al. (2019) collected the photometry and modeled the SED shown in Figure 2.3(d). Although there is no gap, as in the pretransitional LkCa 15 disk, Woitke et al. (2019) find that the data are best fit by two adjacent disk components with different surface density distributions (pictured in the inset of Figure 2.3(d)).

2.1 Signatures of accretion

Accretion signatures help distinguish between Herbig Ae/Be disks, Stage II/T-Tauri disks, and transitional disks. Lack of accretion also differentiates Stage III/weak-line T-Tauri stars, which are newly formed stars whose gaseous disks have dissipated, from classical T-Tauri stars. Figure 2.5 shows the ultraviolet spectrum and Hα line profile (inset) of the M1 star DM Tau, which has mass $0.53M_\odot$ (Piétu et al. 2007) and accretion rate $\dot{M}_* = 6 \times 10^{-9}M_\odot$ yr^{-1} (Ingleby et al. 2009). The labeled lines

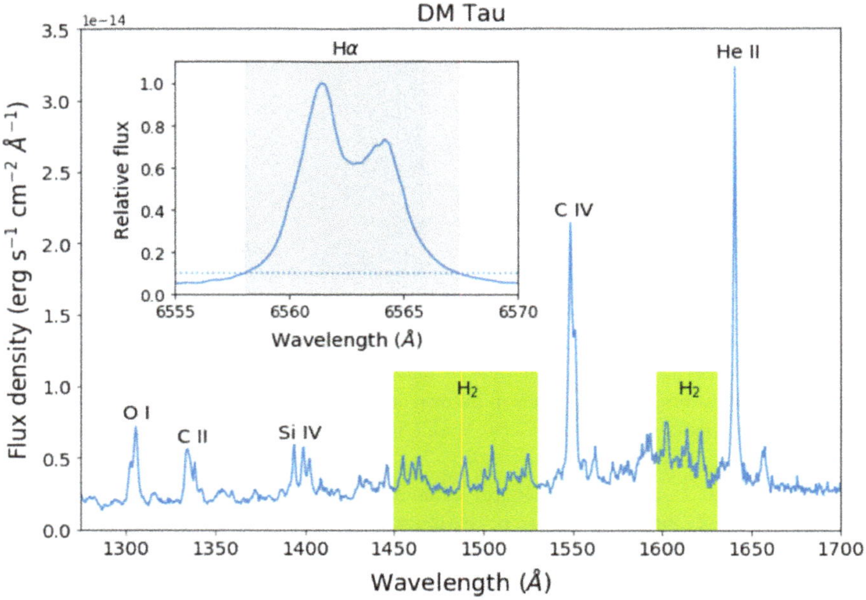

Figure 2.5. Signatures of accretion in Stage II and transitional objects: H_2 fluorescence in the UV spectrum (main plot, highlighted in green) and a wide Hα emission line (inset).

in the UV spectrum may originate from either an accretion shock or an active chromosphere, and so are present in both accreting and nonaccreting T-Tauri stars (Yang et al. 2012). Lyα pumping and collisions with fast free electrons produce the fluorescent H_2 emission lines in the chartreuse-highlighted spectral regions (Herczeg et al. 2002; Bergin et al. 2004; Herczeg et al. 2004; France et al. 2011). While some Stage III T-Tauri stars have remnant cold dust disks that produce detectable far-IR or (sub)millimeter excess and a few even have residual CO gas, H_2 emission is present only if the star is actively accreting (Ingleby et al. 2009) and the disk's main mass reservoir is still intact.

Visible and near-infrared emission lines from atomic hydrogen, which (unlike far-UV H_2 emission) can be observed from the ground, provide the basis for many accretion rate estimates. Model Hα emission-line profiles can be used to estimate $\dot{M}_*$ for stars with known mass and radius (e.g., Muzerolle et al. 2001). White & Basri (2003) suggested distinguishing between Stage II and Stage III stars using the full width of the Hα line at 10% of peak emission, as pictured in Figure 2.5; a velocity width of $\sim$200 km s^{-1} separates accretors from nonaccretors. Finally, one can measure $\dot{M}_*$ by finding the excess continuum luminosity produced by the accretion shock. Excess continuum emission fills in or "veils" absorption lines from the star photosphere; by comparing the spectra of the accreting star and a Stage III star of similar spectral type, one can determine the excess spectrum $F_{\lambda,\mathrm{excess}}$ and the accretion luminosity L_{acc}. The accretion rate can then be determined from

$$L_{\mathrm{acc}} \simeq 0.8\frac{GM_*\dot{M}_*}{R_*}, \tag{2.1}$$

where R_* is the star radius and the factor of 0.8 corrects for the size of the magnetospheric hole in the inner disk (Calvet & Gullbring 1998). Because line veiling-based $\dot{M}_*$ measurements require separation of the photosphere and continuum flux, they are inaccurate for either very low accretion rates (negligible veiling) or very high accretion rates (accretion luminosity dominates the photosphere; absorption lines are completely veiled). Hα and Paβ, by contrast, are well-behaved accretion indicators over a large dynamic range of star mass and accretion rate (Muzerolle et al. 2005; Salyk et al. 2013).

As disk gas starts to dissipate, a Stage II system evolves into a Stage III/weak-line T-Tauri system and a Group I Herbig Ae/Be system moves into Group III. Stage III/Group III stars have no measurable signatures of gas accretion; if disks are detected, they are optically thin remnants of the original planet-forming environment. While terrestrial planet assembly may proceed for more than 100 Myr after the birth of the star, the transition to Stage III signals the end of giant planet formation. In the next section, we provide a simple analytical model for a Stage II disk. Then, in Section 2.4, we present a variety of model extensions, some of which describe the complex surface density profiles observed in the transitional phase.

2.3 A Simple Analytical Disk Model

The goal of a disk model is to describe the density ρ, pressure P, temperature T, and velocity field of the gas, either as static snapshots or with time evolution. Those four quantities are the basis for computations of particle growth rates, gas accretion rates for growing planets, stability analysis of perturbations, and planet migration rates. Even though global, 3D simulations of disk evolution have become tractable in the last decade (Flock et al. 2011, 2015; Lyra et al. 2016; Ruge et al. 2016; Flock et al. 2017; Hord et al. 2017; Liu et al. 2019), analytical and semianalytical models are still vital for interpreting observations (Forgan et al. 2016; Hasegawa et al. 2017; Cridland et al. 2019; Powell et al. 2019) and exploring complex configurations such as warped or circumbinary disks (Foucart & Lai 2014; Zanazzi & Lai 2018; Martin & Lubow 2019; Martin et al. 2019) and disks in infalling clouds (Harsono et al. 2015; Tanaka et al. 2017). (The reference list in this paragraph is by no means exhaustive; it is merely intended to provide the reader with a few examples of each type of disk model mentioned.) Here we define a simple but serviceable analytical model of an axisymmetric disk around a single star. In the next section, we will add extensions to our model for situations that do not meet our simplifying assumptions.

2.3.1 Empirical Constraints on the Disk Model

Observations of astrophysical disks provide insight into the conditions that our analytical disk model must describe. First, we compare disk lifetimes with dynamical times to justify decoupling the radial and vertical structure of our model disk. Using a census of disks in nearby star-forming regions of different ages, one can fit an exponential function to a plot of the fraction of protostars with disks as a function of age (Mamajek 2009):

$$f(t) = Ae^{-t/\tau_l} + C, \tag{2.2}$$

where $f(t)$ is the fraction of stars with disks, t is the age, and A and C are fitted constants. τ_l is then the mean lifetime of disks identified at a specific infrared wavelength. For infrared excesses at 2–4 μm, the mean lifetime is usually 2–3 Myr (million years; Haisch et al. 2001; Hernández et al. 2007, 2008; Mamajek 2009; Fedele et al. 2010; Murphy et al. 2013). In a meta-analysis of 2340 young stars and brown dwarfs in nearby star-forming regions, Ribas et al. (2014) found that the mean lifetime of disks identified at 3.4–4.6 μm (corresponding to dust at 0.01–1 au from the host star or brown dwarf) is 2.3 Myr. Moving to 22–24 μm, which traces dust between 0.3 and 60 au from the objects in their sample, Ribas et al. found a mean lifetime of 5.5 Myr for primordial disks, which have leftover dust and (presumably) gas from the molecular cloud core.[4] Figure 2.6 shows the fraction of stars with disks for clusters with estimated ages <15Myr, broken down by infrared excess wavelength, compiled by Ribas et al. (2014). The mean lifetimes of astrophysical disks match well with estimates of the solar nebula lifetime: >4 Myr based on the lack of solar wind particles, which propagate freely through today's heliosphere but could not penetrate deeply into the solar nebula, embedded in unbrecciated chondritic meteorites (Alexander 2018, Section 4.2).

How does the disk lifetime compare with its longest dynamical time, given by the angular speed at the outer edge? The Keplerian angular speed is

$$\Omega_K = \left(\frac{GM_*}{R^3}\right)^{1/2}, \tag{2.3}$$

where G is the gravitational constant. For a solar-mass protostar with a disk radius of 100 au, the longest dynamical time is $1/\Omega_K = 160$ years. Because disk lifetimes are long compared with dynamical times, we may use a static (nonevolving) model for initial or "background" conditions when we explore rapid processes such as grain growth and turbulence. Then, because most star-forming clouds lack strong magnetic pressure (e.g., Goodman et al. 1993; Caselli et al. 2002) and are cold enough ($T \sim 10$ K) to provide little thermal pressure support against collapse in the direction parallel to the angular momentum vector (e.g., Willacy & Millar 1998), the collapsing cloud forms a vertically thin disk (see the inset disk image in the bottom panel of Figure 2.2), with density scale height $H \ll R$. Because large-scale hydrodynamic motion would destroy the long-lived thin disk morphology, we assume hydrostatic equilibrium for our analytical model (see box 2.3 on hydrostatic

[4] A molecular cloud core is an immediate star formation site; with size ~0.02–0.1 pc (depending on the "crowdedness" of the star-forming cloud), the core is the densest structure on the molecular cloud scale, is gravitationally bound, and will collapse to form a single star or multiple system (Blitz & Williams 1999; Lada et al. 2004).

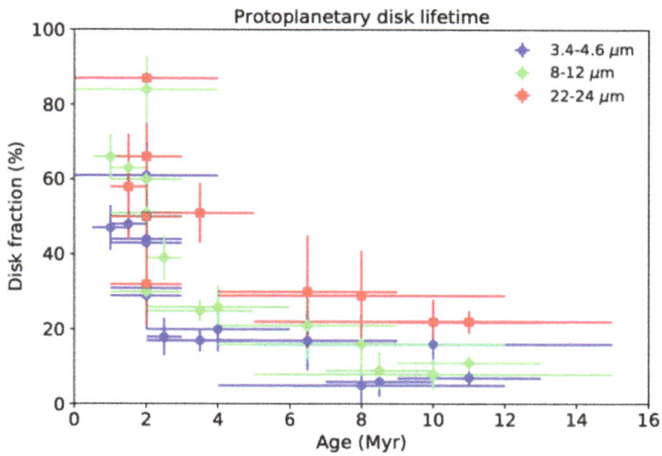

Figure 2.6. Disk fraction as a function of age for nearby star-forming regions and moving groups, broken down by the wavelength of infrared excess. For primordial disks detected at 22–24 μm, the mean lifetime is 5.5 Myr. Data tabulated by Ribas et al. (2014).

equilibrium). Short-timescale, small-scale (magneto)hydrodynamic processes can usually be modeled as perturbations on hydrostatic equilibrium. Finally, the disk's vertical thinness coupled with rotational support in the radial direction ensures that the radial transport time of a parcel of gas $R/v_R \sim \tau_l$ (where v_R is the radial velocity) is much longer than the vertical transport time $H/c_s \sim 1/\Omega_K$ (c_s is the sound speed; we will derive this equation in Section 2.3.2). We use this fact to decouple the radial and vertical directions and treat the radial motion of gas as if it takes place only at the midplane (Pringle 1981). Rewriting the 2D (R, z) structure in 1+1D allows us to solve partial differential equations that describe vertical disk structure as if they were ordinary differential equations: $\partial/\partial z \rightarrow d/dz$.

Since the vertical component of stellar gravity g_z concentrates most of the disk mass near the midplane (see the dark lane in the "flying saucer" disk in the inset of Figure 2.2), it is often useful to describe the disk in terms of mass per unit area or surface density $\Sigma(R)$. A sensible prescription for $\Sigma(R)$ allows us to find volume density:

$$\Sigma(R) = \int_{-\infty}^{\infty} \rho(R, z)dz. \tag{2.4}$$

Non model-dependent measurements of $\Sigma(R)$ in astrophysical disks are rare because models with prescribed surface density profiles are often used to interpret observations (e.g., Andrews et al. 2009), but order-of-magnitude estimates are possible when one knows the total disk mass M_d and the outer radius of the disk $R_{\rm out}$. Unfortunately, a multitude of factors complicate measurements of those two most basic quantities: the main mass component, H_2, has no permanent dipole moment and therefore very weak rotational emission lines, rendering it virtually unobservable

(Bitner et al. 2008) except in fluorescence (see the inset box 2.1); CO gas, which is abundant and chemically stable in molecular clouds, has complex chemistry in disks that limits its use as a mass tracer (e.g., Miotello et al. 2017; Yu et al. 2017a); observations of continuum emission from dust lose sensitivity to particles with radii $>3\lambda$, where λ is the observing wavelength (Draine 2006); and the gas and dust components of disks have different radial extents (e.g., Birnstiel & Andrews 2014). The net effect is that all disk-mass measurements from CO gas and single-wavelength dust continuum observations are lower limits, a viewpoint shared by Dunham et al. (2014), Kratter & Lodato (2016), Galván-Madrid et al. (2018), and Nixon et al. (2018).

Two types of observations offer promising, less biased information about M_d and $\Sigma(R)$. One is the detection of HD (hydrogen deuteride) fundamental rotational emission ($J = 1 \rightarrow 0$, where J is the rotational quantum number) in the far-infrared. Because HD stays in the gas phase and the D/H ratio is well characterized in the local ~100 pc, HD is a good mass tracer provided the density-weighted average gas temperature is known (Trapman et al. 2017). Herschel observations by Bergin et al. (2013) showed that the disk surrounding TW Hya, which is already 10 ± 3 Myr old (Bell et al. 2015), has mass $M_d \geqslant 0.05M_\odot$, or $M_d/M_* \geqslant 0.06$ (stellar mass $0.8M_\odot$; Qi et al. 2013). (We discuss the significance of disk/star mass ratios further in Section 2.4.7.) The radius of the disk, traced by millimeter-size grains and HD gas with $T > 20$ K, is $R_{\rm out} \sim 60$au (Andrews et al. 2012; Schwarz et al. 2016). Herschel HD observations of disks surrounding DM Tau and GM Aur indicate $M_d/M_* = 0.015$–0.072 and 0.022–0.19, respectively (McClure et al. 2016).

Another way to reconstruct the surface density profile of a disk, pioneered by Powell et al. (2017), is to resolve the outermost distances R occupied by dust grains of different sizes. Because dust grains lack the radial pressure support that forces gas to orbit at slightly sub-Keplerian speed, they experience an angular momentum-reducing headwind and drift inward (Weidenschilling 1977a). The inspiral speed depends on both gas density and grain size, with larger particles concentrated closest to the star (e.g., Birnstiel et al. 2010, 2012; Birnstiel & Andrews 2014). Observations at a given wavelength are most sensitive to grains of radius $\lambda/2\pi$, so mapping the outer radius of the disk at several (continuum) wavelengths gives empirical benchmarks of $\rho(R, z = 0)$, which can be converted to $\Sigma(R)$ using $\rho(R, z = 0) \approx \Sigma(R)/2H(R)$ for Gaussian $\rho(z)$. Powell et al. (2019) used surface density profiles of the form

$$\Sigma(R) = \Sigma_0\left(\frac{R}{R_0}\right)^{-\xi} \exp\left[-\left(\frac{R}{R_0}\right)^{2-\xi}\right], \tag{2.5}$$

shallow power laws with steep exponential cutoffs that give well-defined values of $R_{\rm out}$ in observations. (For more on Equation (2.5), see Section 2.3.3.) They found $M_d/M_* = 0.09$–0.27 for a sample of seven disks and suggested that even HD observations give lower limits to disk mass because most of the gas is too cold for rotational emission, a caveat discussed by Bergin et al. (2013) and McClure et al. (2016). The power-law index is usually $\xi \sim 1$ when measured from dust continuum emission but is sometimes shallower when measured from gaseous CO (Guilloteau et al. 2011; Rosenfeld et al. 2012; Guidi et al. 2016; Huang et al. 2016; Tazzari et al. 2016; Williams & McPartland 2016). (Both measurement methods assume optically thin emission, which may not be physically realistic). Following Powell et al. (2019), we adopt $\xi = 1$ for the model-based calculations presented in this chapter. Because the exponential tail in Equation (2.5) accounts for very little of the total disk mass, it is acceptable to leave it out unless one is particularly interested in edge effects. Based on the limited sample of protoplanetary disks with masses measured either from HD or dust lines, we suggest that physically realistic disk models should have $M_d/M_* \gtrsim 0.02$, and that the often-used mmsn model (see the inset below) does not apply to planet formation.

2.2 Minimum-mass solar nebula

The mmsn is constructed by spreading the solid mass contained in the solar system planets into a distribution with $\Sigma \propto R^{-1.5}$ and augmenting it with the hydrogen and helium that were present in the solar nebula, but that the terrestrial planets did not capture (Weidenschilling 1977b; Hayashi 1981). While mmsn-based disk models have led to important insights in disk physics and may provide useful descriptions of the mass distribution once planet formation is nearly complete (e.g., Stage III; see Section 2.2), they do not describe realistic planet-forming conditions. Using the mmsn as a

Figure 2.7. An ALMA image of dust emission from the disk surrounding HL Tau reveals gaps caused by the depletion of large dust grains. The dust may have already been incorporated into planets (Dipierro et al. 2015; Tamayo et al. 2015) or grown to sizes too large to be observed (i.e., in regions where molecules are freezing onto grains, Zhang et al. 2015). For more on ALMA, see the inset box 2.4 Image credit: ALMA (NRAO/ESO/NAOJ), C. Brogan, and B. Saxton (NRAO/AUI/NSF).

planet-forming model requires assuming 100% efficient conversion of solid mass into planets, which we know was not true for the solar nebula given the fact that the Oort cloud—composed of planetesimals ejected during giant-planet formation—has 4–50$M_\oplus$ (Earth masses; Weissman 1990; Neslušan 2007). Other mass sinks are the migration of planets into the host star (see review by Kley & Nelson 2012), planet ejection (e.g., Ford et al. 2005), and radial drift of pebbles (Birnstiel et al. 2012; Lenz et al. 2020; observational evidence presented by Facchini et al. 2019). The models of Pinte et al. (2016) suggest that forming the gaps in the dust distribution surrounding HL Tau (ALMA Partnership et al. 2015; Figure 2.7) required removing up to 40$M_\oplus$ ($1.2 \times 10^{-4} M_\odot$) of dust if the disk started with a smooth radial surface density distribution. Even if the dark rings are produced by grain growth rather than surface density depletion (Zhang et al. 2015), or if the bright rings are produced by sintering of icy aggregates (Okuzumi & Momose 2016; see Section 2.4.2), the leftover dust disk around HL Tau still has estimated mass $5 \times 10^{-4} M_\odot$—five times the solid mass in the mmsn. Models of planet formation should account for the inefficiency of converting mass into dynamically stable planets. Instead of the mmsn, we advocate adopting the "maximum-mass solar nebula," styled MMSN by Nixon et al. (2018) and referred to in this book as MAXSN. From measurements of disk sizes and star accretion rates, Isella et al. (2009) found that the disks orbiting their sample of 14 low- and intermediate-mass protostars formed with masses of 0.05–0.4$M_\odot$, suggesting that the MAXSN is a likely outcome of molecular cloud collapse. The disk model presented in this chapter, which will be used throughout the book, is based on the MAXSN.

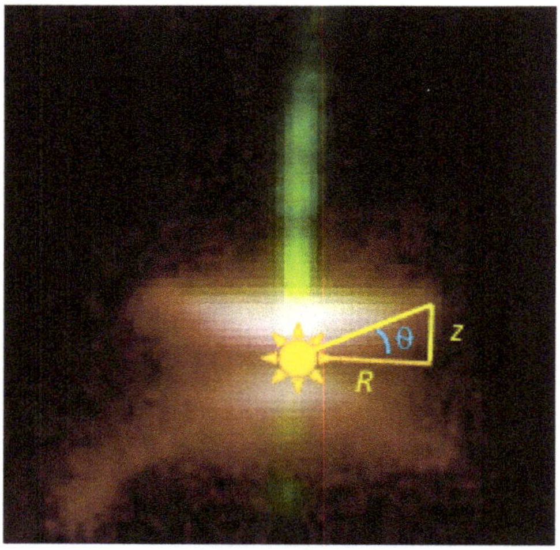

Figure 2.8. HST observation of the flaring disk surrounding HH 30. The jet, rendered green and detected in line emission, carries angular momentum away from the disk (Burrows et al. 1996; see Section 2.4.6 for more on angular momentum transport.) Overlaid is the coordinate system we will use to describe the disk: R is the radial distance from the star in cylindrical coordinates, $r = \sqrt{R^2 + z^2}$ is the actual distance from the star, z is the height above the midplane, and θ is the small angle between vectors $\vec{R}$ and $\vec{r}$. For an axisymmetric, midplane-symmetric disk, we need only model the top half of the (R, z) cross section on one side of the star. Image credit: Chris Burrows (STScI), the WFPC2 Science Team, and NASA; cartoon coordinate system added by author.

2.3.2 Static Model equations

Here we present the equations for our thin, nonevolving disk with $\Sigma \propto R^{-1}$. We simplify the problem by assuming that the disk is symmetric about its rotation axis and midplane, so that $f(R, z) = f(R, -z)$; this way we are describing only a single quadrant of a 1+1D disk (see the coordinate system pictured in Figure 2.8). For $M_d/M_* \gtrsim 0.1$, the disk's self-gravity causes the growth of nonaxisymmetric structures (Kratter & Lodato 2016). We restrict our analytical model to disks with $M_d/M_* \leqslant 0.1$, where disk self-gravity may be ignored, but consider modifications to the model for more massive disks in Section 2.4.7. We assume the velocity field is set by the turbulent viscosity, which also drives accretion; see Section 2.3.3 for more information. (Note that there is no longer a consensus that turbulence is the main mass and angular momentum transport mechanism. Magnetic disk winds may drive accretion and dissipate disks, forcing the transition from Stage II to Stage III, in which case the viscosity prescription presented here does not completely describe disk evolution or velocity fields. See Chambers 2019, for an analytical model combining viscosity and magnetic winds).

2.3 Hydrostatic equilibrium

Fluid in hydrostatic equilibrium has no vertical motion. Figure 2.9 shows the forces on a fluid element at height z in a plane-parallel atmosphere: pressure from gas above the fluid element gives the downward force $P(z + dz)\,dx\,dy$, pressure from gas below the fluid element gives the upward force $P(z)\,dx\,dy$, and the element's own weight is $W = \rho\, g_z\, dx\, dy\, dz$ (where ρ is the density and g_z is the vertical component of the gravitational acceleration). Balancing the forces, we find

$$\begin{aligned} &P(z)\,dx\,dy - P(z + dz)\,dx\,dy - \rho\, g_z\, dx\, dy\, dz \\ &= -\,dP\,dx\,dy - \rho\, g_z\, dx\, dy\, dz \\ &= 0. \end{aligned} \tag{2.6}$$

We rearrange Equation (2.6) to get Equation (2.7), which describes the pressure gradient.

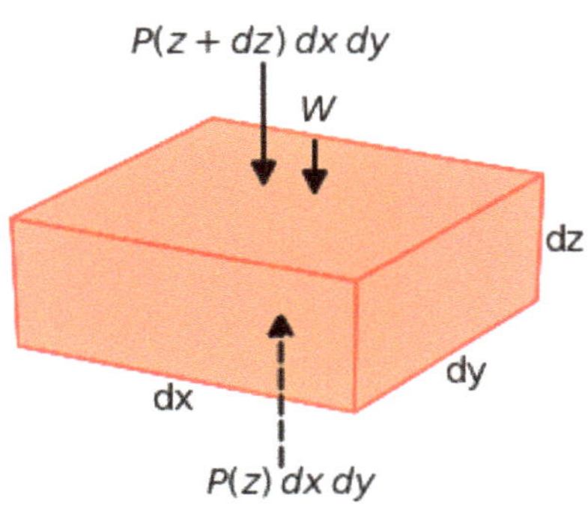

Figure 2.9. Forces on a parcel of fluid in hydrostatic equilibrium.

In hydrostatic equilibrium, the vertical pressure gradient is

$$\frac{\partial P}{\partial z} = -\rho g_z, \tag{2.7}$$

where g_z is the vertical component of the star gravity (see the derivation of Equation (2.7) in the "Hydrostatic equilibrium" inset). Because the disk is thin, $r^2 \simeq R^2$ (where r is the true distance from the star) and the small-angle approximation applies. The vertical component of the stellar gravity is then

$$g_z = \frac{GM_*}{r^2 + z^2} \sin\theta \simeq \frac{GM_* z}{R^3} \simeq \Omega_K^2 z. \tag{2.8}$$

We will make the simplifying and sometimes reasonable assumption that the disk is vertically isothermal: $T(R, z) = T(R, z = 0)$, except for a heated surface layer where starlight is absorbed (Chiang & Goldreich 1997). The equation of state is

$$P = c_s^2 \rho, \tag{2.9}$$

where

$$c_s^2 = \frac{R_g}{\mu} T, \tag{2.10}$$

$R_g = 8.31 \times 10^7$ erg K^{-1} mol^{-1} is the ideal gas constant, and $\mu = 2.33$ g mol^{-1} is the mean molar mass of a solar-composition H_2/He mixture. Solving Equation (2.9) for ρ and plugging into Equation (2.7) along with our approximation for g_z, we find

$$\frac{\partial P}{\partial z} = \left(\frac{P}{c_s^2}\right)\left(\frac{GM_* z}{R^3}\right), \tag{2.11}$$

which is separable and integrable within the isothermal assumption, as $c_s^2(z)=$ constant. The vertical density profile is Gaussian:

$$\rho(R, z) = \rho_0(R) \exp\left[-\frac{1}{2}\frac{\Omega_K^2}{c_s^2} z^2\right] = \rho_0(R) \exp\left[-\frac{1}{2}\frac{z^2}{H^2}\right], \tag{2.12}$$

where

$$H = c_s/\Omega_K \tag{2.13}$$

is the scale height and $\rho_0(R)$ is the density at the midplane. Using Equation (2.4), we find

$$\rho_0(R) = \frac{\Sigma(R)}{\sqrt{2\pi} H(R)}. \tag{2.14}$$

To find $T(R)$, we follow the model of Chiang & Goldreich (1997). We assume the disk is heated passively by starlight with negligible internal heating (Hirose & Turner 2011). Light from the pre-main-sequence star, which peaks in the visible or near-infrared, is absorbed at the disk surface by dust grains with sizes ~0.1–2 μm (e.g., van Boekel et al. 2005) and reradiated at longer wavelengths. Because the disk's optical depth is smallest in the vertical direction, we assume reradiated light mainly propagates vertically. We caution that this 1+1D approach to radiative transfer may underestimate the midplane temperature: in reality, multiple rounds of absorption and reradiation between the disk surface and midplane keep reddening the peak wavelength of dust emission until the light can propagate radially. Furthermore, the purely vertical assumption ignores the fact that the disk may have a puffed-up inner rim that casts a shadow on the outer disk, blocking starlight absorption at some radii (e.g., Natta et al. 2001; see Section 2.4).

Consider the small angle φ at which starlight hits the disk surface at height H_S, where $H_S \ll R$. (Note: φ is often denoted α in the literature; here we use φ to distinguish it from the turbulent efficiency parameter α that appears in Section 2.3.3.) From Equations (2.10) and (2.13), $H \propto R^{3/2}T^{1/2}$. If H_S/H is approximately constant (e.g., Chiang & Goldreich 1997; Chiang et al. 2001), we would need $T \propto R^{-1}$ to make a "wedge" disk with constant φ for all R. But the equilibrium temperature at the disk surface exposed to starlight is given by

$$T^4_{\text{eq, surf}} = \frac{L_*}{4\pi\sigma_{\text{SB}}R^2}\sin\varphi, \tag{2.15}$$

where L_* is the star luminosity, σ_{SB} is the Stefan–Boltzmann constant, and $\sin\varphi$ picks off the component of the incident flux $L_*/(4\pi R^2)$ perpendicular to the disk surface. (For now we will ignore the fact that the disk surface superheats because the dust grains lofted above the midplane are too small to radiate efficiently at mid-IR and longer wavelengths.) Still assuming the disk is isothermal, Equation (2.15) gives $T(R) \propto R^{-1/2}$, a shallower temperature decline than what is required for a wedge. We therefore expect the disk to flare: H_S/R, H/R, and φ should all increase with R (Kenyon & Hartmann 1987). Figure 2.8 shows the flared disk surrounding HH 30 (Burrows et al. 1996). We can see the flaring $H_S(R)$ in HH 30 by following the boundary between the dark interior and the illuminated, optically thin upper layers, where starlight reflecting off small grains can travel freely.

To find H/R and $T(R)$, we need to know $\sin\varphi$. Examining Figure 2.10, which shows the geometry of a vertically stretched, flared disk, we see that the "flare line," which is tangent to the disk surface at (R, H_S) and has slope dH_S/dR, intersects the disk midplane at radius R_f such that $R - R_f = H_S/(dH_S/dR)$. For small angles $\theta \simeq H_S/R$ and $\eta \simeq H_S/(R - R_f)$, we get $\beta \simeq \pi/2 - H_S/R$ and $\zeta \simeq \pi/2 - H_S/(R - R_f)$, so that

$$\sin\varphi \simeq \varphi = \beta - \zeta \simeq H_S/(R - R_f) - H_S/R \simeq \frac{dH_S}{dR} - \frac{H_S}{R}. \tag{2.16}$$

Again assuming approximately constant $H_S/H = C$ and substituting

$$\frac{dH}{dR} - \frac{H}{R} = R\frac{d}{dR}\left[\frac{H}{R}\right] \tag{2.17}$$

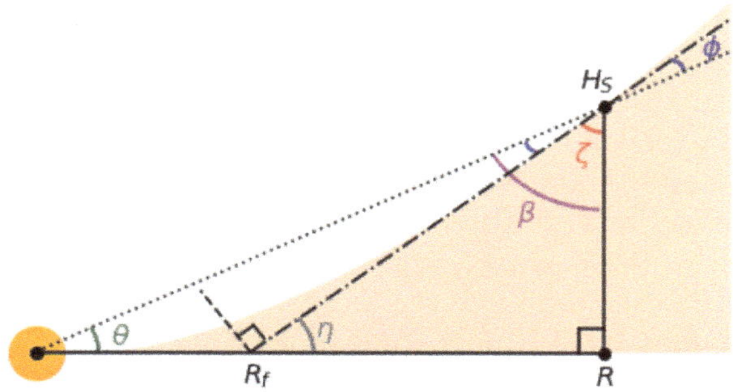

Figure 2.10. The geometry we will use to find the flaring angle φ (blue) and the surface height $H_S(R)$. The dashed–dotted "flare line" is tangent to the curved disk surface at point (R, H_S). The vertical scale is stretched to emphasize the flaring.

into Equations (2.16) and then (2.15), we find

$$\sigma_{SB} T^4 = C \frac{L_*}{4\pi R^2} R \frac{d}{dR}\left[\frac{H}{R}\right]. \tag{2.18}$$

With Equations (2.10), (2.13), and (2.3) and some algebra, we can rewrite Equation (2.18) as a differential equation in H/R:

$$\frac{d}{dR}\left[\frac{H}{R}\right] = \left(\frac{\mu G M_*}{R_g}\right)^4 \frac{4\pi\sigma_{SB}}{CL_*}\left(\frac{H}{R}\right)^8 R^{-3}. \tag{2.19}$$

Equation (2.19) has a power-law solution

$$\frac{H}{R} = \left(\frac{CL_*}{14\pi\sigma_{SB}}\right)^{1/7}\left(\frac{R_g}{GM_*\mu}\right)^{4/7} R^{2/7}, \tag{2.20}$$

where all quantities are in cgs units (as in Equation (2.10)). The scale height varies as $H \propto R^{9/7}$; the disk does indeed flare. The temperature profile is

$$T = \left(\frac{CL_*}{14\pi\sigma_{SB}}\right)^{2/7}\left(\frac{R_g}{GM_*\mu}\right)^{1/7} R^{-3/7}, \tag{2.21}$$

again in cgs units. Empirically measured values of C are between 1 and 5 (Chiang et al. 2001); we expect higher values of C from disks with abundant small dust grains that can easily be lofted above the midplane by turbulence (e.g., Dubrulle et al. 1995). Figure 2.11 shows the interior densities and temperatures of the inner 20 au of a disk with $M_d/M_* = 0.1$ (where $R_{out} = 100$ au) orbiting a solar-mass star with age 1.7 Myr. From the evolutionary tracks of Landin et al. (2010), the star luminosity is 1 $4L_\odot$. Here we use $C = 4$. The scale height and approximate height of starlight absorption are marked. Equation (2.21) and closely related radial power laws have provided temperature profiles for investigations of planet-forming vortices (Klahr & Bodenheimer 2003; Richard et al. 2016), simulations of magnetorotational

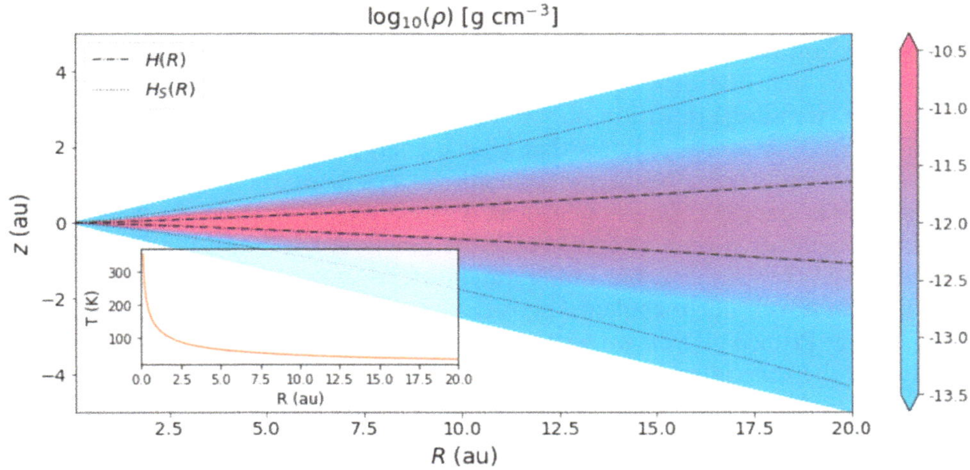

Figure 2.11. The density structure of the inner 20 au of a model disk with $M_d = 0.1M_*$, $M_* = M_\odot$, and $R_{\text{out}} = 100$ au. The young star, age 1.7 Myr, has luminosity $1.4L_\odot$. The black dashed–dotted line marks the scale height H and the dotted line marks $H_S = 4H$, approximately the height above the midplane at which starlight is absorbed. Both axes have the same scale to correctly render the aspect ratio. The inset shows the temperature distribution.

instability-driven accretion (Bai 2011), and measurements of carbon monoxide mass in nearby disks (Piétu et al. 2007). However, they are not adequate for many applications, including (but not limited to) molecular spectroscopy, chemical modeling, and planet trapping. In reality, disks are warmer near their irradiated surfaces than their shielded midplanes (Section 2.4.4). For example, observations with the Spitzer Space Telescope revealed rovibrational emission from gaseous H_2O in warm molecular layers above the midplanes of 22 disks (Pontoppidan et al. 2010), layers that do not exist in our analytical disk model. Planet embryos that might otherwise migrate into the star (see Chapter 7) can become trapped at a transition between disk heating primarily by turbulent viscosity, which this model does not include, and by starlight (Hasegawa & Pudritz 2011). The temperature difference of 19 K versus 20 K affects whether or not N_2H^+, a tracer of ionization rate (Walsh et al. 2012) and nitrogen abundance (Schwarz & Bergin 2014) that is abundant only when CO gas is heavily depleted (Qi et al. 2013), is observable. We describe how to add viscous heating and construct more sophisticated models of stellar heating in Section 2.4.

2.3.3 Viscous Diffusion, Long-Term Evolution, and Velocity Field

For models of the disk's long-term evolution, we require a function $\Sigma(R, t)$; then, time snapshots $\rho(R, z, t)$, $T(R, z, t)$, $P(R, z, t)$, and (optionally) the velocity field can be computed using the same physics as in the static model. We follow Lynden-Bell & Pringle (1974) and assume that turbulent viscosity transfers angular momentum outward, allowing most of the mass to accrete onto the star. (Turbulent viscosity also heats the disk; see Section 2.4.5.) Younger Stage 0/I disks typically have smaller outer radii than older Stage II disks, consistent with viscous spreading (Najita & Bergin 2018). Because most of the mass is concentrated near the midplane, we treat the disk as if it were infinitely thin; retaining the axisymmetric assumption, we model diffusion only in the radial dimension. We caution that our turbulent viscosity model requires that kinetic energy extracted from the Keplerian shear must cascade from the largest eddies down to a dissipation scale of size $\ll H$—in other words, energy dissipation is local. There are large-scale methods of angular momentum transport such as spiral structure (Section 2.4.7) and the Rossby wave instability (RWI; Section 2.4.6; Finite amplitude-initiated, vortex-forming instabilities) that do not fit our turbulent viscosity prescription.

Following Pringle (1981), we write an equation of continuity for the mass in an annulus: in the time interval Δt,

$$\begin{aligned}\Delta M(R, t) &= 2\pi R\, \Delta R\, \Delta\Sigma(R, t)\\ &= 2\pi[R + \Delta R]\Sigma(R + \Delta R)v_R(R + \Delta R)\Delta t - 2\pi R\Sigma(R)v_R(R)\Delta t.\end{aligned} \tag{2.22}$$

In Equation (2.22), ΔM is the mass change in the annulus between R and $R + \Delta R$ and mass moves a distance $v_R\Delta t$, where v_R is the gas radial velocity. In the limit $\Delta R \to 0$ and $\Delta t \to 0$, we find

$$R\frac{\partial\Sigma}{\partial t} = \frac{\partial}{\partial R}(R\Sigma v_R). \tag{2.23}$$

Next we move to the momentum equation. In a plane-parallel flow (Figure 2.12), viscosity—the friction between adjacent layers of gas moving with different speeds—provides shear stress (force per unit area) that is proportional to the velocity gradient: $F_{\rm visc}/A = |\rho\nu\ du/dy|$, where ν is the kinematic viscosity (units cm^2 s^{-1}), u is the horizontal velocity, A is the area of the shearing layers, and y is the vertical coordinate. In our thin disk, each annulus exerts a force per unit *length* on its neighbor:

$$\frac{F_{\rm visc}}{l} = \Sigma\nu R\frac{\partial\Omega_K}{\partial R}, \tag{2.24}$$

where $R(\partial\Omega_K/\partial R)$ is the shear, analogous to du/dy in the plane-parallel case. To find the angular momentum change $\Delta J(R)$ in the time unit Δt, we need to account for (a) the advection of angular momentum into and out of the annulus, and (b) the net torque from both neighboring rings of gas. Given $J(R) = 2\pi R\Delta R\Sigma R^2\Omega_K$,

$$\begin{aligned}\frac{\Delta J}{\Delta t} &= -\,2\pi R\, v_R(R)\,\Sigma(R)\ \cdot\ R^2\Omega_K(R)\\ &\quad + 2\pi[R + \Delta R]\, v_R(R + \Delta R)\,\Sigma(R + \Delta R)\cdot[R + \Delta R]^2\Omega_K(R + \Delta R)\\ &\quad + \tau(R + \Delta R) - \tau(R),\end{aligned} \tag{2.25}$$

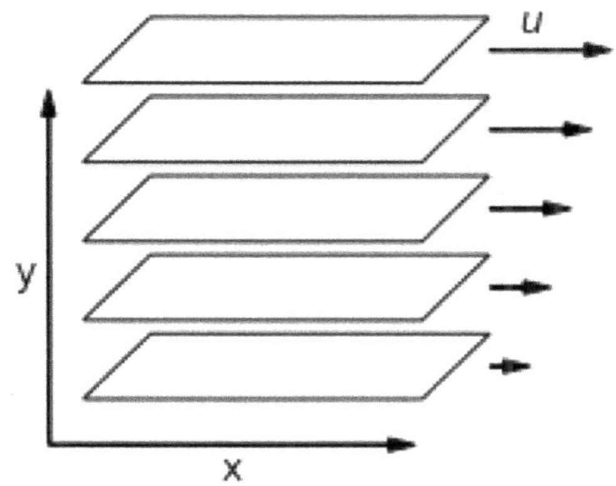

Figure 2.12. A plane-parallel viscous flow.

where τ is the torque an annulus exerts on its exterior neighbor. Using $\vec{\tau} = \vec{R} \times \vec{F}_{\rm visc}$ and $l = 2\pi R$, we find

$$\frac{\partial \tau}{\partial R} = \frac{\partial}{\partial R}\left(2\pi R^3 \Sigma \nu \frac{\partial \Omega_K}{\partial R}\right). \tag{2.26}$$

To find the momentum conservation equation, we substitute Equation (2.26) into Equation (2.25) and take the limit as ΔR, $\Delta t \rightarrow 0$:

$$R\frac{\partial}{\partial t}(\Sigma R^2 \Omega_K) + \frac{\partial}{\partial R}(\Sigma R^3 v_R \Omega_K) = \frac{\partial}{\partial R}\left(\Sigma R^3 \nu \frac{\partial \Omega_K}{\partial R}\right). \tag{2.27}$$

Combining Equation (2.23) with Equation (2.27) to eliminate v_R gives the diffusion equation

$$\frac{\partial \Sigma}{\partial t} = \frac{3}{R}\frac{\partial}{\partial R}\left[R^{1/2}\frac{\partial}{\partial R}(\Sigma \nu R^{1/2})\right]. \tag{2.28}$$

One may simulate the evolution of the (R, z) cross section by advancing the surface density profile forward in time using Equation (2.28), then computing independent $T(z)$ and $\rho(z)$ profiles at each radial gridpoint and timestep (or a subset of the timesteps) using either a radiative transfer model (see Section 2.4) or Equations (2.7), (2.14), and (2.21). This method was suggested by Pringle (1981) and used by, e.g., Dodson-Robinson et al. (2009) and Landry et al. (2013). Equation (2.28) has similar properties to the parabolic heat equation and can be solved numerically by the same algorithms—either BTCS (backward in time, centered in space) differentiation, which is unconditionally stable but with only first-order accuracy in time and second order in space, or the more accurate Crank-Nicholson method (Crank & Nicholson 1947). The disk diffusion equation can be solved analytically if $\nu(R)$ is a time-independent power law in radius (Pringle 1981). Equation (2.5), the power-law surface density profile with exponential taper, is a self-similar solution to Equation (2.28) (Lynden-Bell & Pringle 1974; Hartmann et al. 1998). We capture the viscous evolution of a disk with vertically nonuniform $\nu(z)$ (i.e., realistic turbulence) by substituting a mass-weighted vertical average viscosity $\bar{\nu}$ in for ν in Equation (2.28):

$$\bar{\nu} = \frac{2}{\Sigma} \int_{z=0}^{\infty} \nu \rho dz. \tag{2.29}$$

Gas velocities are required for modeling grain growth and molecular emission lines. A description of the velocity field follows from the turbulent viscosity ν. Following the mixing-length theory (Prandtl 1925), which is also used to parameterize convection in stellar interiors (e.g., Spiegel 1971) but which Prandtl himself described as "only a rough approximation" (Bradshaw 1974), we write the turbulent viscosity (also called eddy viscosity) as the product of a length scale ℓ and a characteristic turbulent velocity v_t: $\nu = \ell v_t$. Going back to Newton's law of viscosity (Equation (2.24) for disks), we assume the velocity scale v_t depends linearly on the length scale ℓ and shear rate $R(\partial\Omega_K/\partial R)$, so $\nu \propto |\ell^2 R\ (\partial\Omega_K/\partial R)| \implies \nu \propto (3/2)\ell^2\Omega_K$. Following Shakura & Sunyaev 1973; we let our velocity scale be the sound speed c_s multiplied by some constant that describes the efficiency of energy transfer from shear into turbulence. The associated length scale, which is also the size of the largest eddy, is $\ell = (\text{constant}) \times c_s/\Omega_K = (\text{constant}) \times H$ (our constant has subsumed the factor of 2/3). We now have our viscosity prescription:

$$\nu = \alpha c_s H, \tag{2.30}$$

where α is the dimensionless constant. We find the characteristic turbulent speed v_t by stipulating that the largest allowable eddy turns over on the orbital timescale $1/\Omega_K$; because the eddy turnover time is ℓ/v_t, we derive

$$\begin{aligned} \nu &= \frac{v_t^2}{\Omega_K} \\ v_t &= \sqrt{\alpha}\, c_s. \end{aligned} \tag{2.31}$$

Although models with isotropic turbulence and a Maxwellian speed distribution around v_t can lead to useful constraints on disk structure (e.g., Flaherty et al. 2015, 2017; Yu et al. 2017b), Keplerian shear is likely to extend the eddies in the azimuthal direction, so that $\ell_\phi > \ell_R, \ell_z$ (e.g., Flock et al. 2012; Romanova et al. 2012). (The vertical shear instability, VSI, however, creates tall, columnar vortices; see Section 2.4.6.)

It is important to note that the α viscosity prescription, the $\xi = 1$ surface density power law, and the Chiang & Goldreich (1997) temperature profile with $T \propto R^{-3/7}$ (Equation (2.21)) do not combine to form a steady-state disk, in which the mass flow rate $\dot{M}$ through each annulus is the same everywhere. With

$$\begin{aligned} v_R &= -\frac{3}{\Sigma R^{1/2}} \frac{\partial}{\partial R}(\Sigma \nu R^{1/2}), \\ \dot{M} &= 2\pi R \Sigma v_R, \end{aligned} \tag{2.32}$$

one achieves steady state by either retaining $\xi = 1$ and setting $T \propto R^{-1/2}$, or retaining $T \propto R^{-3/7}$ and setting $\xi = 15/14$. If one were to use Equation (2.28) to simulate the diffusion of a disk with $\Sigma \propto R^{-1}$ at $t = 0$, while setting $T \propto R^{-3/7}$, the surface density profile would evolve. Although it is not necessary (or even possible, given no exterior mass source) for a disk to be in steady state, a truly self-consistent simulation of disk diffusion requires including accretion heating and iteratively solving for the temperature, scale height, and viscosity (see Section 2.4.5).

2.4 Atacama Large Millimeter Array

Gas and dust near the midplane of a T-Tauri protoplanetary disk has characteristic temperature $T \sim 20$ K (e.g., D'Alessio et al. 1998), so is detectable only in the far-infrared or radio. The far-infrared is out of reach of ground-based telescopes because of atmospheric absorption, but the atmosphere is partially clear at radio wavelengths $\lambda \gtrsim 0.3$mm. To maximize transparency and minimize noise from the atmosphere, the Atacama Large Millimeter Array, or ALMA, was built on the Chajnantor plateau in the Atacama desert, where the precipitable water vapor is <1 mm. The 43-antenna interferometer, which has the resolution of a 16 km telescope in its most extended configuration, delivers high-resolution images of planet-forming gas and dust (Figure 2.13).

Figure 2.13. ALMA antennas seen in silhouette against the sunset over the Chajnantor Plateau in February 2018. Image credit: H. Calderón—ALMA (ESO/NRAO/NAOJ).

As with disk mass, observational constraints on α and v_t are limited and model dependent. Because turbulence is expected to be subsonic, very high signal-to-noise (S/N) observations that yield well-resolved molecular emission-line profiles from which the rotational, thermal, and turbulent components can be disentangled are required. Teague et al. (2016) show that turbulent speeds traced by CO, CS, and CN molecules in the TW Hya disk may range from zero to $\approx$130 m s^{-1}, with the highest speeds at the inner edge of the spatially resolved region. Measurement precision is limited by the absolute flux calibration of the ALMA interferometer, lack of knowledge of the spatial distribution of the observed molecules, and uncertainties in the disk's temperature distribution. The fastest inferred turbulent speeds are consistent with $v_t \sim 0.2$–$0.4c_s$, for $\alpha \sim 0.04$–0.16—higher than the proposed $\alpha = 0.01$ selected to match observed accretion rates of T-Tauri stars (Hartmann et al. 1998) and the new, lower estimates based on simulations (see review by Lyra & Umurhan 2019). However, a follow-up study of TW Hya by Flaherty et al. (2018) combining ALMA and SMA observations of CO could not separate the thermal and turbulent contributions to line profiles, instead placing an upper limit to v_t between $<0.04c_s$ and $<0.13c_s$, depending on model assumptions. The authors suggest that the disagreement with the Teague et al. (2016) results is due to different assumptions about the vertical temperature structure. Given that many types of turbulence are not isotropic (Section 2.4.6), observers may want to explore different ratios of $v_{t,r}$, $v_{t,\phi}$, and $v_{t,z}$ when interpreting emission-line profiles.

Flaherty et al. (2017) also detect only upper limits on turbulent speeds measured from CO and DCO^+ line profiles in the disk of HD 163296, this time <0.04–$0.06c_s$ ($\alpha \lesssim 3 \times 10^{-3}$). However, Yu et al. (2017b) show that measurements of turbulent speeds are degenerate with the spatial distribution of CO and the disk age: an old, turbulent disk mimics a young, laminar disk. Pinte et al. (2016) again inferred low turbulent speeds and low $\alpha = 10^{-4}$ from modeling the vertical dust distribution (turbulent eddies "kick" dust away from the midplane, where it tends to settle, with stronger turbulence giving faster kicks) in the disk surrounding HL Tau (ALMA Partnership et al. 2015). Despite uncertainties, low turbulent speeds inferred from both simulations and observations suggest that turbulence may be inadequate to drive observed accretion rates and are leading many disk theorists to explore large-scale magnetic winds (e.g., Bai 2013; Gressel et al. 2015; Hasegawa et al. 2017; Takahashi & Muto 2018) and spiral waves (e.g., Anthony & Carlberg 1988; Vorobyov & Basu 2006; Heinemann & Papaloizou 2009; Bae et al. 2016; Kuffmeier et al. 2018) as angular momentum transport mechanisms. For more on turbulence, disk winds, and angular momentum transport, see Section 2.4.6.

2.4 Model Extensions: Adding Complexity

The simple model given here is a good foundation for many planet formation-related projects, but real protoplanetary disks have features we have not yet described. Astrophysical disks often break the assumptions of axisymmetry and midplane symmetry or have gaps or bumps in the surface density distribution. Real disks also have a complex vertical structure with $H_S(R)$ not monotonically increasing, allowing parts of the inner disk to cast shadows on the outer disk. Here we present a list of extensions to our simple model.

2.4.1 Radial Pressure Support

So far we have considered only vertical pressure support, but gas pressure also acts in the radial direction. Each annulus of gas feels a slightly lower radial effective gravity $g_{r,\mathrm{eff}} < g_r$ because of the negative pressure gradient, leading to a sub-Keplerian angular speed $\Omega < \Omega_K$. The equation of motion for a pressure-supported disk is

$$\frac{v_\phi^2}{R} = \frac{GM_*}{R^2} + \frac{1}{\rho_0}\frac{\partial P}{\partial R} \tag{2.33}$$

for the disk midplane. We can recast our equation of motion as

$$v_\phi^2 = \frac{GM_*}{R} + c_s^2\frac{\partial \ln P}{\partial \ln R}, \tag{2.34}$$

where we have used Equation (2.9) to substitute for P/ρ in the second term of the right-hand side of Equation (2.34). Defining the true angular speed of the gas using $v_\phi = R\Omega$ and using Equation (2.13) to replace c_s^2, our final product is

$$\Omega^2 = \Omega_K^2\left[1 + \left(\frac{H}{R}\right)^2\frac{\partial \ln P}{\partial \ln R}\right]. \tag{2.35}$$

Because the departure of Ω from Ω_K is small, it is usually not necessary to consider the radial pressure gradient when modeling just the gas disk (Equations (2.7)–(2.28)). However, the sub-Keplerian Ω has strong effects on the dust dynamics (Weidenschilling 1977a; see Chapter 3). Radial pressure support is the simplest of the model extensions listed here.

2.4.2 Gaps and Holes

Astrophysical disks often have gaps, inner holes, and/or local maxima in their brightness distributions. Such features may result from planet–disk interactions (e.g., Rice et al. 2003; Najita et al. 2007; Dodson-Robinson & Salyk 2011; Zhu et al. 2011; Dong et al. 2015; Isella et al. 2016), grain growth at ice lines where the disk temperature drops below the condensation temperature of an abundant molecule (Zhang et al. 2015; Cieza et al. 2016; van der Marel et al. 2018; Okuzumi & Tazaki 2019), radial variations in magnetic flux (Bai & Stone 2014; Béthune et al. 2016; Suriano et al. 2019), or photoevaporation (Hollenbach et al. 1994; Alexander et al. 2006; Ercolano et al. 2008; Gorti & Hollenbach 2008; Owen et al. 2010). In fact, very few disks observed at the unprecedented angular resolution of the Atacama Large Millimeter Array (ALMA) lack radial substructure. (See the inset in box 2.4 for more on ALMA.) Figure 2.14 shows images of disks from the ALMA DSHARP survey (Andrews et al. 2018), all of which have complex $\Sigma(R)$ profiles. Because dust grains and pebbles drift toward gas pressure maxima (Chapter 3), overdense rings or the outer edges of gaps may aid giant-planet formation by concentrating solids.

Codes that can simulate the magnetohydrodynamic (MHD) effects of disk structure are `ZEUS`[5] (Stone & Norman 1992a, 1992b), `PLUTO`[6] (Mignone et al. 2007), `Athena`[7] (Stone et al. 2008), and PENCIL. Such simulations usually use

[5] https://www.astro.princeton.edu/~jstone/zeus.html

[6] http://plutocode.ph.unito.it/

[7] https://princetonuniversity.github.io/Athena-Cversion/

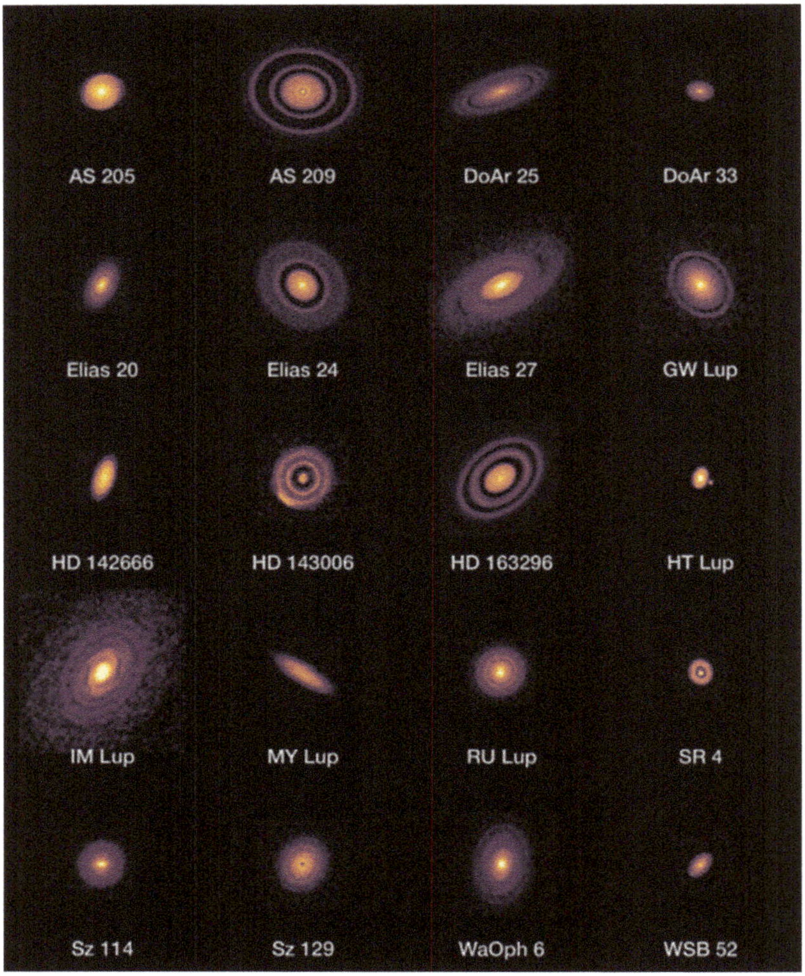

Figure 2.14. ALMA high-resolution images of nearby protoplanetary disks from the Disk Substructures at High Angular Resolution Project (DSHARP; Andrews et al. 2018). Note the complexity of the radial intensity profiles: almost all disks have gaps, bright rings, or spiral structure. Credit: ALMA (ESO/NAOJ/NRAO), S. Andrews et al.; N. Lira.

models like the one in Section 2.3.2 to provide the initial conditions or, in the case of isothermal disks, the nonevolving temperature profile. (For a temperature profile that includes a warm surface layer and accretion heating, see Sections 2.4.4 and 2.4.5.) If MHD-altered $\Sigma(R)$ or $\rho(R, z)$ is required as input to nondynamical computations such as molecular line radiative transfer or chemical reaction rates, the output of a dynamical simulation can be postprocessed.

The structures of gaps opened by planets can be calculated self-consistently from simulations of planet/disk interactions using a fluid dynamics code such as FARGO/FARGO-3D[8] (Masset 2000; Benítez-Llambay & Masset 2016) or PENCIL[9] (Brandenburg & Dobler 2002). In their study of the impact of photoevaporation on planet migration, Wise & Dodson-Robinson (2018) provide an example FARGO input parameter file (see their table 3). For gaps opened by (multiple) planets, the surface density profile can be calculated cheaply, requiring only ~200 orbits of simulation time for the outermost planet. Longer simulations are required to assess the dynamical stability of multiple planets in the disk. There is also the option of using semianalytical formulae fitted to numerical simulations to prescribe the surface density in a planet's gap. From Fung et al. (2014),

$$\begin{aligned} \Sigma_{\text{gap}} &\propto q^{-1}\alpha^{1.3}\left(\frac{H}{R}\right)^{6.1} \quad \text{brown–dwarf mass regime} \\ \Sigma_{\text{gap}} &\propto q^{-2.2}\alpha^{1.4}\left(\frac{H}{R}\right)^{6.6} \quad \text{Jupiter mass regime} \\ \Sigma_{\text{gap}} &\propto q^{-2}\alpha\left(\frac{H}{R}\right)^{5} \quad \text{Neptune mass regime,} \end{aligned} \tag{2.36}$$

where Σ_{gap} is the average surface density in the gap and q is the planet-to-star mass ratio. Crida et al. (2006) give a differential equation for $d\Sigma/dR$ that can be integrated numerically to fully specify $\Sigma(R)$ around the planet (see their equations (11), (13), and (14)). Gas in highly turbulent disks or hot disks rapidly diffuses into the tidal gap, so the surface density contrast between the gap and the disk around it depends on α and H/R as well as planet mass. See Figure 2.15 (top row) for example simulations of gaps opened by Jupiter-mass planets.

It is not always accurate to talk about ice lines creating gaps in the surface density distribution because condensation does not, by itself, remove or redistribute mass—though growth may increase the grains' inward drift speed, "drying" the area beyond the ice line (Ciesla & Cuzzi 2006; Najita et al. 2013). Instead, there is a local maximum or minimum in the radial distribution of dust grains of a particular size. Accumulation of ice mantles on grains may cause the grains to grow beyond the observable size at a given wavelength ($\sim\lambda/3$; Draine 2006), causing dark rings in images. Freezeout also causes opacity changes that can be detected in spectral indices, which measure flux ratios between (sub)millimeter wavelengths (e.g., Cieza et al. 2016; see box 2.5 for more on opacity). Grain growth can be parameterized by smoothly varying the mean grain size or size distribution across the radial width of the ice line. Ros & Johansen (2013) provide an analytical estimate of the maximum particle size achievable by water ice condensation as a function of α, scale height H, (R, z), and a fitted constant (their equations (53) and (54)); additional computational results that apply to other volatiles besides water come from Cridland et al. (2017) and Pinilla et al. (2017). In contrast to

[8] http://fargo.in2p3.fr/

[9] http://pencil-code.nordita.org/

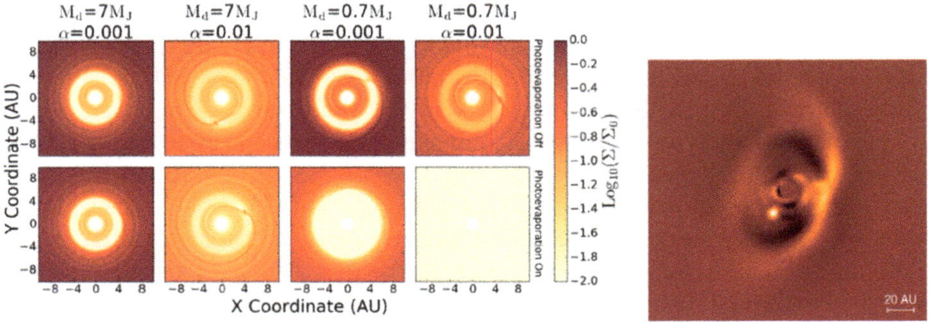

Figure 2.15. Left: Planets in nonphotoevaporating disks (top row) and photoevaporating disks (bottom row). The column headings M_d and α give the disk mass and viscous efficiency, respectively (M_J is the mass of Jupiter). The gap depth depends on the disk viscosity, as high levels of turbulence work to "infill" deep tidal gaps. Photoevaporation can trigger rapid disk dissipation if the disk has already been depleted by viscous accretion or magnetic winds (lower right two panels). Image credit: Reprinted from Wise & Dodson-Robinson (2018). Right: SPHERE coronagraph image of the transitional disk PDS 70 and its "super-Jupiter" planet candidate, which has a radius of 1.4–3.7 R_J (Jupiter radii). The planet candidate appears to be in a near-circular orbit. Image credit: ESO/A. Müller, MPIA.

condensation, sintering—in which fusion of ice particles within the same large aggregate stiffens the aggregate so that grain collisions result in bouncing—frustrates grain growth, causing bright rings just beyond ice lines. Sintering may explain the bright rings in the HL Tau image (Figure 2.7). Okuzumi & Momose (2016) show maps of the representative aggregate radius (i.e., size of large grains) and dust surface density as a function of distance from HL Tau (their figure 8). However, van der Marel et al. (2019) find no relationship between temperature and ring location in a set of 20 disks with radial structure, and so conclude that freezeout and sintering are not the dominant drivers of rings and gaps in submillimeter images.

Blueshifted and non-Keplerian line profiles from He, CO, [Ne II], and H_2O provide evidence that disks are sources of wide-angle, diffuse winds (Edwards et al. 2006; Pascucci & Sterzik 2009; Gorti et al. 2011; Pontoppidan et al. 2011; Salyk et al. 2019). One wind source is photoevaporation, in which UV and X-ray photons from the central star heat the disk surface layers enough so that the sound speed exceeds the gravitational escape speed (Hollenbach et al. 1994). Models of photoevaporation by extreme ultraviolet (EUV) photons suggest that photoevaporation first opens a gap in the disk, which widens into a hole that extends all the way to the star after the material interior to the original gap accretes onto the star (Alexander et al. 2006). X-ray photoevaporation proceeds in much the same way, though the location of the initial gap differs from the EUV case. In contrast, far-ultraviolet (FUV) radiation from either the central star (Gorti & Hollenbach 2009) or nearby, massive stars (Adams et al. 2004) shaves away the disk from the outside in, truncating it rather than opening a gap. Disk dissipation by photoevaporation is a two-timescale problem: for many millions of years, photoevaporation has little effect on the surface density profile, but the gap and hole open rapidly once accretion has depleted the disk to the point that the inward flow cannot keep refilling the nascent gap (Alexander et al. 2014). One can speed up computations of the surface density profile in a photoevaporating gap by using analytical solutions to Equation (2.28) (Pringle 1981) to find the surface density profile of an old disk already depleted by accretion, then turning on a numerical model. For a simple combined 1D model of viscous accretion and photoevaporation, add a wind term to Equation (2.28):

$$\frac{\partial \Sigma}{\partial t} = \frac{3}{R}\frac{\partial}{\partial R}\left[R^{1/2}\frac{\partial}{\partial R}(\Sigma \bar{\nu} R^{1/2})\right] - \dot{\Sigma}_w(R, t), \tag{2.37}$$

where $\dot{\Sigma}_w(R, t)$ is the surface density loss rate due to wind. Time-independent fitting formulae for $\dot{\Sigma}(R)$ based on numerical simulations of the "prehole" epoch (before the material in the gap drains onto the star) are available for EUV photoevaporation (appendix of Alexander & Armitage 2007; originally from models by Font et al. 2004) and X-ray photoevaporation (Owen et al. 2012; appendix B1).

Merín et al. (2010) and Pinilla et al. (2018) state that photoevaporation is probably not responsible for most inner holes in transitional disks, in which the hole size and star accretion rate appear to be unconnected. However, the combination of a gap-opening giant planet and photoevaporation can dissipate a disk that is already severely depleted by accretion onto the star (Alexander & Pascucci 2012; Rosotti et al. 2013). Figure 2.15 (left panel) shows FARGO simulations of gap-opening giant planets in photoevaporating disks (bottom) and nonphotoevaporating disks (top) from Wise & Dodson-Robinson (2018). PDS 70, a gapped disk with a single detected planet candidate (Keppler et al. 2018; Müller et al. 2018; Figure 2.15, right panel), may be a candidate for combined tidal/photoevaporative clearing (though it is certainly possible that smaller, undetectable planets also inhabit the system).

2.4.3 Inner Rim

Because dust provides most of the opacity in protoplanetary disks (see the inset box 2.5 for an explanation of opacity), the hot inner disk where dust-forming molecules are in the gas phase is optically thin to starlight. Dust just outside the sublimation radius is then directly exposed to starlight, as opposed to intercepting light at a small grazing angle like the surface of the flared disk in Section 2.3.2. Such dust acts as a cylindrical inner wall to photons trying to enter the disk midplane (Natta et al. 2001; Dullemond et al. 2001; Dullemond & Dominik 2004a; Isella & Natta 2005; Mulders et al. 2010). The high dust opacity creates a steep temperature gradient across the cylinder where starlight is absorbed (optical depth to starlight $\tau_* = 1$; box 2.5): in models by Flock et al. (2016), the midplane temperature drops from 1500 K to just above 1000 K over only a 0.01 au change in radius. The gas in the inner wall heats up by colliding with hot dust grains, increasing the scale height so that a "puffed-up" inner rim forms. Because gas beyond the inner wall is much cooler, its scale height is lower than predicted by a flared disk model (Section 2.3.2) and the inner rim casts a shadow. With the outer disk only indirectly illuminated by near-infrared emission from the inner wall, rather than directly illuminated by starlight, the disk SED has a high near-IR/far-IR flux ratio.

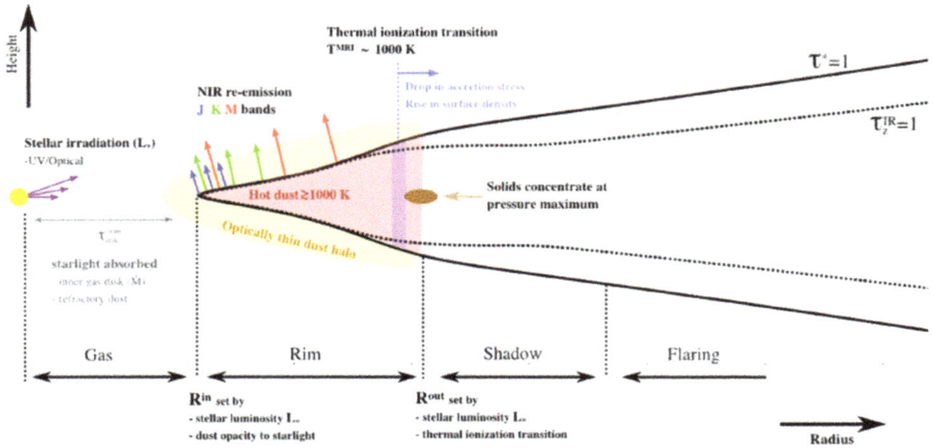

Figure 2.16. The structure of the puffed-up inner rim of a protoplanetary disk. Starlight is absorbed where the optical depth to $\lambda \lesssim 1\ \mu$m is unity. The radial extent of the shadowing rim is determined by the star luminosity, the dust opacity, and the location where the temperature drops too low to thermally ionize the gas, quenching magnetohydrodynamic turbulence. We discuss sources of turbulence in Section 2.4.6; the significance of the pressure maximum marked in the figure will become apparent in Chapter 3. Image credit: Flock et al. (2016).

Models of the inner rim are most important for interpreting star+disk SEDs, but they may also describe planet-forming conditions in the inner disk. Figure 2.16 shows a cartoon of the inner rim structure based on the radiation–hydrodynamic models of Flock et al. (2016). Solids are concentrated just outside the thermally ionized disk interior, possibly accelerating the formation of short-period planets. (The physics of how pebbles drift toward the pressure maximum at the thermal ionization front—where the temperature drops too low for H_2 to be collisionally ionized—is explained in Chapter 3; the reason the ionization front causes a pressure maximum will become clear in Section 2.4.6.) See Ueda et al. (2017) for analytical expressions for the temperature profile and gas-to-dust ratio in the inner rim. Disks surrounding massive HAe/Be stars are more likely to have inner rims than disks around T-Tauri stars (Dullemond et al. 2001).

2.5 Opacity, optical depth, and intensity: A lightning-quick introduction

We quantify the propagation of light through the disk using opacity κ, the cross section per unit disk mass (units cm^2 g^{-1}). The number of mean free paths traversed by a photon traveling between two points is the optical depth, given by

$$\tau_\lambda = \int_{s_0}^{s_f} \kappa_\lambda \rho \, ds, \quad (2.38)$$

where s_0 and s_f are the starting and ending locations of the light path and the λ subscript indicates that both opacity and optical depth are functions of wavelength. At any location in the disk, the mean free path of a photon is

$$\lambda_{\mathrm{mfp},\lambda} = 1/(\kappa_\lambda \rho). \tag{2.39}$$

Optical depth is therefore the approximate length of the light path in units of the mean free path at wavelength λ. We say "approximate" because ρ and κ may change along the light path. Because dust opacities are orders of magnitude higher than gas opacities, temperature calculations may ignore the gas opacity in any part of the disk where dust exists. Except for grains with radius $\sim$1 μm, which yield emission features at 10 μm and 18 μm, silicate dust (with or without an ice coating) is largely "gray" to starlight at wavelengths $\lambda < 2\pi r$, where r is the grain radius. However, at $\lambda \geqslant 2\pi r$, the optical cross section drops precipitously. The left panel of Figure 2.17 shows the absorption opacity $\kappa_{\lambda,\mathrm{abs}}$ for three different grain sizes plus a continuous size distribution. Most of the wavelength dependence comes from grain size, not the material properties of silicates (though adding organic mantles to the dust gives a "red" opacity function).

We next introduce the specific intensity I_λ, units erg s^{-1} cm^{-3} ster^{-1}, which is the power (erg s^{-1}) passing through a surface area (cm^{-2}) in a given wavelength

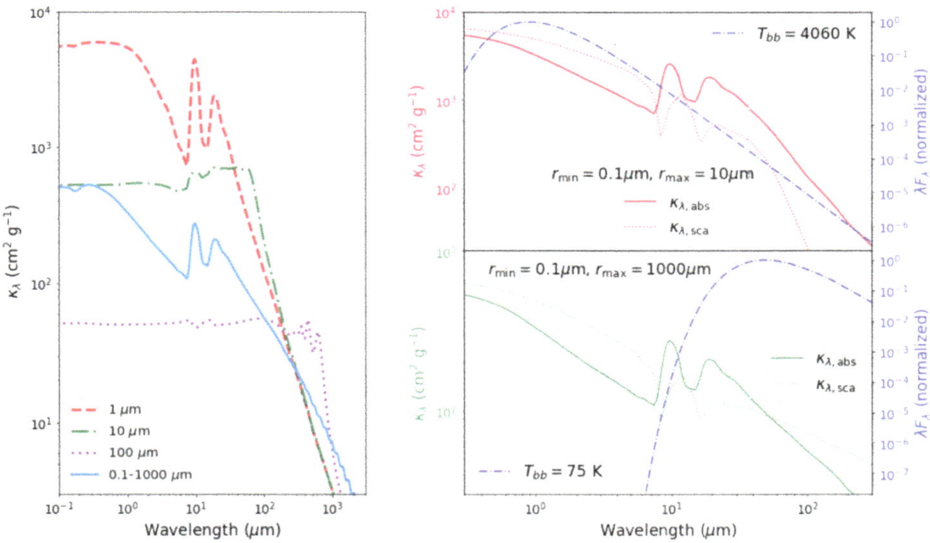

Figure 2.17. Left: wavelength-dependent absorption opacity per gram of dust for three different grain radii, plus a size distribution with 0.1 μm $\leqslant r \leqslant$ 1000 μm and $dn/dr \propto r^{-3.5}$, where n is the grain number density. Following Woitke et al. (2016), the grain composition is 60% amorphous silicate, 15% amorphous carbon, and 25% porosity by volume. To get the opacity per gram of mixed gas and dust, multiply by the dust-to-gas mass ratio (typically ~0.01). Top right: absorption+scattering opacities and incident spectrum above H_S, where only small grains exist ($r_{\max}$ = 10 μm). We approximate the radiation field incident on the disk with a T = 4060 blackbody, the same as the photosphere temperature of RU Lup (Figure 2.3; here we ignore the UV excess). Bottom right: radiation field thermalized at the disk interior temperature, 75 K, at 12 au in the model shown in Figure 2.11 and absorption+scattering opacities of a grain size distribution appropriate to the deep interior ($r_{\max}$ = 1000 μm).

band (cm^{-1}) into a solid angle ($ster^{-1}$). Specific intensity is a useful quantity because in a vacuum, it is conserved along a light ray. However, dust in the disk absorbs and scatters photons. We may break the opacity into separate absorption and scattering terms $\kappa_{\lambda,\text{abs}}$ and $\kappa_{\lambda,\text{sca}}$. In the absence of scattering, we have

$$\frac{dI_\lambda}{I_\lambda} = -\kappa_{\lambda,\text{abs}}\rho\ ds, \tag{2.40}$$

$$I_\lambda = I_{\lambda,0}e^{-\tau_\lambda}. \tag{2.41}$$

On optically thin light paths where $\tau_\lambda \leqslant 1$, most photons of wavelength λ are neither absorbed nor scattered; optically thick paths have $\tau_\lambda > 1$ and most photons get scattered and/or absorbed+reemitted—perhaps many times—along the path. In disks, we can ignore scattering on optically thin light paths because silicate dust grains tend to forward-scatter, which means the path of the scattered photon does not change very much. Equation (2.41) therefore describes radiative transfer above the disk surface at H_S.

The concept of optical depth leads us to a more sophisticated understanding of the disk surface height (Section 2.3.2). H_S is the z-coordinate of the location where starlight entering the disk at grazing angle φ has traveled one mean free path, so that $\tau_* = 1$. In defining τ_*, the disk atmosphere's optical depth to starlight, we are collapsing our wavelength-dependent absorption opacity into a single intensity-weighted mean opacity averaged over the star's spectrum:

$$\kappa_* = \frac{\int_0^\infty B_{\lambda,*}\kappa_{\lambda,\text{abs}}d\lambda}{\int_0^\infty B_{\lambda,*}d\lambda}, \tag{2.42}$$

where $B_{\lambda,*}$ is the Planck function for the star's effective temperature. κ_* is called the Planck mean opacity; the Planck function biases the mean toward the wavelengths carrying the most power and is used for optically thin light paths.

The disk has a different opacity and optical depth to its own thermal radiation, which includes some combination of reprocessed starlight and viscous heating. Where the disk is vertically optically thick, we may approximate the disk's vertical half-plane (midplane to surface) optical depth to its escaping thermal radiation as

$$\tau_{1/2} \approx \kappa_R\Sigma/2, \tag{2.43}$$

where κ_R is the Rosseland mean opacity:

$$\frac{1}{\kappa_R} = \frac{\int_0^\infty (1/\kappa_\lambda)(\partial B_\lambda/\partial T)d\lambda}{\int_0^\infty (\partial B_\lambda/\partial T)d\lambda}. \tag{2.44}$$

In Equation (2.44), $\kappa_\lambda = \kappa_{\lambda,\text{abs}} + \kappa_{\lambda,\text{sca}}$ is the total opacity at a given wavelength and $\partial B_\lambda/\partial T$ is the derivative of the Planck function with respect to the local disk temperature. We replace the Rosseland mean with the Planck mean where the disk becomes optically thin to light at the peak of $B_\lambda[T(R)]$, where $T(R)$ is the local isothermal temperature (Equation (2.21)). The right two panels of Figure 2.17 show radiation fields and wavelength-dependent opacities that would be combined to find κ_* for the surface layers (top) and κ_R for 12 au in our model disk interior (bottom).

Let us consider photons moving through the optically thick disk interior, which is in local thermodynamic equilibrium, or LTE. Here, collisions between gas molecules and dust grains are common enough that all types of matter equilibrate to a single temperature. The complete equation for radiative transfer is[10]

$$\frac{dI_\lambda}{d\tau_\lambda} = S_\lambda - I_\lambda, \tag{2.45}$$

which one derives by combining Equations (2.38) and (2.40) and adding a source function to describe radiation generated in the disk (e.g., by turbulence-driven accretion or reemission of absorbed starlight). In LTE, Kirchoff's law for radiation states that the source function S_λ is the Planck function $B_\lambda[T]$. Using the Rosseland mean opacity κ_R allows us to consider radiative transfer over all wavelengths simultaneously: we can work with I, or bolometric intensity (units erg s^{-1} cm^{-2} $ster^{-1}$), instead of I_λ. We will also want to average I over all solid angles to find the bolometric mean intensity J:

$$J = \langle I \rangle = \frac{1}{4\pi} \int_{4\pi} I d\Omega. \tag{2.46}$$

In radiative equilibrium, $\dot{T}(R, z) = 0$: no net energy may be deposited or removed anywhere in the vertical column as radiation travels up and down, giving $S = J$ for all (R, z). Finally, because our source function S_λ is a blackbody and $\int_0^\infty B_\lambda[T] d\lambda = \sigma_{SB} T^4/\pi$, we find

$$J = \frac{\sigma_{SB}}{\pi} T^4. \tag{2.47}$$

We can use Equation (2.47) to find the temperature the disk would have in the absence of stellar irradiation, if accretion were the only heat source. We will first expand our disk model by using τ_* and κ_* to add a heated surface layer to our disk, then quantify the accretion heating in Section 2.4.5, and finally combine the two for disks with both heat sources.

2.4.4 Warm Surface Layer

Planet-forming regions have $\tau_* \gg 1$ (see the inset in box 2.5), meaning they are optically thick to starlight: incident photons are absorbed and reemitted many times before reach the midplane. Each time a photon is absorbed by a dust grain (as opposed to scattered), some of its energy goes into heating the grain, so the reemitted photon is redder than the absorbed photon. Even at the $\tau_* = 1$ surface, the incident flux has already been attenuated by a factor of $1/e$. In the absence of significant accretion heating (Section 2.4.5), an optically thick disk will be warmer at the surface than at the midplane. Accurate temperature profiles are critical for mapping

[10] In our 1+1D disk model, we are working in a plane-parallel framework and assuming that radiative transfer in the optically thick interior is always vertical. The factor of $\cos\theta$ (where θ is the angle between the normal vector to the disk midplane and the direction of light propagation) therefore does not appear in Equation (2.45): $\cos\theta = 1$. Vertical radiative transport is generally not a bad approximation because the radial optical depth far exceeds the vertical optical depth at all wavelengths, but sometimes (sub)millimeter radiation is able to propagate radially.

the spatial distribution of molecules observed in disks (Long et al. 2018; Pinte et al. 2018; Schwarz et al. 2019) and determining the ice inventory available for planet formation (e.g., Dodson-Robinson et al. 2009; Qi et al. 2013; Salinas et al. 2016). Here we define a 1+1D analytical prescription for the temperature profile of a passively heated disk (negligible viscous heating) with a superheated atmosphere. This model was developed by Dullemond et al. (2001) and applied by Ricci et al. (2010) to determine the outer radii and dust properties of disks in the Taurus-Auriga star-forming region.

We will break the disk into two layers: a surface layer directly irradiated by starlight and an upper interior that sees some reradiated, reddened light. We expect our two disk layers to have different opacities because they see different radiation fields and have different grain size distributions (large grains settle toward the midplane; see Chapter 3). Our upper interior opacity is the Planck mean opacity for dust at the disk's local isothermal temperature (Equation (2.21)):

$$\kappa_d = \frac{\int_0^\infty B_\lambda[T(R)]\kappa_{\lambda,\mathrm{abs}}d\lambda}{\int_0^\infty B_\lambda[T(R)]d\lambda}, \tag{2.48}$$

which we multiply by the dust-to-gas mass ratio to get the optical cross section per gram of mixed gas and dust (box 2.5). Note that we are using the Planck mean opacity for the disk's upper interior because we are confining this analysis to optically thin light paths.

We can adapt Equation (2.38) to find $\tau_*(z)$, the Planck mean optical depth to starlight at any distance above the midplane:[11]

$$\tau_*(z) = \frac{1}{\sin\varphi}\int_z^\infty \rho(z)\kappa_* dz, \tag{2.49}$$

where $1/\sin\varphi$ is the length ratio of the slant path of incoming starlight to the height z. A blackbody grain lofted high above the $\tau_* = 1$ surface at height H_S has equilibrium temperature[12]

$$T_{bb}(R) = \left(\frac{L_*}{16\pi R^2\sigma_{\mathrm{SB}}}\right)^{1/4}. \tag{2.50}$$

[11] Here we are following Hillenbrand et al. (1992), Gullbring et al. (1998), and Dullemond et al. (2001) and assuming that the disk has an inner hole from (e.g.) dust sublimation or magnetospheric truncation. Evidence for sublimation-induced inner holes comes from the correlation between L_* and inner-disk radius measured from near-IR interferometry (Monnier & Millan-Gabet 2002). With the inner hole, the illuminated disk surface "sees" the entire star surface, as opposed to half the star surface if the disk were to extend all the way to the star.

[12] To derive Equation (2.50), set the energy absorbed by the grain equal to the energy reemitted by the grain: $\pi r^2 L_*/4\pi R^2 = 4\pi r^2\sigma_{\mathrm{SB}}T_{bb}^4$, where r is the grain radius.

But grains high above the surface are not blackbodies: they are too small to emit efficiently in the mid- and far-infrared. For example, Equation (2.21) predicts $T = 75$ K at 12 au for $C = 4$ (Figure 2.11), giving a peak wavelength of 39 μm for blackbody reemission. But for grains with $r = 1$ μm, the opacity at 39 μm, and with it the emissivity, is an order of magnitude lower than its maximum value at shorter wavelengths, because a grain cannot absorb or emit efficiently at wavelengths larger than its circumference (see Figure 2.17). In order to maintain thermal equilibrium with the incident starlight, the grain must superheat to $T > 75$ K so it can emit at shorter wavelengths. This superheating can be parameterized by the opacity ratio κ_*/κ_d:

$$T_{H>H_S} = \frac{\kappa_*}{\kappa_d} T_{bb}. \tag{2.51}$$

We adjust the "near-surface" equilibrium temperature by attenuating the stellar flux $L_*/4\pi R^2$ incident at H_S by a factor of $e^{-\tau_*}$:

$$T_{\rm eq,surf}(R,\, z) = \left(e^{-\tau_*(z)} \frac{\kappa_*}{\kappa_d} \frac{L_*}{16\pi R^2 \sigma_{\rm SB}}\right)^{1/4}. \tag{2.52}$$

Equation (2.52) does not apply deep in the disk interior, where the propagating radiation field thermalizes at the disk temperature and no longer resembles the star photosphere. Near the disk midplane, τ_* is so high that $T_{\rm eq,surf} \to 0$. But recall that we already have Equation (2.21) to describe the isothermal interior temperature $T(R)$. We may use a flux sum to smoothly transition from the superheated surface temperature to the more blackbody-like interior, where larger grains reside:

$$T(R,\, z)^4 = T_{\rm eq,surf}(R,\, z)^4 + T(R)^4. \tag{2.53}$$

Figure 2.18 shows the vertical temperature profile of the disk specified in Section 2.3.2 at 1 au, 5 au, and 10 au, plus the 2D $\log_{10}[T(R,\, z)]$ distribution in the inner 10 au. The "snow line" (also called ice line or condensation front) where H_2O gas freezes onto grain surfaces is shown in cyan. The wavelength-dependent opacities used to compute κ_* and κ_d are those plotted on the right-hand side of Figure 2.17. The warm layer required to explain observations of volatile gases at large radii, where midplane temperatures are $\lesssim$50 K (e.g., Aikawa et al. 2002; Glassgold et al. 2004; Schwarz et al. 2019), is now present. Besides the Ricci et al. 2010 study of Taurus-Auriga disks, Equation (2.53) has been used in estimates of grain size distributions and disk masses surrounding intermediate-mass stars (Testi et al. 2003; Natta et al. 2004) and models of the dust emission and CO velocity field of the HD 163296 disk (Isella et al. 2007).

Equation (2.53) still does not fully capture the complexity of passive heating. Our simple approach to radiative transfer does not account for the fact that the superheated surface layer can change the angle of starlight incidence φ. Equation (2.53) can also fail because the disk mean opacity κ_d treats each photon as if its free path were $\lambda_{\rm mfp}$, while light on the Rayleigh–Jeans tail of $B_\lambda[T(R)]$ and $B_{\lambda,*}$ can propagate much farther than $\lambda_{\rm mfp}$. Photons near the midplane may be able to travel macroscopic distances in the radial direction, while Equation (2.53) is built on the

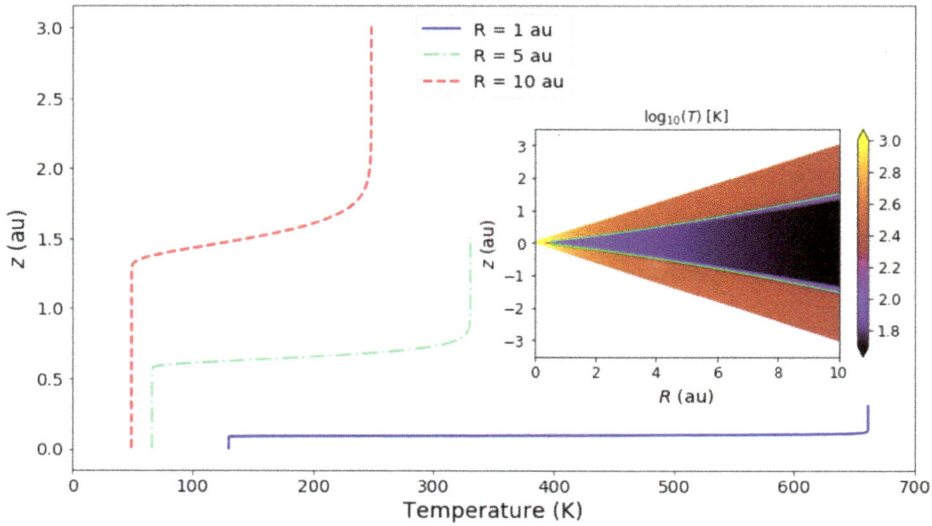

Figure 2.18. Vertical temperature profiles at 1 au, 5 au, and 10 au (line plot), plus the 2D distribution of $\log_{10}(T)$ in the inner 10 au of the disk (inset) with the ice line (T = 160 K) marked in cyan. Here we set $T_* = 4060$ K, the observed temperature of RU Lup (Dionatos et al. 2019). We assume that only small grains with $0.1 < r < 10$ μm contribute to κ_*, while the interior grains that contribute to κ_d have grown to millimeter radii. Both grain size distributions have $\log_{10}(dn/dr) = -3.5$.

assumption that they cannot. A temperature underestimate is likely in regions with $\tau_* \gg 1$ but $\tau_{1/2} \sim 1$. On the other hand, the disk midplane can also cool beyond the isothermal temperature "floor" set by Equation (2.21) as photons propagating deep within the disk get reddened with each round of absorption and reemission. By the time radiation reaches the disk midplane, it has much less energy than the incident starlight. For example, the Flying Saucer disk shown in Figure 2.2 has midplane temperature $T \leqslant 10$ K at 100 au (Guilloteau et al. 2016; Dutrey et al. 2017), much colder than predicted by data-informed models such as those of Isella et al. (2009) and Andrews et al. (2013). Equation (2.53) may overestimate temperatures where surface densities are high and the disk is optically thick to both starlight and thermalized radiation trying to escape the midplane ($\tau_* > \tau_{1/2} \gg 1$). More sophisticated but still computationally cheap models can be computed with the publicly available code DISKSTRUCT[13] (Dullemond 2011; based on the models of Dullemond et al. 2002), which uses 1+1D geometry with the flaring angle prescription of Chiang et al. (2001) but adds frequency-dependent opacities. Other upgrades to Equation (2.53) come from D'Alessio et al. (1998, 1999), and D'Alessio et al. (2006), who use Planck and Rosseland mean opacities for the star and disk radiation, respectively, but who include scattering and viscous heating and simultaneously and self-consistently solve for H_S, $T(R, z)$, and $\rho(R, z)$.

[13] https://www.ita.uni-heidelberg.de/~dullemond/software/diskstruct/index.shtml

Without the type of tomography data that was used to reconstruct the midplane temperature profile of the Flying Saucer—which was available only because the disk is edge on and seen against a bright background of CO-emitting clouds—radiative transfer software that uses frequency-dependent dust opacities provides the most realistic temperature calculations. We recommend the publicly available code RADMC-3D[14] (Dullemond et al. 2012) and its predecessor RADMC (Dullemond & Dominik 2004a), which are extensively documented and have been benchmark-tested for continuum radiative transfer in protoplanetary disks (Pascucci et al. 2004). RADMC-3D will iterate over the vertical structure to make sure $H(R)$, $H_S(R)$, $T(R, z)$, and $\rho(R, z)$ are self-consistent. HOCHUNK3D[15] (Whitney et al. 2003, 2013) is also publicly available and has a wide range of built-in disk structure options, including gaps, inner rims, and spirals, while MCMax and MCFOST (Pinte et al. 2006) are available by collaboration with the developers.[16] While radiative transfer software is ideal for modeling the internal structure of a user-specified star and disk, it is most often used to solve the inverse problem: given a SED, find a disk structure (or set of structures) that fits (e.g., Andrews et al. 2013; Woitke et al. 2016, 2019; Ballering & Eisner 2019).

It is often possible to model gas emission lines originating from within ~2 scale heights of the midplane by solving for the dust temperature only and assuming that frequent gas–grain collisions cause gas to assume the dust temperature (e.g., Andrews et al. 2009; Banzatti et al. 2012; Yu et al. 2017a). RADMC-3D can model molecular or atomic emission in LTE. However, in tenuous regions where gas–grain collisions are infrequent, the gas temperature exceeds the dust temperature (Jonkheid et al. 2004; Kamp & Dullemond 2004; Nomura & Millar 2005; Meijerink et al. 2009; Blevins et al. 2016) and non-LTE models are required. Line Modeling Engine (Brinch & Hogerheijde 2010), or LIME,[17] is a publicly available gas radiative transfer code with both a fast LTE ray tracer and a non-LTE module. As with SEDs, line radiative transfer software is often used to find the best-fit mass, emitting area (i.e., location in the disk), and characteristic temperature for a molecular line data set (Keane et al. 2014; McClure et al. 2016; Kama et al. 2016; Miotello et al. 2017). Finally, we have built our disk models by assuming that the star and dust are in thermal equilibrium (i.e., starlight heats the disk, dust reradiates the energy from starlight, and the two processes are in balance), but emission lines can also be important cooling agents (Gorti & Hollenbach 2008; Morris et al. 2009; Woitke et al. 2009; Aresu et al. 2014). Thermochemical models such as ProDiMo (Woitke et al. 2009), which is available by registration and application,[18] will simultaneously solve for the gas and dust temperature, density, and molecular abundances, but are computationally expensive and not suited to large model grids.

[14] https://www.ita.uni-heidelberg.de/~dullemond/software/radmc-3d/
[15] https://gemelli.colorado.edu/~bwhitney/codes/codes.html
[16] https://dianaproject.wp.st-andrews.ac.uk/data-results-downloads/main-modelling-software/
[17] https://github.com/lime-rt/lime
[18] https://dianaproject.wp.st-andrews.ac.uk/data-results-downloads/main-modelling-software/

2.4.5 Viscous Heating

Turbulent viscosity channels kinetic energy from the Keplerian shear into the largest eddies and then down to the smallest eddies at the dissipation scale, where the energy is deposited as heat. Although the SEDs of most disks around T-Tauri stars are consistent with "passive" models where starlight provides the only heating (e.g., Woitke et al. 2019), disks around more massive Herbig Ae/Be stars have non-negligible accretion heating (Manoj et al. 2006; Mendigutía et al. 2012). Furthermore, viscous heating can alter the midplane temperatures in the turbulent inner few astronomical units of T-Tauri disks (D'Alessio et al. 1998) and the surface temperatures of magnetically active T-Tauri disks (Turner et al. 2006; Hirose & Turner 2011; Simon et al. 2013). Here we present a prescription for viscous heating that can be used wherever energy dissipation is local (i.e., a true turbulent cascade); it should not be used with nonturbulent angular momentum transport mechanisms such as spiral waves.

The viscous heating rate per unit mass (units of erg g^{-1} s^{-1}) is proportional to the square of the shear:

$$q_{+,\,m} = \nu\left(R\frac{\partial\Omega_K}{\partial R}\right)^2. \tag{2.54}$$

We substitute in $(R\partial\Omega_K/\partial R)^2 = (9/4)\Omega_K^2$ and rewrite Equation (2.54) as a heating rate per unit volume (units of erg cm^{-3} s^{-1}; Pringle 1981):

$$q_{+,\,v} = \frac{9}{4}\Omega_K^2\nu\rho. \tag{2.55}$$

To keep the disk in thermal equilibrium, the viscously generated luminosity from the interior must propagate to the surface and escape. We integrate $q_{+,\,v}$ upward from the midplane to find the accretion flux (units of erg s^{-1} cm^{-2}) through the top surface of a given vertical column:

$$F_{\rm acc} = \int_{z=0}^{\infty}\frac{9}{4}\Omega_K^2\nu\rho\, dz \approx \frac{9}{8}\Omega_K^2\bar{\nu}\Sigma, \tag{2.56}$$

If the disk is symmetric about the midplane, an equal amount of flux propagates downward, leaving the bottom of the disk. The accretion flux exits at the height $z_{\rm acc}$ at which the disk becomes optically thin to its own thermal radiation; usually $z_{\rm acc} < H_S$. Here we are assuming that most of the turbulent energy dissipation takes place near the midplane so that $F_{\rm acc}(z)$ is approximately constant: escaping radiation does not pick up "extra" flux as it propagates vertically. However, many types of turbulence are stronger at the surface than the midplane (Section 2.4.6), in which case one must use Equations (2.63) and (2.65).

In the absence of stellar irradiation, the temperature at $z_{\rm acc}$ would be given by

$$F_{\rm acc} = \sigma_{\rm SB}T_{\rm acc,surf}^4 \tag{2.57}$$

$$T^4_{\text{acc,surf}} = \frac{9}{8}\frac{\Omega_K^2 \bar{\nu}\Sigma}{\sigma_{\text{SB}}}. \tag{2.58}$$

Knowing that $T_{\text{acc,surf}}$ is approximately uniform above z_{acc}, we use the flux sum of the accretion temperature and the passively heated temperature $T_{\text{eq,surf}}$ (Equation (2.52)) to find the true temperature at the disk surface:

$$T^4_{\text{surf}} = T^4_{\text{acc,surf}} + T^4_{\text{eq,surf}}. \tag{2.59}$$

Unless $T_{\text{acc,surf}} \ll T_{\text{eq,surf}}$, which may happen if α is very small ($\sim 10^{-4}$ or lower), the accretion heating will increase the scale height and change the angle at which starlight hits the disk surface. Equations (2.7), (2.9), and (2.59) must then be solved iteratively until a self-consistent value of T_{surf} is reached. For small α, one may ignore the fact that accretion heating alters the vertical structure and simply use Equation (2.59) without an iterative model.

The last piece of the puzzle is the temperature in the deep interior, where the disk is optically thick to its own radiation ($\tau_{1/2} > 1$). We start by finding the volume density of energy carried by photons:

$$u = \frac{4\pi}{c}J, \tag{2.60}$$

which we derive by integrating J over all 4π steradians and dividing by the speed of light c at which photons exit our volume. In a photon gas, energy density and radiation pressure are related by $P_{\text{rad}} = u/3$, giving

$$P_{\text{rad}} = \frac{4\pi}{3c}J. \tag{2.61}$$

We can relate the radiation pressure, optical depth, and flux:

$$\frac{dP_{\text{rad}}}{d\tau_R} = -\frac{F_{\text{acc}}}{c}, \tag{2.62}$$

where $d\tau_R = \kappa_R\, \rho\, dz$ is the differential Rosseland mean optical depth (Equation (2.38), with $\kappa_\lambda \to \kappa_R$ and $ds \to dz$ because radiation moves vertically).[19] Combining Equations (2.61) and (2.62) gives

$$\frac{dJ}{dz} = -\frac{3\kappa_R \rho}{4\pi}F_{\text{acc}}. \tag{2.63}$$

Substituting in Equations (2.57) and (2.47) gives

$$\frac{d}{dz}(T^4_{\text{acc}}) \approx -\frac{27}{32}\frac{\kappa_R \rho}{\sigma_{\text{SB}}}\Omega_K^2 \bar{\nu}\Sigma, \tag{2.64}$$

[19] To understand Equation (2.62), recall that F_{acc} is the rate of vertical energy transfer through a horizontal area element in the disk; the area element divides two adjacent volume elements through which photons carry energy at speed c, and τ_R keeps track of the photons that are lost to absorption or scattering on their path between the volume elements.

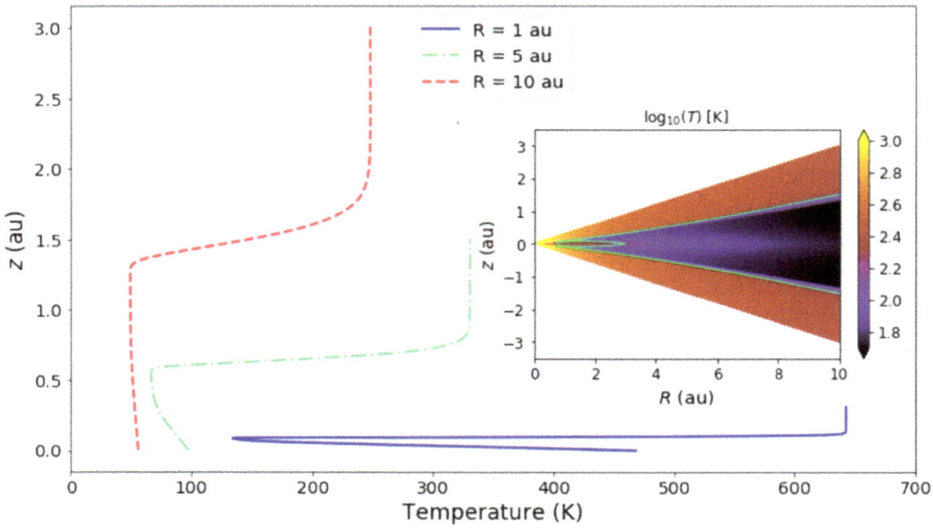

Figure 2.19. Temperature profiles $z(T)$ at 1 au, 5 au, and 10 au with both viscous heating (Equations (2.58) and (2.64)) and passive heating (Equation (2.52)). We use constant $\nu(z)$ and $\alpha(R, z)$, with $\alpha = 2 \times 10^{-3}$ (higher than implied by some observations, as discussed in Section 2.3.3, but the thermal effects of weaker turbulence are limited to $\lesssim$1 au and are difficult to visualize) with the same opacities as in Sections 2.4.3 and 2.4.4. Plots are on the same temperature scale as in Figure 2.18, and the 2D structure of the ice line is shown in cyan.

which can be integrated inward from the surface where $T_{\mathrm{acc,surf}}$ is known. The flux sum also works in the disk interior to find the true temperature at any point (R, z): $T_{\mathrm{true}}^4 = T_{\mathrm{acc}}^4 + T^4$, where T^4 is given by Equation (2.53). Figure 2.19 shows the temperature structure in the inner 10 au of our model disk when we combine viscous heating with $\alpha = 2 \times 10^{-3}$ and our analytical warm surface layer model (Section 2.4.4; Figure 2.18). In regions that are vertically optically thin to incident starlight, $T(z)$ is given by Equation (2.53) on its own.

If $\nu(z)$ and $\alpha(z)$ are not constant, as in magnetorotational turbulence (Section 2.4.6, Gammie 1996), Equations (2.56), (2.58), and (2.64) cannot be used as written. Instead, Equation (2.59) still applies, but one must find $T_{\mathrm{acc}}(R, z)$, $\rho(R, z)$, and $F_{\mathrm{acc}}(R, z)$ in optically thick regions by numerically integrating a set of coupled differential equations:

$$\begin{aligned} \frac{\partial F_{\mathrm{acc}}}{\partial z} &= q_{+,\,v} \\ \frac{\partial T_{\mathrm{acc}}}{\partial z} &= \nabla_{\mathrm{rad}} \frac{T}{P} \frac{\partial P}{\partial z}, \end{aligned} \tag{2.65}$$

and Equation (2.7). In Equation (2.65), the thermodynamic gradient $\nabla_{\rm rad} = d \ln T / d \ln P$ is[20]

$$\nabla_{\rm rad} = \frac{3}{16} \frac{\kappa_R P F_{\rm acc}}{\sigma_{\rm SB} g_z T^4}. \qquad (2.66)$$

Equations (2.4), (2.9), and (2.59), and the boundary condition $F_{\rm acc}(z = 0) = 0$, which applies to a midplane-symmetric disk, close the system. In the 1+1D approximation, $\partial / \partial z \rightarrow d / dz$ and one may use numerical methods for solving ordinary differential equations to find the temperature profile.

2.4.6 Turbulence, Vortices, and Angular Momentum Transport

Transfer of angular momentum from star-forming material to planet-forming material begins right at the onset of star formation. To become part of a star, a parcel of gas from a collapsing cloud core must reduce its specific angular momentum by six to seven orders of magnitude (Goodman et al. 1993; Caselli et al. 2002; see Figure 2.20.). Rotationally supported disks that remove angular momentum from accreting gas surround some of the youngest, most embedded protostars (Jørgensen et al. 2009; Choi et al. 2010; Enoch et al. 2011; Murillo et al. 2013; Ohashi et al. 2014; Tobin et al. 2015; Yen et al. 2017; Lee et al. 2018; Segura-Cox et al. 2018; Tobin et al. 2018). Although specific angular momentum appears to be conserved during infall from a protostellar envelope onto a disk,[21] the disk itself is "lossy:" specific angular momentum declines by a factor of 100 as material moves through the disk to the star surface (Hennebelle 2013; Li et al. 2014). The angular momentum transport mechanism affects all stages of planet formation. For example, in Section 2.3.3 we assumed disk evolution was driven by viscosity: strong turbulence inhibits the growth of solid bodies (Dubrulle et al. 1995; Dullemond & Dominik 2004b; Chapter 3) but may stall planet migration (Crida & Morbidelli 2007), preserving planets from plunging into their host stars.

Modern simulations of magnetorotational turbulence and analysis of molecular emission-line profiles usually suggest low values of viscous efficiency α that might not be adequate to drive the observed accretion rates (Section 2.3.3), stimulating interest in spiraling magnetic wind as an angular momentum transport mechanism (Bai 2013; Bai & Stone 2013; Gressel et al. 2015; Bai 2017). The wind removes mass and angular momentum from the inner few astronomical units of the disk, sustaining accretion rates of $10^{-8} M_\odot$ yr^{-1} (Wang et al. 2019) in agreement with observations of Stage II disks (Hartmann et al. 1998). ALMA CO ($J = 2 \rightarrow 1$) observations of the Herbig Ae disk HD 163296 reveal a disk wind with a double corkscrew shape that is consistent with the magnetocentrifugal wind model

[20] Here we are assuming energy is transported by radiation, which is true throughout most of the disk, but convection is more efficient if $\nabla_{\rm rad} > \nabla_{\rm ad}$, where $\nabla_{\rm ad} = (\gamma - 1)/\gamma$ is the adiabatic thermodynamic gradient and γ is the ratio of specific heats. In convective regions, substitute $\nabla_{\rm ad}$ for $\nabla_{\rm rad}$ in Equation (2.65).

[21] Note that the first round of spindown reduces specific angular momentum by a factor of 10 during the transition from the molecular cloud core to the infalling envelope and is probably driven by magnetic braking (Belloche 2013).

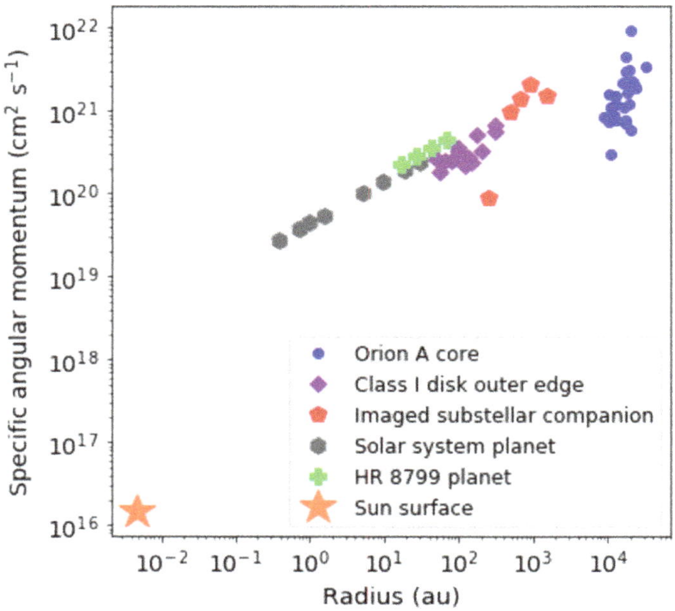

Figure 2.20. The angular momentum "ladder": specific angular momentum of star- and planet-forming structures, imaged substellar companions, and known planets as a function of radius (cloud cores) or distance from the star (disks, companions, planets). Data sources are Tatematsu et al. (2016; Orion A cloud cores), Najita & Bergin (2018; Class I disk radii), De Rosa et al. (2014; directly imaged companion ζ Del b); Close et al. (2007; directly imaged companion 11 Oph b); Metchev & Hillenbrand (2006; directly imaged companion HD 203030 B); Burgasser et al. (2000; directly imaged companion GJ 570 D); Bailey et al. (2014; directly imaged planet HD 106906 b); and Marois et al. (2008, 2010; planets orbiting HR 8799).

(Klaassen et al. 2013). Chambers (2019) presents an analytical model for a magnetically driven disk wind and explores the parameter space of slow versus fast winds and turbulent versus laminar disks. Magnetic angular momentum transport—whether magnetorotational instability-driven (Section 2.4.6, MRI) or magnetocentrifugal—offers promising possibilities for planet formation: it allows accreting disks to be laminar rather than turbulent, creates surface density gradients that halt planet migration (Suzuki & Muto 2010; Ogihara et al. 2015), and may even trigger the formation of close-in super-Earths (Ogihara et al. 2018). See Pudritz et al. (2007) for a review of disk outflow physics.

Yet turbulence in disks certainly exists: the disk spreading between Stage 0/1 and Stage II documented by Najita & Bergin (2018) is strong evidence of a local angular momentum transport mechanism, and Guilloteau et al. (2012) detect transonic turbulence (Mach number ~0.4–0.5) in the warm molecular layer of the DM Tau disk. Furthermore, fitting the emission-line profile of the CO overtone ($\Delta v = 2$, where v is the vibrational quantum number) from 51 Oph requires turbulent motion (Berthoud et al. 2007), while polarized emission from micron-size dust grains in the AB Aur outer disk (Figure 2.3) suggests turbulent lofting above the midplane (Li et al. 2016). Although the magnetorotational instability (MRI) was considered the leading turbulence driver (Balbus & Hawley 1991, 1998; Hawley 2001) until many simulations showed that the ionization fraction may be too low for sustained turbulence in the inner disk (Wardle 2007; Bai 2011, 2013; Pandey & Wardle 2012; Kunz & Lesur 2013; Bai 2014; Simon et al. 2015), recent research into purely hydrodynamic instabilities suggests that there are many ways to "wake up" turbulence in the dead zone where ionization is too low for MRI (e.g., Klahr & Bodenheimer 2003; Lesur & Papaloizou 2010; Lyra & Klahr 2011; Nelson et al. 2013); see review by Lyra & Umurhan (2019). (Note that a revitalized dead zone may actually impede planet formation by stirring dust grains out of the midplane and removing the local density maximum caused by low turbulent viscosity; Isella et al. 2016). Large-scale vortices also transport angular momentum and have been observed in at least seven disks (van der Marel et al. 2013; Pérez et al. 2014; Baruteau et al. 2019). Here we briefly describe candidate turbulence- and vortex--causing instabilities, which operate on dynamical timescales, and offer ways to simulate or parameterize their effects on the disk structure.

• **MRI:** H_2/He gas in protoplanetary disks is partially ionized by cosmic rays from nearby stars or the interstellar medium (Glassgold et al. 2012), X-rays from the central star's corona (Glassgold et al. 1997), UV from the central star and nearby stars (Perez-Becker & Chiang 2011), and decay of short-lived radionuclides (Umebayashi & Nakano 2009). Furthermore, Herbig Ae/Be stars may have inner disks hot enough ($\sim 10^4$ K) for thermal ionization of sodium and potassium atomic gas (Umebayashi & Nakano 1988; Fromang et al. 2002; see Section 2.4.3). If the disk is ionized, magnetic fields will couple parts of the disk that lie along the same field line. When a field line is bent, it exerts a restoring tension

$$\vec{F_t} = \frac{(\vec{B} \cdot \vec{\nabla})\vec{B}}{4\pi} \tag{2.67}$$

($\vec{B}$ is magnetic flux density; cgs units of $\vec{F_t}$ are dyn cm^{-3}) that acts to straighten the field line. But the Keplerian shear makes it inevitable that all field lines, except purely azimuthal ones, will be bent or stretched. Consider the simple case shown in Figure 2.21. A vertical field line (blue solid curve) couples a parcel of gas at the midplane (blue box) with parcels of gas above and below the midplane (green boxes). There is a slight angular velocity gradient over the vertical column: while Ω_K is the midplane angular speed, the true angular speed is a function of z:

$$\Omega(z) = \sqrt{\frac{GM_*}{(R^2 + z^2)^{3/2}}} \tag{2.68}$$

(Figure 2.21, red arrows). The gas parcel at the midplane runs ahead of its off-midplane counterparts, but doing so perturbs the connecting magnetic field line.[22] The magnetic tension F_t (green arrows) tries to straighten out the field line, slowing

[22] MRI simulations usually seed the turbulence with random perturbations on top of the Keplerian velocity field rather than horizontal shear created by $\Omega(z)$; we use horizontal shear as a pedagogical example but note that any type of perturbation (δR, $\delta\phi$, or δz) will generate magnetic tension.

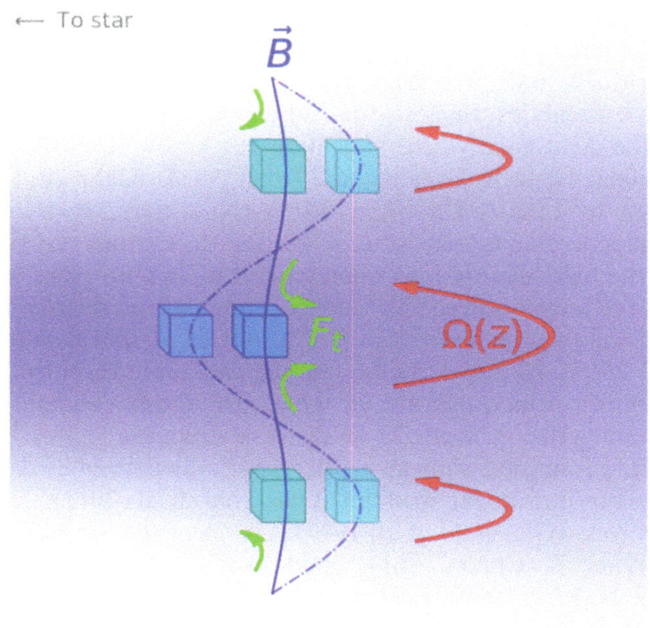

Figure 2.21. An initially vertical field line $\vec{B}$ (blue curve) gets deformed by vertical Keplerian shear (varying $\Omega(z)$, red) as midplane gas (blue box) orbits slightly faster than off-midplane gas (green boxes). The restoring tension F_t (bright green arrows) tries to return the field line to vertical by slowing down the midplane gas and speeding up the off-midplane gas. However, a Keplerian disk has a negative heat capacity: slowing down the orbiting midplane gas causes it to lose angular momentum, drop to a smaller orbital radius, and speed up, while the opposite happens to the off-midplane gas. The dashed–dotted blue curve and the lighter-colored boxes show the field line and its connected gas parcels after a small timestep $\Delta t \ll 1/\Omega_K$. A positive feedback loop develops with perturbations growing exponentially.

the midplane gas parcel and speeding up the off-midplane gas parcels. But a Keplerian disk has a negative heat capacity: the slowed-down midplane gas loses angular momentum and drops closer to the star, $R \to R - \Delta R$, where it speeds up. The off-midplane, initially sped-up gas parcels gain angular momentum, move outward, and slow down. After a small time interval $\Delta t \ll 1/\Omega_K$, the field line is bent even more (dashed–dotted blue curve and lighter-colored boxes). The magnetic field's attempt to straighten itself out leads to further bending, which increases the magnetic tension yet again in a positive feedback loop. Displacements grow at a rate $\sim\Omega_K$ (Balbus & Hawley 1991) until the turbulence is fully developed.

The MRI is robust in the ideal magnetohydrodynamics (MHD) limit, where field lines are frozen in to the gas (i.e., charged particles cannot cross field lines). To reach ideal MHD, the neutral molecules must inherit the velocity field of the charged particles, which is set by the combination of bent field lines and Keplerian shear.

In the absence of purely hydrodynamic instabilities, there is a nonideal, non-turbulent "dead zone" in the inner disk where the surface density is high enough to screen out ionizing cosmic rays and X-rays (Gammie 1996; Glassgold et al. 1997; Sano et al. 2000; Fleming & Stone 2003). The combination of low ionization fraction and high density means that charged particles receive too many random velocity "kicks" from collisions with neutrals for the frozen-field approximation to be true, an effect called Ohmic dissipation. The dead zone (if it exists) may have important consequences for planet formation: its low turbulence allows efficient dust grain growth (Dubrulle et al. 1995; Dullemond & Dominik 2004b), while its outer edge can halt migrating planets and drifting pebbles (Armitage et al. 2001; Masset et al. 2006; Hasegawa & Pudritz 2011; Mohanty et al. 2018) or spin off grain-trapping vortices (Lyra et al. 2009). Estimates of the dead zone radius range from 15 to 20 au in disks with surface densities similar to the mmsn (Matsumura & Pudritz 2006; Dzyurkevich et al. 2013; Landry et al. 2013; box 2.2) all the way to 60 au in a system with $M_d = 0.085M_\odot$ and $M_* = 0.5M_\odot$ (Flock et al. 2015), though results strongly depend on the size and abundance of dust grains (Ilgner & Nelson 2006). Other nonideal effects are ambipolar diffusion, which halts the MRI in tenuous regions where ion–neutral collisions are too rare for the ions to "lead" the neutrals into turbulence, and the Hall current, which comes from the relative motion of electrons and ions when electrons are coupled to the magnetic field but ions are coupled to the neutrals. The Hall current can revive MRI in zones with $\vec{\Omega} \cdot \vec{B} > 0$ that would otherwise be dead due to Ohmic dissipation (Wardle 2007; Lesur et al. 2014). Figure 2.22 shows $\nu(R, z)$ from the models of Landry et al. (2013) in a system with $M_* = M_\odot$, $M_d = 0.03M_*$, star age 1 Myr, and $R_{\text{out}} = 70\text{au}$. The models include Ohmic dissipation and ambipolar diffusion, but not the Hall current.

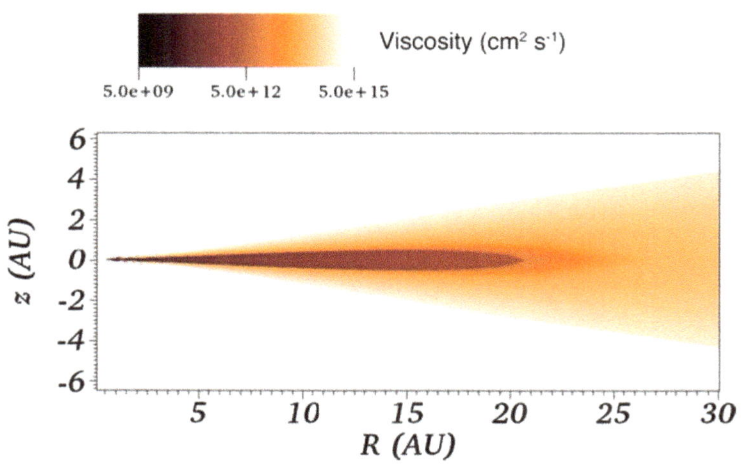

Figure 2.22. The viscosity profile of a disk orbiting a young sunlike star of age 1 Myr. The disk has mass $0.03M_\odot$ contained within 70 au of the star. The dead zone is the dark red region of low viscosity extending out to 20 au. Image: Reprinted from Landry et al. (2013).

To include the Ohmic dead zone in a global disk model, use the criterion $\Lambda_z > 1$ in turbulent regions, where the dimensionless Elsässer number Λ_z is the ratio of the length scale of the most rapidly growing unstable vertical mode to the diffusive length scale:

$$\Lambda_z \simeq \frac{v_A^2}{10\eta\Omega_K} \tag{2.69}$$

In Equation (2.69), $v_A = B/\sqrt{4\pi\rho}$ is the Alfvén speed of disturbances traveling along magnetic field lines, η is the magnetic diffusivity (units of $cm^2\ s^{-1}$; analogous to kinematic viscosity), and the factor of 10 accounts for the fact that the vertical magnetic pressure is $\sim 1/10$ of the total magnetic pressure of the mainly toroidal field produced by fully developed turbulence (Turner & Sano 2008). Exact computations of magnetic diffusivity η require chemical reaction networks that treat ionization and recombination in the gas phase and on grain surfaces (e.g., Semenov et al. 2004; Ilgner & Nelson 2006; Xu et al. 2019); from Lesur et al. (2014) and Lyra & Umurhan (2019), an approximate fitting formula is

$$\eta \approx 234\frac{T^{1/2}}{x}, \tag{2.70}$$

where x is the ionization fraction. To calculate x, follow the prescription in Section 2.4 of Lesur et al. (2014).

Observational limits on B in the TW Hydrae disk come from the polarimetry of Vlemmings et al. (2019), who find $|B_z| < 0.8$ mG (note that the stellar wind may exclude cosmic rays; Cleeves et al. 2013)—much lower than $B = 0.1$–1 G ($B_z \approx 0.03$–0.3 G) for the solar nebula, as measured by remnant magnetism in meteorites (Levy & Sonett 1978). AB Aur dust polarimetry reveals an ordered, probably poloidal magnetic field with $B = 1$–10 mG (Li et al. 2016). Theorists usually assume $\beta_0 = 10^3$–10^5, where

$$\beta = \frac{8\pi P}{B^2} \tag{2.71}$$

is the ratio of gas pressure to magnetic pressure and β_0 is the midplane value, from which B can be recovered (Fromang & Nelson 2006; Simon et al. 2018). If $\Lambda_z > 1$, turbulence develops with viscous efficiency (Bai & Stone 2011)

$$\alpha(R, z) = \frac{1}{2\beta(R, z)}. \tag{2.72}$$

With few observational constraints on magnetic field strengths in disks, parameter studies exploring a range of B and β are often required. To simulate the MRI turbulence itself, use one of the magnetohydrodynamic codes mentioned in Section 2.4.2: `ZEUS`, `PENCIL`, `PLUTO`, or `Athena`.

- **Baroclinic family:** Vertical shear instability (VSI; Nelson et al. 2013) and convective overstability (Petersen et al. 2007; Lesur & Papaloizou 2010) are products of baroclinic flows (Klahr & Bodenheimer 2003), where

$$\vec{\nabla} P \times \vec{\nabla}\rho \neq 0. \tag{2.73}$$

In the inviscid limit, the Navier–Stokes momentum equation is

$$\frac{d\vec{v}}{dt} + (\vec{v} \cdot \vec{\nabla})\vec{v} = -\frac{1}{\rho}\vec{\nabla} P + \vec{g}, \tag{2.74}$$

where $\vec{v}$ is the fluid velocity and $\vec{g}$ is the acceleration due to gravity. Focusing on the gravity term, $g_\phi = 0$ and g_R is balanced by the Keplerian flow. The disk gas would be fully hydrostatic in the vertical direction if $g_z = (1/\rho)(\partial P/\partial z)$ (as derived in box 2.3), with $\vec{\nabla} P \| \vec{\nabla}\rho$. However, in our vertically isothermal model (Section 2.3.2) the temperature contour lines cannot under any circumstances be parallel to the density contour lines given that there is a vertical density gradient. The isothermal equation of state (Equation (2.9)) tilts the isobars (pressure contour lines) with respect to the density contours, ensuring that g_z causes a small amount of vertical motion. Even nonisothermal passively heated disks (see Section 2.4.4) are likely to be baroclinic. Figure 2.23 shows crossing isobars and density contours from 10 to 14 au in the model disk of Yu et al. (2016), with the temperature structure computed using RADMC (Section 2.4.4). Since the disk's MRI dead zone extends out to 15 au, Figure 2.23 shows how Ohmically dead regions may meet the necessary conditions for baroclinic turbulence.

While in Section 2.3.3 we assumed the radial Keplerian shear was the energy source for the turbulent cascade from large eddies to small eddies, the VSI extracts energy from the baroclinic vertical shear. The VSI requires rapid cooling: a bubble

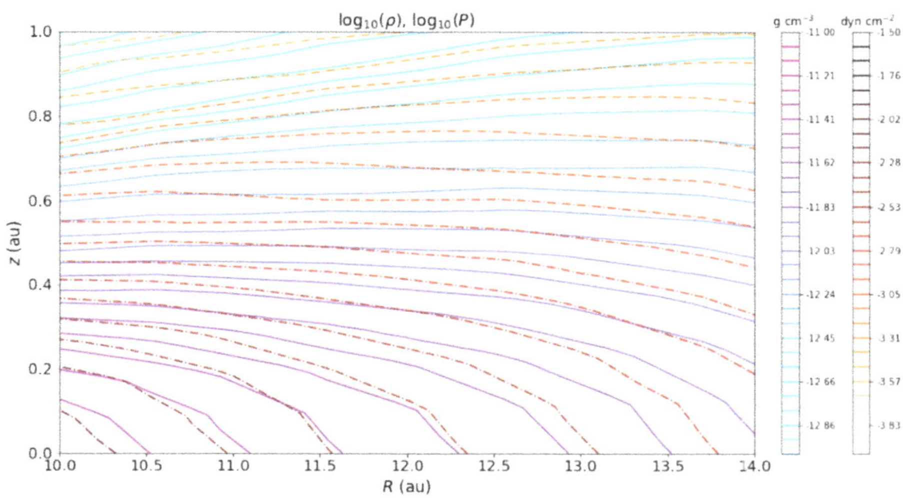

Figure 2.23. Pressure and density contours in a small section of the (R, z) plane from the Yu et al. (2016) disk model. Contours of $\log_{10}(\rho)$ are on the blue/purple color scale and contours of P are on the yellow/red/black color scale. The fact that the pressure and density contours cross means $\vec{\nabla} P \times \vec{\nabla}\rho \neq 0$, which can lead to baroclinic growth of vortices. Yu et al. (2016) used RADMC (Dullemond & Dominik 2004a) to compute the temperature profile.

of hot gas must transport out its excess heat in $\lesssim 1/10$ of an orbit (Nelson et al. 2013; Richard et al. 2016), or else buoyancy will restore vertical hydrostatic equilibrium. From Lin & Youdin (2015), the cooling criterion is

$$\Omega_K t_c < \frac{H}{R}\left|\frac{\partial \ln \Omega}{\partial \ln R}\right| \frac{1}{\gamma - 1}, \tag{2.75}$$

where t_c is the cooling time and $\gamma = 7/5$ is the ratio of specific heats of a diatomic gas. In optically thick regions of the disk, the cooling time is the radiative diffusion time:

$$\begin{aligned} \chi &= \frac{16\sigma_{\rm SB}T^3}{3\kappa_R\rho^2 c_V} \\ t_c &= \frac{\ell_p^2}{\chi}, \end{aligned} \tag{2.76}$$

where χ is the thermal diffusivity (units cm^2 s^{-1}; analog of kinematic viscosity), ℓ_p is the length scale of the perturbation, and c_V is the specific heat at constant volume in erg g^{-1} K^{-1} (Lyra & Umurhan 2019; note that we used molar units in Section 2.3.2: the c_V we use here has different units from the R_g in Section 2.3.2). The VSI forms tall, columnar flows of alternating positive and negative δv_ϕ, where the angular speed is $v_\phi = R\Omega_K + \delta v_\phi$ (Nelson et al. 2013; Stoll & Kley 2014). From Richard et al. (2016) and Flock et al. (2017), the turbulent speeds are a few percent of the local sound speed; one could parameterize VSI turbulence by setting, for example, $v_t = 0.03c_s$ and recovering the nonuniform $\alpha(R, z)$ using Equation (2.31) in the regions where Equation (2.75) is true. This parameterization would be adequate to model angular momentum transport from VSI, but one should not use it to decompose gas velocity into vector components (e.g., in modeling molecular line emission) because the VSI is not isotropic. Lyra & Umurhan (2019) note that the inertial range of the VSI has not been fully characterized. It is possible that VSI generates vortices but does not fully develop turbulence that cascades down to the molecular viscosity dissipation scale, in which case VSI would not be a source of accretion heating (Section 2.4.5). Long-lived vortices could aid planet formation by concentrating solids.

Convective overstability comes from radial convection, though vertical convection may be important in some regions of the disk (Klahr et al. 1999; Lesur & Ogilvie 2010) and could improve chances for planet formation by direct gravitational collapse (Boss 2002). The key ingredient in convective overstability is a negative radial entropy gradient (Klahr & Bodenheimer 2003). A hot, underdense gas blob moves outward because it is buoyant in the radial pressure gradient, and the entropy gradient amplifies its outward motion. The cycle completes when the blob cools, sinking back inward as its density increases. The Coriolis deflection forces the gas blob to oscillate azimuthally as well as radially, and as the blob cools it generates an azimuthal temperature and entropy gradient in the surrounding gas. A vortex grows as the azimuthal temperature gradient combines with the radial pressure/entropy gradient to satisfy Equation (2.73): the flow is now baroclinic. Overstability means each in–out cycle, or epicycle, is slightly larger than the preceding cycle. See Latter's (2016) Figure 2 for a cartoon illustrating the overstable oscillations. Convective overstability is not turbulence: rather than a cascade of energy from large to small eddies, small perturbations grow into large-scale vortices (Lyra 2014; Klahr & Hubbard 2014), which excite angular-momentum-transporting spiral waves. Heinemann & Papaloizou (2009) and Heinemann & Papaloizou (2012) find an angular momentum transport efficiency equivalent to $\alpha = 10^{-3}$, which can be put into Equations (2.30) and (2.28) for long-timescale disk evolution models. Because spiral waves are large-scale and do not dissipate energy locally, convective overstability should not be used as a source for accretion heating (Section 2.4.5). From Lyra & Umurhan (2019), convective overstability operates at the midplane when

$$\xi < 3\left(q_T - \frac{1}{2}\right), \tag{2.77}$$

where q_T is the power-law index of the radial temperature profile ($T \propto R^{-q_T}$). In our fiducial model disk with $\xi = 1$ and $q_T = 3/7$, convective overstability would not generate vortices.

• **Finite amplitude-initiated, vortex-forming instabilities:** While the MRI, VSI, and convective overstability grow from low-amplitude perturbations ($\delta v \ll c_s, v_\phi$), some instabilities are initiated by finite-amplitude disturbances. These instabilities form vortices that may be very large, even of order the disk radius. Some vortices decay into a (partial) turbulent cascade (Lesur & Papaloizou 2010) or migrate into the star (Paardekooper et al. 2010), while others live long enough for their pressure gradients to completely size-segregate the disk's dust grains. We begin with the Rossby wave instability (RWI; Lovelace et al. 1999; Li et al. 2000), which is a radial relative of the baroclinic instabilities described above. The RWI begins when a vertically thin disk modeled in 2D (R, ϕ) has a local extremum or strong gradient in a vorticity-related ratio

$$\mathcal{L} = \frac{\Sigma}{2(\vec{\nabla} \times \vec{v})_z} S^{2/\Gamma}, \tag{2.78}$$

where S is the entropy

$$S(R) = \frac{\int_{-H}^{H} P(z)dz}{\Sigma(R)^{\Gamma}}, \tag{2.79}$$

Γ is an effective adiabatic index (see Section 3 of Li et al. 2000), and $\vec{\nabla} \times \vec{v}$ is the vorticity, the z-component of which records velocity curl in the (R, ϕ) plane. One may achieve an extremum of $\mathcal{L}$ by maximizing or minimizing Σ, P, or $(\vec{\nabla} \times \vec{v})_z$: a large enough nonaxisymmetric perturbation creates surface density contours and isobars that are not parallel ($\vec{\nabla}\Sigma \times \vec{\nabla}P \neq 0$) so that pressure generates azimuthal flow. However, Lovelace et al. (1999) suggest that the pressure maximum is the most effective instability generator. Lovelace et al. (1999) point out that a local pressure maximum can be achieved at the centrifugal radius of a Stage I envelope (Section 2.2), where gas is deposited at the outer edge of the disk: mass and pressure buildup will naturally occur if the infall rate from the cloud exceeds the accretion rate through the disk. The steep pressure gradients at the edges of gaps produced by planets are also natural candidates for Rossby vortices (Koller et al. 2003; Huang et al. 2018; Baruteau et al. 2019), as are the edges of dead zones (Varnière & Tagger 2006; Pierens & Lin 2018), where the sudden change in inward gas radial velocity v_R due to an abrupt jump in ν changes the pressure according to Bernoulli's principle.

The pressure/surface density fluctuations that initiate the RWI must have a finite amplitude because the perturbations that grow are those that are "trapped" at the extremum or gradient of $\mathcal{L}(R)$: small-scale disturbances can damp out as they cross the region with strongly varying $\mathcal{L}(R)$. (Indeed, the Lovelace et al. 1999 linear stability analysis is built on the assumption that $\ell_p > H$.) Li et al. (2000) state that a 10%–20% change in $\Sigma(R)$ over length scale $\ell_p \gtrsim H$ will result in instability. Perturbations may spin into long-lived vortices if the pressure or surface density extremum that formed them is semipermanent, such as a dead-zone edge or planet gap; vortices created at a transitory disk structure will be destroyed by the Keplerian shear. An example long-lived Rossby vortex is Jupiter's great red spot. Many disks have large "crescent" structures that could be Rossby vortices, most notably Oph-IRS 48 (van der Marel et al. 2013; Figure 2.24), SAO 206462, SR 21 (Pérez et al. 2014), and MWC 758 (Baruteau et al. 2019). The anticyclonic vortices (i.e., spinning clockwise in a counterclockwise flow) are visible because they trap millimeter to centimeter-size dust grains, creating asymmetry in (sub)millimeter continuum images.

To parameterize the RWI in an analytical model, one could use the Li et al. (2000) condition (10%–20% $\Sigma(R)$ variation over length H) to check whether vortices could form. Guidance on the number of vortices in the unstable ring (i.e., the mode number of the instability) and the vortex size comes from Meheut et al. (2010, 2012),

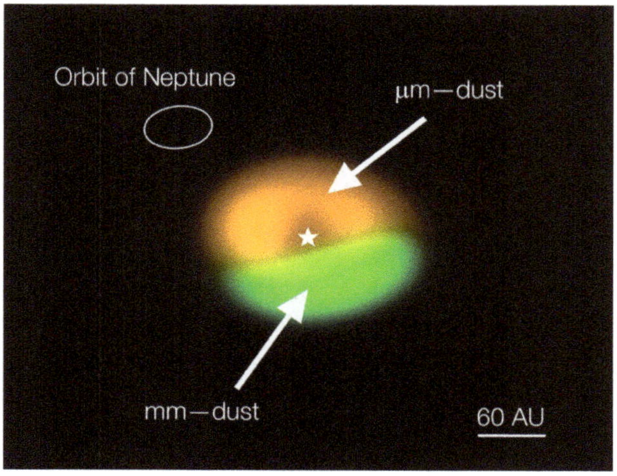

Figure 2.24. A vortex in the disk surrounding Oph-IRS 48 traps millimeter-size dust grains on one side of the disk, creating a safe zone for planet-forming pebbles (Chapter 3). Image credit: ALMA (ESO/NAOJ/NRAO)/ Nienke van der Marel.

who used the hydrodynamics module of the publicly available code `MPI-AMRVAC`[23] (Keppens et al. 2012) to simulate the RWI in 3D. `MPI-AMRVAC` also has a magnetohydrodynamic module and could be used to self-consistently simulate the growth of the RWI at dead-zone edges. Rossby waves can initiate a series of inward- and outward-traveling vortices that transport mass and angular momentum, but it is not clear whether the vortices break down into turbulence. We do not recommend using the RWI as a turbulent viscosity source within the α framework (either Section 2.3.3 or 2.4.5), which assumes that energy is dissipated locally. Instead, the RWI is most relevant to the disk's large-scale structure. Lyra & Lin (2013) present an analytical prescription for the gas density in a vortex (their equations 36 and 37) and also specify in-vortex dust distributions that can be used to infer vortex properties from images such as Figure 2.24.

Another finite amplitude-initiated turbulence candidate is the zombie vortex instability (ZVI), discovered by Marcus et al. (2013). Once a vortex is formed (e.g., by convective overstability or the RWI), it gets azimuthally stretched into a filament by the Keplerian flow and forms a strong gradient in $\mathcal{L}(R)$: the vortex is now a Rossby wave source in its own right. Because the vortex is orbiting the star, its Rossby waves originate from a moving source and so get Doppler-shifted. If the Doppler-shifted waves resonate with the buoyancy frequency (the Brunt–Väisälä frequency, on which displaced gas parcels oscillate from their hydrostatic equilibrium positions) at any disk location, small buoyancy perturbations will grow and get stretched into filaments and the cycle begins anew.[24] From Lesur & Latter (2016) and Malygin et al. (2017); the ZVI is active under the following conditions:

$$N^2 t_c^2 \gg 1 \quad \text{and} \quad Pe > 10^4, \tag{2.80}$$

where t_c is the radiative diffusion time given by Equation (2.76), N is the Brunt–Väisälä frequency, and Pe, the Peclet number, is the ratio of the advection time to the thermal diffusion time:

$$\begin{aligned} Pe &= \frac{\Omega L^2}{\chi} \\ N &= \sqrt{-\frac{g_z}{\rho_0}\frac{\partial \rho}{\partial z}}. \end{aligned} \tag{2.81}$$

In Equation (2.81), g_z is the vertical component of the stellar gravity and ρ_0 is the midplane density.

Because ZVI growth rates are slow (timescale $\lesssim 0.1/\Omega_K$), filament cooling times must also be slow: $t_c > 10$ orbits (Lesur & Latter 2016). The requirement that the

[23] http://amrvac.org/

[24] Lyra & Umurhan (2019) liken the ZVI to resonances in stringed instruments: a violinist or violist knows that playing a perfectly in-tune "A" on the D string will make the open A string vibrate as well, creating a full, ringing sound. Conversely, one can induce vibrato when playing an open string by fingering (with vibrato) the same note on another string; the resulting sound has the carrying power of the open string, but without its stridency.

cooling time be long means the growing modes are large: small-scale perturbations have time to radiate away excess heat before the instability can grow. Umurhan et al. (2016) and Marcus et al. (2016) suggest that the instability propagates into the midplane and the filaments truly break down into turbulence. If so, it is acceptable to use the ZVI as a source of α viscosity for angular momentum transport (Section 2.3.3; Equation (2.28)) and accretion heating (Section 2.4.5). According to Marcus et al. (2016), $\alpha \approx 4 \times 10^{-4}$ for ZVI-initiated turbulence. However, Barranco et al. (2018) find that the ZVI spawns large-scale zonal flows—successive rings of gas with perturbed velocities δv in opposite directions. More research is needed to determine whether the ZVI fits the α framework. Figure 2.25 shows self-propagating zombie vortices in the fiducial model of Lesur & Latter (2016; their figure 1).

For guidance on which purely hydrodynamic instabilities operate in which parts of the disk, see figure 12 of Malygin et al. (2017) and figure 15 of Lyra & Umurhan (2019). Note that both of these papers use the mmsn (box 2.2) as their base model; we assert that the mmsn represents the end stages of planet formation, not the beginning, and recommend a more massive model disk.

Our last type of turbulence arises not from (magneto)hydrodynamic instabilities, but from the disk's self-gravity. In the next section, we develop analytical tools for assessing the disk's gravitational stability and examine possible outcomes of instability, including gravitoturbulence, spiral waves, and fragmentation.

2.4.7 Self-Gravity: Gravitoturbulence, Spirals, and Fragmentation

So far we have assumed the disk's self-gravity is negligible compared to the star's gravity. At $M_d = 0.1 M_\odot$, our fiducial model (Sections 2.3.2, 2.4.4, and 2.4.5) is at the uppermost end of the mass ratio range M_d/M_* over which that assumption is realistic. We begin our analysis of self-gravity by deriving the Toomre (1964) Q criterion, which determines whether an overdense "blob" will collapse under its own

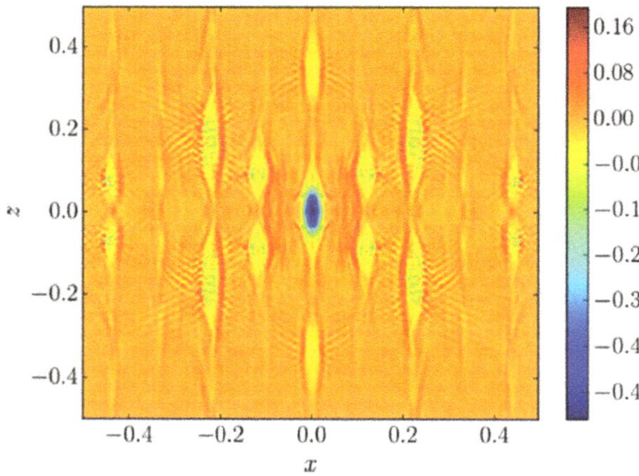

Figure 2.25. An (R, z) slice of the shearing box simulations of Lesur & Latter (2016) showing the formation of zombie vortices ($x = R$ in the figure label). The color scale shows the z-component of the vorticity, $\vec{\nabla} \times \vec{v}$ (where $\vec{v}$ is the gas velocity). Image: Reproduced from Lesur & Latter (2016), copyright of OUP, copyright 2016.

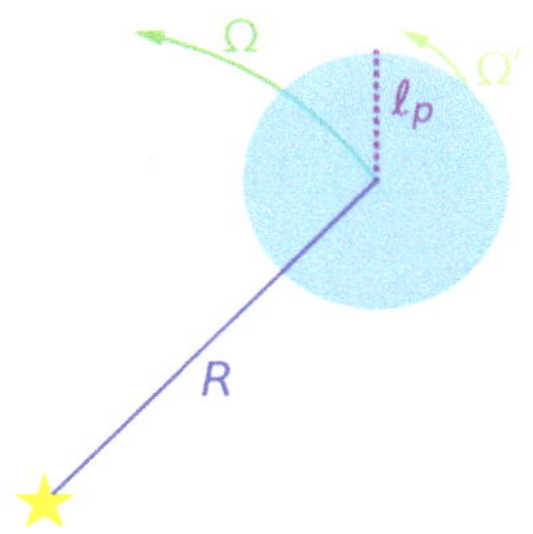

Figure 2.26. Surface density perturbation for stability analysis: an overdense gas blob with radius ℓ_p is located at distance R from the central star and has angular speed Ω at its center. The outermost edge of the blob has angular speed Ω'.

self-gravity or get smoothed out. As in Section 2.3.3, we model an infinitely thin disk. Figure 2.26 shows our blob—a circular perturbation with uniform surface density Σ, radius ℓ_p, distance from the central star R, and angular speeds Ω and Ω' at its center and outermost edge, respectively.

First we test whether gas pressure can stabilize the blob. If sound waves traveling at speed c_s can traverse the blob before it has time to collapse under its own gravity, the perturbation will damp out. We find the maximum stable perturbation size or Jeans length L_J (Jeans 1902; Carroll & Ostlie 2017, chapter 12) by setting the sound-crossing time L_J/c_s equal to the free-fall time $\tau_{\rm ff}$. An annulus at the edge of our perturbation feels gravitational acceleration $a_{\rm blob} = GM_{\rm blob}/\ell_p^2 = \pi G\Sigma$ toward the blob center, and the annulus will fall a distance of $d = (1/2)a_{\rm blob}t^2$ in time t.[25] By setting $d = \ell_p$, we derive

$$\begin{aligned} \tau_{\rm ff} &= \sqrt{\frac{2\ell_p}{\pi G\Sigma}} \\ L_J &= \frac{2c_s^2}{\pi G\Sigma}. \end{aligned} \tag{2.82}$$

The other stabilizing influence is the Keplerian shear. Without self-gravity, the blob's outer edge orbits the star more slowly than its center ($\Omega' < \Omega$) and the perturbation stretches out and eventually dissipates. However, the blob may be massive enough for self-gravity to defeat Keplerian shear. We find the minimum blob size L_s that can be destroyed by shear by setting $\tau_s = \tau_{\rm ff}$, where τ_s is the shear time:

$$\tau_s = \frac{\ell_p}{v_\phi - v_\phi'} = \frac{\ell_p}{R\Omega - (R + \ell_p)\Omega'}. \tag{2.83}$$

[25] Here we are assuming an "inside-out" collapse so that no falling annulus of gas gets stopped by a collision with its inner neighbor.

Given $v'_\phi = (GM_*)^{1/2}(R + \ell_p)^{-1/2}$ and assuming $\ell_p \ll R$, we can rewrite $R + \ell_p = R(1 + \ell_p/R)$ and use a binomial expansion to find a first-order approximation $(R + \ell_p)^{-1/2} \approx R^{-1/2}(1 - \ell_p/2R)$. The denominator of Equation (2.83) is then $\ell_p v_\phi/2R$, which yields

$$\begin{aligned} \tau_s &= \frac{2}{\Omega} \\ L_s &= \frac{2\pi G\Sigma}{\Omega^2}. \end{aligned} \tag{2.84}$$

Because the shear time does not formally depend on the perturbation size, Keplerian shear is best able to stabilize large-scale perturbations with long free-fall times. On the other hand, pressure is best at stabilizing small-scale perturbations because the sound-crossing time increases linearly with ℓ_p, while $\tau_{\rm ff} \propto \sqrt{\ell_p}$. A growing perturbation therefore has $L_J < \ell_p < L_s$. Wherever $L_s < L_J$, the disk is gravitationally stable to all surface density perturbations. We define the stability criterion as $Q = \sqrt{L_J/L_s}$, or

$$Q = \frac{c_s\Omega}{\pi G\Sigma}. \tag{2.85}$$

$Q = 1$ is the stability threshold at which $L_J = L_s$; with increasing Q comes increasing gravitational stability.

If $Q \lesssim 1$, what happens to the disk? One possibility is gravitoturbulence. Our derivation of the Toomre Q assumes a perfectly isothermal disk, but in reality, a contracting blob of gas will experience adiabatic heating. The heat produced by contraction must be radiated away or the perturbation cannot fully collapse. Instead, it dissolves into turbulence, as in the MRI and VSI (Section 2.4.6). Because optically thin regions are effectively isothermal and cool almost instantaneously, we usually expect gravitoturbulence in optically thick regions. From Gammie (2001), Rice et al. (2005), Rafikov (2009), and Lin & Kratter (2016), gravitoturbulence has nearly uniform efficiency:

$$\alpha = \frac{4}{9}\frac{\left(\sqrt{\gamma} - 1\right)^{1/2}}{\gamma\left(\sqrt{\gamma} + 1\right)} \approx 0.06. \tag{2.86}$$

The conditions for gravitoturbulence are $\tau_c > \tau_{\rm ff}$, where τ_c is the cooling time, and $Q < Q_{\rm crit}$. τ_c differs from t_c from Equation (2.76) in that it covers both optically thin and optically thick regions:

$$\tau_c = \frac{4}{9\gamma(\gamma - 1)}\frac{\Sigma c_s^2}{\sigma_{\rm SB}T^4}\tau_R \tag{2.87}$$

(Kratter & Lodato 2016). The critical Q value is $0.7 \lesssim Q_{\rm crit} \lesssim 1.7$ (Kim & Ostriker 2002; Boss 2002), depending on which physical complexities are included in the analysis. For example, vertical pressure support stabilizes disks, requiring $Q < 1$ for the onset of gravitoturbulence (Rice et al. 2003), while turbulent angular momentum transport (e.g., by MRI or hydrodynamic instabilities) removes rotational support and has a destabilizing influence (Lin & Kratter 2016). It is usually safe to assume that gravitoturbulence dissipates energy on small-enough scales to fit our prescription for accretion heating (Section 2.4.5).

Another possible outcome when $Q \lesssim 1$ is a large-scale spiral structure. While our "circular blob" model builds useful physical insight about gravitational instability, large-scale surface density disturbances actually take the form

$$\Sigma(R,\, t) = \Sigma_0 \exp[i(kR + \omega t)], \tag{2.88}$$

where $k = 2\pi/\ell_p$ is the radial wavenumber and ω is the growth rate of the perturbation. Equation (2.88) comes from a Fourier series solution to Poisson's equation for the gravitational potential Φ from the disk gas:

$$\nabla^2\Phi = 4\pi G\Sigma\delta(z), \tag{2.89}$$

where the delta function encodes our assumption that all the mass is in the $(R,\, \phi)$ plane. In Equation (2.88), both k and ω are complex. We can rewrite the equation as

$$\Sigma(R,\, t) = \Sigma_0 \exp(ikR)[\cos(\omega_r t) + i\,\sin(\omega_r t)]\exp(-\omega_i t), \tag{2.90}$$

where ω_r and ω_i are the real and imaginary parts of ω. The angle of ω in the complex plane indicates the wave-launching direction. From inspecting the final term on the right-hand side of Equation (2.90), we see that a negative value of ω_i causes the perturbation to grow, destabilizing the disk. From Toomre (1964) and Lin & Shu (1964), the dispersion relation for our density waves is

$$[\omega - m\Omega(R)]^2 = c_s^2 k^2 - 2\pi G\Sigma|k| + \kappa_e^2, \tag{2.91}$$

where m is the number of spiral arms and κ_e is the epicyclic frequency of radial density oscillations. Setting $m = 0$ to specify axisymmetric perturbations and recalling that $\kappa_e = kc_s = \Omega$ for a Keplerian disk, we recover the Toomre Q stability criterion from Equation (2.85)!

Globally, a spiral structure occurs when $M_d/M_* \gtrsim 0.1$ (Kratter & Lodato 2016). Dong et al. (2018) show that gravitational instability, as opposed to wakes from forming giant planets, is the best explanation for the spiral structure observed in a set of Herbig Ae/Be disks, which tend to have higher M_d/M_* than T-Tauri disks. Locally, spiral structure and gravitoturbulence appear in almost identical conditions. From Gammie (2001) and Durisen et al. (2007), spirals form when $Q \lesssim Q_{\rm crit}$ and $\beta_{\rm cool} = \Omega_K \tau_c \gtrsim 3$. When $\beta_{\rm cool} \lesssim 3$, perturbations can truly detach from the surrounding disk and form fragments—possible precursors to stars, brown dwarfs, or super-Jupiters (Chapter 6). Longer cooling times saturate the instability so that spiral waves reach a maximum surface density contrast that scales as

$$\delta\Sigma/\Sigma \propto \frac{1}{\sqrt{\beta_{\rm cool}}}, \tag{2.92}$$

then stop growing (Cossins et al. 2009; Kratter & Lodato 2016). Figure 2.27 shows a $\lambda = 1.3$mm ALMA image of the $m = 2$ spiral in the Elias 2-27 disk (Pérez et al. 2016) along with the molecular cloud where the disk resides. If the continuum emission is optically thin (a big assumption), the dust surface density contrast is $\delta\Sigma/\Sigma \gtrsim 2$. The logarithmic spiral pitch angle is 7.9°, which is actually too loose to satisfy the "tight-winding" approximation used to derive Equation (2.91). Elias 2-27 is nonetheless our best observational benchmark for models of spiral structure.

Torques from spiral waves provide large-scale angular momentum transport (Lin & Pringle 1987). Kratter et al. (2008) show that one can shoehorn this type of disk evolution into the diffusion formalism of Section 2.3.3 and Equation (2.28) despite the fact that momentum exchange is not local. We use Equation (2.30) to define a spiral-wave "viscosity" with

$$\alpha = \max\left[\frac{1.4 \times 10^{-3}(2 - Q)}{\mu_d^{5/4} Q^{1/2}}, 0\right], \tag{2.93}$$

where $\mu_d = M_d/(M_d + M_*)$. Equation (2.93) should not be used to describe accretion heating. Usually the $m = 2$ mode dominates mass and angular momentum transport by spiral waves (e.g., Mejía et al. 2005).

It is not always obvious whether self-gravity should lead to spiral structure or gravitoturbulence. The outcome is determined by the extent to which waves with different values of k interact (mode–mode coupling): gravitoturbulence occurs when

Figure 2.27. The Ophiuchus star-forming region (left) is home to Elias 2-27, an unusually massive disk ($M_d \sim 0.1 M_\odot$) that may be gravitationally unstable. On the right is an unsharp-masked ALMA image showing a spiral structure. Image credit: L. Pérez (MPIfR), B. Saxton (NRAO/AUI/NSF), ALMA (ESO/NAOJ/NRAO), NASA/JPL Caltech/WISE Team.

energy from large perturbations can cascade to smaller ones (Laughlin et al. 1997). Mode–mode coupling only appears in second-order (or higher) perturbation analyses or numerical simulations, and its manifestations often depend on the viscosity prescription (Durisen et al. 2007). However, spiral structure and gravitoturbulence yield very different planet formation prospects: while the overdense spiral arms can spur planetesimal formation or even fragment into gas giants if grain growth improves their cooling efficiency, gravitoturbulence stirs the dust grains and inhibits planet formation.

2.4.8 Disks in Binary Systems

So far we ha/ve developed an extensive disk-modeling toolbox including turbulence, passive and active heating, self-gravity, inner holes, and more. Yet we have restricted our analysis to single stars, even though the multiplicity fraction is 0.46 for F and G dwarfs (Raghavan et al. 2010; Tokovinin 2014) and rises for higher-mass stars! Focusing on single stars has the obvious benefit of simplifying our disk models and is perhaps forgivable if our goal is to connect with observations. Planets in known binary systems number in the dozens (Martin et al. 2019), whereas thousands of confirmed exoplanets orbit single stars.[26] Radial-velocity (RV) planet searchers have either avoided tight binary targets for fear of spectral contamination from the secondary or have come up empty when searching for circumbinary planets in systems with low-mass, low-flux secondaries (Konacki et al. 2009; Martin et al. 2019). Yet RV searches for planets orbiting stars in wide-separation systems are starting to pay off, exemplified by the discovery of Proxima Centauri b—a potentially habitable near neighbor (Anglada-Escudé et al. 2016). Furthermore, the Kepler and K2 missions have detected circumbinary "Tatooine" planets (e.g., Kostov et al. 2016), including one located inside the theoretical semimajor axis limit for stable orbits (Doyle et al. 2011). Both data and models suggest that while disks around single stars may be the most fertile ground for growing planets (Paardekooper et al. 2008; Meschiari 2012), there is still a need for models of planet-forming disks in binary systems. Boss (2006) even suggests that a marginally gravitationally stable disk (Section 2.4.7) may be pushed over the edge into fragmentation by a companion star, perhaps creating a pathway toward the formation of super-Jupiters.

The observational biases against disks in young binary systems—either circumbinary or individual to each star—are less severe than the biases against planets in mature binary systems. For example, in ALMA observations of Stage II stars with either infrared excess or evidence of accretion (see Section 2.2 and box 2.1), Akeson et al. (2019) detected millimeter emission from at least one star in every binary pair, signaling the existence of a (hopefully) planet-forming disk. The sample of disks in young binaries is large enough for us to draw tentative conclusions about the underlying population. Here are some significant findings from observations of disks in binaries:

[26] http://exoplanet.eu/catalog/

- **Disks in close binary systems do not last long.** When the binary separation is $\lesssim$40 au, the disk(s) (either a single circumbinary or individual truncated disks surrounding each star) have a mean life of only 1 Myr (Kraus et al. 2012). In contrast, we saw in Section 2.3.1 that the ensemble of disks in nearby star-forming regions has mean lifetimes of 2–5.5 Myr, depending on observing wavelength (Ribas et al. 2014). We will see in Chapters 3 and 4 how much planet formation is a race against the clock—especially for giant planets. The paucity of planets detected in close binary systems may reflect physical reality as well as observational bias.
- **The prospects for disk survival improve dramatically with increasing binary separation.** Kraus et al. (2012) found that wide binaries (projected separation $\gtrsim$40 au) in Taurus-Auriga, age ~2 Myr, have the same disk occurrence rate as single stars, while Cieza et al. (2009) found that binary systems with projected separations between 40 and 400 au were twice as likely to host at least one disk than binaries with separation $\lesssim$40 au. Duchêne (2010) even asserts that disks in binaries with $a \gtrsim$ 100 au (where a is the semimajor axis) are identical to disks surrounding single stars (Figure 2.28; top panel), though Fragner & Nelson (2010) suggest that such disks may still develop warps as they are slowly torqued toward alignment with the binary orbit. Companion stars separated by $\gtrsim$100 au probably formed in separate fragments of the collapsing cloud core; the resulting disks are expected to be misaligned (Bate & Bonnell 1997).
- **In wide binaries, the circumprimary disk is not necessarily more massive than the circumsecondary disk.** M_d/M_* appears not to depend on the mass ratio of the host stars (Akeson & Jensen 2014). If both disks are accreting from an envelope, as in Stage I, it is even possible that the circumsecondary disk will grow fastest because envelope gas has to lose more angular momentum to join the circumprimary disk rather than the circumsecondary disk. In main-sequence systems, the primary star should not necessarily be the preferred planet-search target, even in the absence of practical constraints such as spectral type and star radius.
- **The inner holes of circumbinary disks are not fully empty.** We saw in Section 2.4.2 that gaps surrounding growing planets are depleted, but not fully cleared. Physical intuition and Equation (2.36) tell us that the bigger the mass of the companion, the bigger the surface density contrast between the gap and the rest of the disk. However, even binary orbits might not completely sweep away the surrounding gas. While some circumbinary disks no longer feed star accretion (Ireland & Kraus 2008), others show evidence for hot gas surrounding the star orbits (Najita et al. 2003). Artymowicz & Lubow (1996) demonstrated that channeled flows could cross the gap between a star surface and the disk's inner wall without either being choked off or refilling the gap. The bottom panel of Figure 2.28 shows the circumbinary disk L1551 NE, the inner hole of which hosts channel flows that feed each individual star's subdisk (Takakuwa et al. 2014).

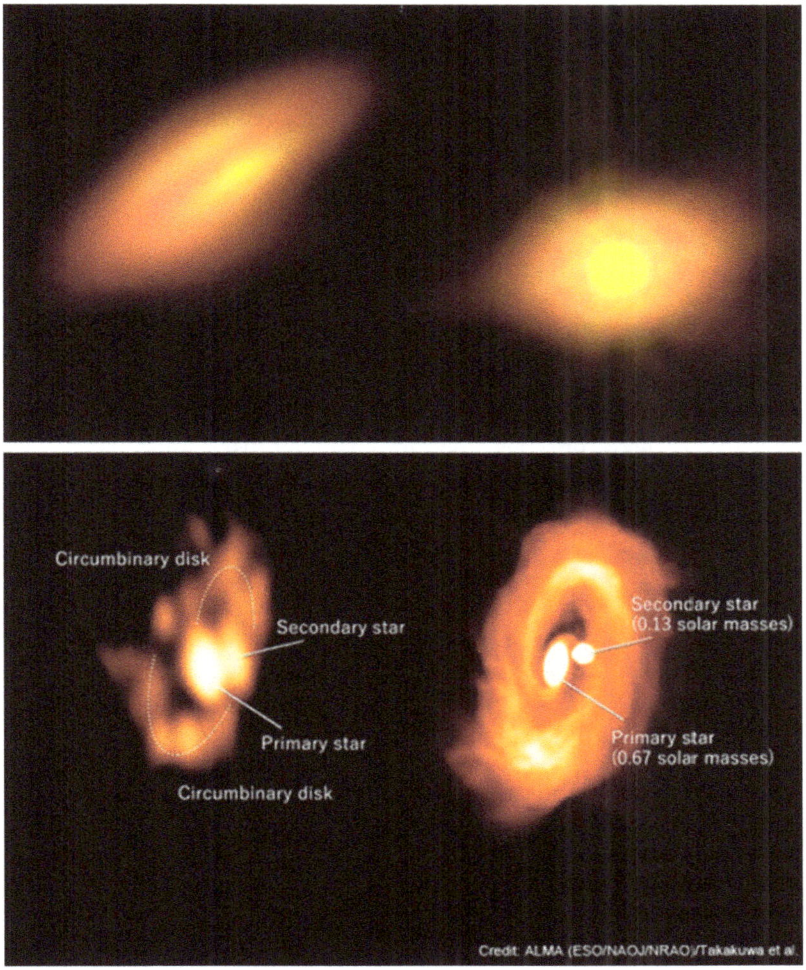

Figure 2.28. Top: ALMA and Hubble Space Telescope composite image of the young binary system HK Tau (Stapelfeldt et al. 1998; Jensen & Akeson 2014). The two disks are mutually misaligned by 60°, meaning at least one of the disks is misaligned with the orbital plane by a minimum of 30°. The disks will likely stay misaligned with each other and the binary orbit given the large projected separation of 386 au. HK Tau is an example of the situation where each disk could be modeled as if it orbited a single star (Duchêne 2010). Image credit: B. Saxton (NRAO/AUI/NSF), K. Stapelfeldt et al. (NASA/ESA Hubble). Bottom: ALMA image (left) and simulation (right) of the binary protostar L1551 NE and its circumbinary disk (Takakuwa et al. 2014). The stars have opened an inner hole in the larger disk, but each is surrounded by its own subdisk.

From a disk-modeling point of view, we are primarily concerned with three consequences of binarity: tidal truncation, warps/spirals, and eccentricity. Truncation, which takes place on the dynamical timescale, can take the form of either a wide inner hole in a circumbinary disk or small outer radii for the circumstellar disks in a binary system. For example, Wagner et al. (2018) measured an outer radius of only $R_{\mathrm{out}} = 40$ au for the disk surrounding HD 100435 A, whereas some disks have $R_{out} > 300$ au (e.g., CoKu Tau 1 and DM Tau; Padgett et al. 1999; Simon et al. 2000). HD 100435 is, of course, a binary with semimajor axis $a = 105$ au. FARGO-3D and PLUTO can be extended to simulate the dynamical evolution of circumbinary disks, with embedded planets if so desired (Thun & Kley 2018). In such models, the two stars are not part of an N-body simulation with the planets; instead, their gravitational effects are captured by prescribed potentials. In a numerical model, disk truncation—including the shapes of inner/outer edges—is computed self-consistently. To parameterize truncation, use the results of Artymowicz & Lubow (1994) as a guide: in circumbinary disks,

$$\begin{aligned} R_{in} &> 1.8a \quad \text{if } e = 0 \\ R_{in} &> 2.6a \quad \text{if } e = 0.25, \end{aligned} \tag{2.94}$$

where R_{in} is the radius of the inner-disk edge, a is the binary semimajor axis, and e is the eccentricity. In wide binaries, the circumstellar disks must have $R_{out} < 0.15$–$0.45a$, depending on the exact orbital parameters. Note that Artymowicz & Lubow (1994) only considered disks that are coplanar with the binary orbit; observations show that disks are typically slightly, but not wildly, misaligned with respect to the binary orbit ($\lesssim 20°$, Jensen et al. 2004).

The discussion of disk/orbit (mis)alignment leads us to consider warps and spiral arms. Warps are the means by which disks are slowly brought into alignment with the binary orbit. Pontoppidan et al. (2008) presented spectroastrometry of TW Hya that suggested an inner-disk warp, which could be caused by a massive, embedded companion (i.e., giant planet or brown dwarf). The inner-disk gap later discovered by Andrews et al. (2016) is consistent with the embedded companion hypothesis. The disk surrounding AS 205 N has spiral arms that may result from a close encounter with its companion AS 205 S, which is itself a tight binary encircled by a disk (Kurtovic et al. 2018). MacFadyen & Milosavljević (2008) suggest that spiral density waves are stationary in reference frames corotating with the binary. Binary-induced warps and spirals present a much more formidable modeling challenge than truncation; warps, especially, must be simulated in 3D, and a cylindrical coordinate system is not appropriate when the simulation must include both disks. Price et al. (2018) used the smoothed particle hydrodynamics code PHANTOM[27] with radiative transfer postprocessing by MCFOST to model the myriad features—warp, spiral, inner hole, dust horseshoe, and fast radial flows—in the HD 142527 circumbinary disk. They were able to confirm that all could be produced by torques from the eccentric binary as it approaches periastron. On the analytical side, Foucart & Lai (2014) offer steady-state solutions to the warp and twist profiles in a circumbinary disk misaligned to the orbital plane. Facchini et al. (2013) do the same, but focus specifically on low-viscosity disks; we note that all of the viscosity-generating mechanisms discussed in this chapter fall into their "low viscosity" regime.

Eccentricity within the disk has profound consequences for planet formation. Collisions between pebbles and planetesimals are least destructive when the objects'

[27] https://phantomsph.bitbucket.io/

velocity dispersion is low; increasing both the orbital speed (as at the periastron of an eccentric orbit) and the random velocity component (as with repeated passes through a spiral wave while drifting inward) could severely frustrate planet growth (Meschiari 2012). The MWC 758 disk has spirals plus an outer edge that is best fitted by an ellipse with eccentricity 0.1 rather than a circle, suggesting a companion beyond 100 au (Dong et al. 2018). By transporting angular momentum outward, spiral waves tend to increase the disk eccentricity (Papaloizou et al. 2001); the disk then precesses slowly (MacFadyen & Milosavljević 2008). The apoastra of individual eccentric orbits line up to form the spiral. While binary companions typically force higher-order modes, the $m = 1$ mode in which the entire disk is eccentric and rigidly precessing is the longest lasting: if the $m = 1$ mode is excited, the eccentricity could persist for the entire disk lifetime. Equation (1) of Lee et al. (2019) can be solved numerically to find the surface density distribution of an eccentric disk.

Having read this chapter, you now have the tools to build your own model of a planet-forming environment. In the next chapter, we will investigate the growth of solid bodies from grains into pebbles—the first step in a growth process that will eventually churn out giant planets.

References

Adams, F. C., Hollenbach, D., Laughlin, G., et al. 2004, ApJ, 611, 360
Adams, F. C., Lada, C. J., & Shu, F. H. 1987, ApJ, 312, 788
Aikawa, Y., van Zadelhoff, G. J., van Dishoeck, E. F., et al. 2002, A&A, 386, 622
Akeson, R. L., & Jensen, E. L. N. 2014, ApJ, 784, 62
Akeson, R. L., Jensen, E. L. N., Carpenter, J., et al. 2019, ApJ, 872, 158
Alexander, C. M. O., McKeegan, K. D., & Altwegg, K. 2018, SSRv, 214, 36
Alexander, R. D., & Armitage, P. J. 2007, MNRAS, 375, 500
Alexander, R. D., Clarke, C. J., & Pringle, J. E. 2006, MNRAS, 369, 229
Alexander, R. D., & Pascucci, I. 2012, MNRAS, 422, L82
Alexander, R., Pascucci, I., Andrews, S., et al. 2014, in Protostars and Planets VI (Tucson, AZ: Arizona Univ. Press), 475
ALMA Partnership,, Brogan, C. L., & Pérez, L. M. 2015, APJL, 808, L3
André, P., Ward-Thompson, D., & Barsony, M. 2000, in Protostars and Planets IV (Tucson, AZ: Arizona Univ. Press), 59
Andrews, S. M., Huang, J., Pérez, L. M., et al. 2018, ApJ, 869, L41
Andrews, S. M., Rosenfeld, K. A., Kraus, A. L., et al. 2013, ApJ, 771, 129
Andrews, S. M., Wilner, D. J., Espaillat, C., et al. 2011, ApJ, 732, 42
Andrews, S. M., Wilner, D. J., Hughes, A. M., et al. 2009, ApJ, 700, 1502
Andrews, S. M., Wilner, D. J., Hughes, A. M., et al. 2012, ApJ, 744, 162
Andrews, S. M., Wilner, D. J., Zhu, Z., et al. 2016, ApJL, 820, L40
Anglada-Escudé, G., Amado, P. J., Barnes, J., et al. 2016, Natur, 536, 437
Anthony, D. M., & Carlberg, R. G. 1988, ApJ, 332, 637
Aresu, G., Kamp, I., Meijerink, R., et al. 2014, A&A, 566, A14
Armitage, P. J., Livio, M., & Pringle, J. E. 2001, MNRAS, 324, 705
Artymowicz, P., & Lubow, S. H. 1994, ApJ, 421, 651
Artymowicz, P., & Lubow, S. H. 1996, ApJL, 467, L77

Arun, R., Mathew, B., Manoj, P., et al. 2019, AJ, 157, 159
Aso, Y., Ohashi, N., Saigo, K., et al. 2015, ApJ, 812, 27
Bae, J., Nelson, R. P., Hartmann, L., et al. 2016, ApJ, 829, 13
Bai, X.-N. 2011, ApJ, 739, 50
Bai, X.-N. 2013, ApJ, 772, 96
Bai, X.-N. 2014, ApJ, 791, 137
Bai, X.-N. 2017, ApJ, 845, 75
Bai, X.-N., & Stone, J. M. 2011, ApJ, 736, 144
Bai, X.-N., & Stone, J. M. 2013, ApJ, 769, 76
Bai, X.-N., & Stone, J. M. 2014, ApJ, 796, 31
Bailey, V., Meshkat, T., Reiter, M., et al. 2014, ApJ, 780, L4
Balbus, S. A., & Hawley, J. F. 1991, ApJ, 376, 214
Balbus, S. A., & Hawley, J. F. 1998, RvMP, 70, 1
Ballering, N. P., & Eisner, J. A. 2019, AJ, 157, 144
Banzatti, A., Meyer, M. R., Bruderer, S., et al. 2012, ApJ, 745, 90
Barranco, J. A., Pei, S., & Marcus, P. S. 2018, ApJ, 869, 127
Baruteau, C., Barraza, M., Pérez, S., et al. 2019, MNRAS, 486, 304
Bate, M. R., & Bonnell, I. A. 1997, MNRAS, 285, 33
Beichman, C. A., Myers, P. C., Emerson, J. P., et al. 1986, ApJ, 307, 337
Bell, C. P. M., Mamajek, E. E., & Naylor, T. 2015, MNRAS, 454, 593
Belloche, A. 2013, EAS Publications Series, 25
Benisty, M., Juhász, A., Facchini, S., et al. 2018, A&A, 619, A171
Benisty, M., Stolker, T., Pohl, A., et al. 2017, A&A, 597, A42
Benítez-Llambay, P., & Masset, F. S. 2016, ApJS, 223, 11
Benson, P. J., & Myers, P. C. 1989, ApJS, 71, 89
Bergin, E., Calvet, N., Sitko, M. L., et al. 2004, ApJL, 614, L133
Bergin, E. A., Cleeves, L. I., Gorti, U., et al. 2013, Natur, 493, 644
Bergin, E. A., Du, F., Cleeves, L. I., et al. 2016, ApJ, 831, 101
Berthoud, M. G., Keller, L. D., Herter, T. L., et al. 2007, ApJ, 660, 461
Béthune, W., Lesur, G., & Ferreira, J. 2016, A&A, 589, A87
Birnstiel, T., & Andrews, S. M. 2014, ApJ, 780, 153
Birnstiel, T., Klahr, H., & Ercolano, B. 2012, A&A, 539, A148
Birnstiel, T., Ricci, L., Trotta, F., et al. 2010, A&A, 516, L14
Bitner, M. A., Richter, M. J., Lacy, J. H., et al. 2008, ApJ, 688, 1326
Blevins, S. M., Pontoppidan, K. M., Banzatti, A., et al. 2016, ApJ, 818, 22
Blitz, L., & Williams, J. P. 1999, arXiv:astro-ph/9903382
Bodenheimer, P. 2011, Principles of Star Formation (Berlin: Springer)
Boss, A. P. 2002, ApJ, 576, 462
Boss, A. P. 2006, ApJ, 641, 1148
Bouwman, J., Henning, T., Hillenbrand, L. A., et al. 2008, ApJ, 683, 479
Bradshaw, P. 1974, Natur, 249, 135
Brandenburg, A., & Dobler, W. 2002, CoPhC, 147, 471
Brinch, C., & Hogerheijde, M. R. 2010, A&A, 523, A25
Burgasser, A., Kirkpatrick, J., Cutri, R., et al. 2000, ApJ, 531, L57
Burrows, C. J., Stapelfeldt, K. R., Watson, A. M., et al. 1996, ApJ, 473, 437
Calvet, N., D'Alessio, P., Watson, D. M., et al. 2005, ApJ, 630, L185

Calvet, N., & Gullbring, E. 1998, ApJ, 509, 802
Calvet, N., Hartmann, L., Kenyon, S. J., et al. 1994, ApJ, 434, 330
Carroll, B. A., & Ostlie, D. A. 2017, An Introduction to Modern Astrophysics (Cambridge: Cambridge Univ. Press)
Caselli, P., Benson, P. J., Myers, P. C., & Tafalla, M. 2002, ApJ, 572, 238
Chambers, J. 2019, ApJ, 879, 98
Chiang, E. I., & Goldreich, P. 1997, ApJ, 490, 368
Chiang, E. I., Joung, M. K., Creech-Eakman, M. J., et al. 2001, ApJ, 547, 1077
Chiang, H.-F., Looney, L. W., & Tobin, J. J. 2012, ApJ, 756, 168
Choi, M., Tatematsu, K., & Kang, M. 2010, ApJ, 723, L34
Cieza, L. A., Padgett, D. L., Allen, L. E., et al. 2009, ApJL, 696, L84
Clarke, C. J., Gendrin, A., & Sotomayor, M. 2001, MNRAS, 328, 485
Ciesla, F. J., & Cuzzi, J. N. 2006, Icar, 181, 178
Cieza, L. A., Casassus, S., Tobin, J., et al. 2016, Natur, 535, 258
Cleeves, L. I., Adams, F. C., & Bergin, E. A. 2013, ApJ, 772, 5
Close, L. M., Zuckerman, B., Song, I., et al. 2007, ApJ, 660, 1492
Cossins, P., Lodato, G., & Clarke, C. J. 2009, MNRAS, 393, 1157
Cox, E. G., Harris, R. J., Looney, L. W., et al. 2015, ApJ, 814, 28
Crank, J., & Nicholson, P. 1947, PCPS, 43, 50
Crida, A., & Morbidelli, A. 2007, MNRAS, 377, 1324
Crida, A., Morbidelli, A., & Masset, F. 2006, Icar, 181, 587
Cridland, A. J., Pudritz, R. E., & Alessi, M. 2019, MNRAS, 484, 345
Cridland, A. J., Pudritz, R. E., & Birnstiel, T. 2017, MNRAS, 465, 3865
Currie, T., Marois, C., Cieza, L., et al. 2019, ApJ, 877, L3
D'Alessio, P., Calvet, N., Hartmann, L., et al. 1999, ApJ, 527, 893
D'Alessio, P., Calvet, N., Hartmann, L., et al. 2006, ApJ, 638, 314
D'Alessio, P., Cantö, J., Calvet, N., et al. 1998, ApJ, 500, 411
D'Antona, F., & Mazzitelli, I. 1994, ApJS, 90, 467
De Rosa, R. J., Patience, J., Ward-Duong, K., et al. 2014, MNRAS, 445, 3694
Dionatos, O., Woitke, P., Güdel, M., et al. 2019, A&A, 625, A66
Dipierro, G., Price, D., Laibe, G., et al. 2015, MNRAS, 453, L73
Dodson-Robinson, S. E., & Salyk, C. 2011, ApJ, 738, 131
Dodson-Robinson, S. E., Willacy, K., Bodenheimer, P., et al. 2009, Icar, 200, 672
Dong, R., Liu, S.-y., Eisner, J., et al. 2018, ApJ, 860, 124
Dong, R., Najita, J. R., & Brittain, S. 2018, ApJ, 862, 103
Dong, R., Zhu, Z., & Whitney, B. 2015, ApJ, 809, 93
Doyle, L. R., Carter, J. A., Fabrycky, D. C., et al. 2011, Sci, 333, 1602
Drabek-Maunder, E., Mohanty, S., Greaves, J., et al. 2016, ApJ, 833, 260
Draine, B. T. 2006, ApJ, 636, 1114
Dubrulle, B., Morfill, G., & Sterzik, M. 1995, Icar, 114, 237
Duchêne, G. 2010, ApJL, 709, L114
Dullemond, C. P., & Dominik, C. 2004a, A&A, 417, 159
Dullemond, C. P., & Dominik, C. 2004b, A&A, 421, 1075
Dullemond, C. P., & Dominik, C. 2005, A&A, 434, 971
Dullemond, C. P. 2011, DISKSTRUCT: A simple 1+1-D disk structure code, Astrophysics Source Code Library, ascl:1108.015

Dullemond, C. P., Dominik, C., & Natta, A. 2001, ApJ, 560, 957
Dullemond, C. P., Juhasz, A., Pohl, A., et al. 2012, RADMC-3D: A multi-purpose radiative transfer tool, Astrophysics Source Code Library, ascl:1202.015
Dullemond, C. P., van Zadelhoff, G. J., & Natta, A. 2002, A&A, 389, 464
Dunham, M. M., Vorobyov, E. I., & Arce, H. G. 2014, MNRAS, 444, 887
Durisen, R. H., Boss, A. P., Mayer, L., et al. 2007, in Protostars and Planets V (Tucson, AZ: Arizona Univ. Press), 607
Dutrey, A., Guilloteau, S., Piétu, V., et al. 2017, A&A, 607, A130
Dutrey, A., Semenov, D., Chapillon, E., et al. 2014, in Protostars and Planets VI (Tucson, AZ: Arizona Univ. Press), 317
Dzyurkevich, N., Turner, N. J., Henning, T., et al. 2013, ApJ, 765, 114
Edwards, S., Fischer, W., Hillenbrand, L., et al. 2006, ApJ, 646, 319
Eisner, J. A., Hillenbrand, L. A., Carpenter, J. M., et al. 2005, ApJ, 635, 396
Enoch, M. L., Corder, S., Duchêne, G., et al. 2011, ApJS, 195, 21
Enoch, M. L., Evans, N. J., Sargent, A. I., et al. 2009, ApJ, 692, 973
Ercolano, B., Drake, J. J., Raymond, J. C., et al. 2008, ApJ, 688, 398
Espaillat, C., Calvet, N., D'Alessio, P., et al. 2007, ApJ, 670, L135
Espaillat, C., Calvet, N., Luhman, K. L., Muzerolle, J., & D'Alessio, P. 2008, ApJL, 682, L125
Evans, N., Calvet, N., Cieza, L., et al. 2009, arXiv:0901.1691
Evans, N. J., Dunham, M. M., Jørgensen, J. K., et al. 2009, ApJS, 181, 321
Facchini, S., Lodato, G., & Price, D. J. 2013, MNRAS, 433, 2142
Facchini, S., van Dishoeck, E. F., Manara, C. F., et al. 2019, A&A, 626, L2
Fedele, D., van den Ancker, M. E., Henning, T., et al. 2010, A&A, 510, A72
Fernandes, R. B., Mulders, G. D., Pascucci, I., Mordasini, C., & Emsenhuber, A. 2019, ApJ, 874, 81
Fischer, D. A., & Valenti, J. 2005, ApJ, 622, 1102
Flaherty, K. M., Hughes, A. M., Rose, S. C., et al. 2017, ApJ, 843, 150
Flaherty, K. M., Hughes, A. M., Rosenfeld, K. A., et al. 2015, ApJ, 813, 99
Flaherty, K. M., Hughes, A. M., Teague, R., et al. 2018, ApJ, 856, 117
Fleming, T., & Stone, J. M. 2003, ApJ, 585, 908
Flock, M., Dzyurkevich, N., Klahr, H., et al. 2011, ApJ, 735, 122
Flock, M., Dzyurkevich, N., Klahr, H., et al. 2012, ApJ, 744, 144
Flock, M., Fromang, S., Turner, N. J., et al. 2016, ApJ, 827, 144
Flock, M., Nelson, R. P., Turner, N. J., et al. 2017, ApJ, 850, 131
Flock, M., Ruge, J. P., Dzyurkevich, N., et al. 2015, A&A, 574, A68
Font, A. S., McCarthy, I. G., Johnstone, D., et al. 2004, ApJ, 607, 890
Ford, E. B., Lystad, V., & Rasio, F. A. 2005, Natur, 434, 873
Forgan, D. H., Ilee, J. D., Cyganowski, C. J., et al. 2016, MNRAS, 463, 957
Foucart, F., & Lai, D. 2014, MNRAS, 445, 1731
Fragner, M. M., & Nelson, R. P. 2010, A&A, 511, A77
France, K., Schindhelm, E., Burgh, E. B., et al. 2011, ApJ, 734, 31
Fromang, S., & Nelson, R. P. 2006, A&A, 457, 343
Fromang, S., Terquem, C., & Balbus, S. A. 2002, MNRAS, 329, 18
Fung, J., Shi, J.-M., & Chiang, E. 2014, ApJ, 782, 88
Galván-Madrid, R., Liu, H. B., Izquierdo, A. F., et al. 2018, ApJ, 868, 39
Gammie, C. F. 1996, ApJ, 457, 355

Gammie, C. F. 2001, ApJ, 553, 174
Glassgold, A. E., Galli, D., & Padovani, M. 2012, ApJ, 756, 157
Glassgold, A. E., Najita, J., & Igea, J. 1997, ApJ, 480, 344
Glassgold, A. E., Najita, J., & Igea, J. 2004, ApJ, 615, 972
Goodman, A. A., Benson, P. J., Fuller, G. A., & Myers, P. C. 1993, ApJ, 406, 528
Gorti, U., & Hollenbach, D. 2008, ApJ, 683, 287
Gorti, U., & Hollenbach, D. 2009, ApJ, 690, 1539
Gorti, U., Hollenbach, D., Najita, J., et al. 2011, ApJ, 735, 90
Grady, C. A., Muto, T., Hashimoto, J., et al. 2013, ApJ, 762, 48
Gressel, O., Nelson, R. P., & Turner, N. J. 2011, MNRAS, 415, 3291
Gressel, O., Turner, N. J., Nelson, R. P., et al. 2015, ApJ, 801, 84
Guidi, G., Tazzari, M., Testi, L., et al. 2016, A&A, 588, A112
Gullbring, E., Hartmann, L., Briceño, C., et al. 1998, ApJ, 492, 323
Guilloteau, S., Dutrey, A., Piétu, V., et al. 2011, A&A, 529, A105
Guilloteau, S., Dutrey, A., Wakelam, V., et al. 2012, A&A, 548, A70
Guilloteau, S., Piétu, V., Chapillon, E., et al. 2016, A&A, 586, L1
Haisch, K. E., Lada, E. A., & Lada, C. J. 2001, ApJ, 553, L153
Harsono, D., van Dishoeck, E. F., Bruderer, S., et al. 2015, A&A, 577, A22
Harsono, D., Jørgensen, J. K., van Dishoeck, E. F., et al. 2014, A&A, 562, A77
Hartmann, L., Calvet, N., Gullbring, E., et al. 1998, ApJ, 495, 385
Hasegawa, Y., Okuzumi, S., Flock, M., et al. 2017, ApJ, 845, 31
Hasegawa, Y., & Pudritz, R. E. 2011, MNRAS, 417, 1236
Hawley, J. F. 2001, ApJ, 554, 534
Hayashi, C. 1981, PThPS, 70, 35
Heinemann, T., & Papaloizou, J. C. B. 2009, MNRAS, 397, 64
Heinemann, T., & Papaloizou, J. C. B. 2012, MNRAS, 419, 1085
Herbig, G. H. 1960, ApJS, 4, 337
Herczeg, G. J., Linsky, J. L., Valenti, J. A., et al. 2002, ApJ, 572, 310
Herczeg, G. J., Wood, B. E., Linsky, J. L., et al. 2004, ApJ, 607, 369
Hernández, J., Calvet, N., Briceño, C., et al. 2007, ApJ, 671, 1784
Hernández, J., Hartmann, L., Calvet, N., et al. 2008, ApJ, 686, 1195
Hennebelle, P. 2013, EAS Publications Series, 67
Hillenbrand, L. A., Strom, S. E., Vrba, F. J., et al. 1992, ApJ, 397, 613
Hirose, S., & Turner, N. J. 2011, ApJ, 732, L30
Hollenbach, D., Johnstone, D., Lizano, S., et al. 1994, ApJ, 428, 654
Hord, B., Lyra, W., Flock, M., et al. 2017, ApJ, 849, 164
Howard, A. W., Marcy, G. W., Bryson, S. T., et al. 2012, ApJS, 201, 15
Huang, J., Öberg, K. I., & Andrews, S. M. 2016, ApJ, 823, L18
Huang, P., Isella, A., Li, H., et al. 2018, ApJ, 867, 3
Hueso, R., & Guillot, T. 2005, A&A, 442, 703
Ilgner, M., & Nelson, R. P. 2006, A&A, 445, 205
Ingleby, L., Calvet, N., Bergin, E., et al. 2009, ApJL, 703, L137
Ireland, M. J., & Kraus, A. L. 2008, ApJL, 678, L59
Isella, A., Carpenter, J. M., & Sargent, A. I. 2009, ApJ, 701, 260
Isella, A., Guidi, G., Testi, L., et al. 2016, PhRvL, 117, 251101
Isella, A., & Natta, A. 2005, A&A, 438, 899

Isella, A., Natta, A., Wilner, D., et al. 2010, ApJ, 725, 1735
Isella, A., Tatulli, E., Natta, A., et al. 2008, A&A, 483, L13
Isella, A., Testi, L., Natta, A., et al. 2007, A&A, 469, 213
Jeans, J. H. 1902, RSPTA, 199, 1
Jensen, E. L. N., & Akeson, R. 2014, Natur, 511, 567
Jensen, E. L. N., & Mathieu, R. D. 1997, AJ, 114, 301
Jensen, E. L. N., Mathieu, R. D., Donar, A. X., et al. 2004, ApJ, 600, 789
Johnson, J. A., Aller, K. M., Howard, A. W., et al. 2010, PASP, 122, 905
Jonkheid, B., Faas, F. G. A., van Zadelhoff, G.-J., et al. 2004, A&A, 428, 511
Jørgensen, J. K., van Dishoeck, E. F., Visser, R., et al. 2009, A&A, 507, 861
Kama, M., Bruderer, S., Carney, M., et al. 2016, A&A, 588, A108
Kamp, I., & Dullemond, C. P. 2004, ApJ, 615, 991
Kamp, I., Tilling, I., Woitke, P., et al. 2010, A&A, 510, A18
Kant, I. 1755, Allgemeine Naturgeschichte und Theorie des Himmels (Koenigsberg/Leipzig: Johann Friederich Petersen)
Keane, J. T., Pascucci, I., Espaillat, C., et al. 2014, ApJ, 787, 153
Kenyon, S. J., & Hartmann, L. 1987, ApJ, 323, 714
Keppens, R., Meliani, Z., van Marle, A. J., et al. 2012, JCoPh, 231, 718
Keppler, M., Benisty, M., Müller, A., et al. 2018, A&A, 617, A44
Kim, W.-T., & Ostriker, E. C. 2002, ApJ, 570, 132
Klaassen, P. D., Juhasz, A., Mathews, G. S., et al. 2013, A&A, 555, A73
Klahr, H. H., & Bodenheimer, P. 2003, ApJ, 582, 869
Klahr, H. H., Henning, T., & Kley, W. 1999, ApJ, 514, 325
Klahr, H., & Hubbard, A. 2014, ApJ, 788, 21
Kley, W., & Nelson, R. P. 2012, ARA&A, 50, 211
Koller, J., Li, H., & Lin, D. N. C. 2003, ApJL, 596, L91
Konacki, M., Muterspaugh, M. W., Kulkarni, S. R., et al. 2009, ApJ, 704, 513
Kostov, V. B., Orosz, J. A., Welsh, W. F., et al. 2016, ApJ, 827, 86
Kratter, K., & Lodato, G. 2016, ARA&A, 54, 271
Kratter, K. M., Matzner, C. D., & Krumholz, M. R. 2008, ApJ, 681, 375
Kraus, A. L., & Ireland, M. J. 2012, ApJ, 745, 5
Kraus, A. L., Ireland, M. J., Hillenbrand, L. A., et al. 2012, ApJ, 745, 19
Kristensen, L. E., & Dunham, M. M. 2018, A&A, 618, 158
Kuffmeier, M., Frimann, S., Jensen, S. S., et al. 2018, MNRAS, 475, 2642
Kunz, M. W., & Lesur, G. 2013, MNRAS, 434, 2295
Kurtovic, N. T., Pérez, L. M., Benisty, M., et al. 2018, ApJL, 869, L44
Lada, C. J., Huard, T. L., Crews, L. J., et al. 2004, ApJ, 610, 303
Landin, N. R., Mendes, L. T. S., & Vaz, L. P. R. 2010, A&A, 510, A46
Landry, R., Dodson-Robinson, S. E., Turner, N. J., & Abram, G. 2013, ApJ, 771, 80
Latter, H. N. 2016, MNRAS, 455, 2608
Laughlin, G., Korchagin, V., & Adams, F. C. 1997, ApJ, 477, 410
Lee, C.-F., Li, Z.-Y., Hirano, N., et al. 2018, ApJ, 863, 94
Lee, W.-K., Dempsey, A. M., & Lithwick, Y. 2019, ApJ, 822, 11
Lenz, C. T., Klahr, H., Birnstiel, T., et al. 2020, A&A, 640, A61
Lesur, G. R. J., & Latter, H. 2016, MNRAS, 462, 4549
Lesur, G., Kunz, M. W., & Fromang, S. 2014, A&A, 566, A56

Lesur, G., & Ogilvie, G. I. 2010, MNRAS, 404, L64
Lesur, G., & Papaloizou, J. C. B. 2010, A&A, 513, A60
Levy, E. H., & Sonett, C. P. 1978, in IAU Colloq. 52: Protostars and Planets (Tuscon, AZ: Arizona Univ. Press), 516
Li, D., Pantin, E., Telesco, C. M., et al. 2016, ApJ, 832, 18
Li, H., Finn, J. M., Lovelace, R. V. E., et al. 2000, ApJ, 533, 1023
Li, J. I.-H., Liu, H. B., Hasegawa, Y., et al. 2017, ApJ, 840, 72
Li, Z.-Y., Banerjee, R., Pudritz, R. E., et al. 2014, in Protostars and Planets VI (Tucson, AZ: Arizona Univ. Press), 173
Lin, C. C., & Shu, F. H. 1964, ApJ, 140, 646
Lin, D. N. C., & Pringle, J. E. 1987, MNRAS, 225, 607
Lin, M.-K., & Kratter, K. M. 2016, ApJ, 824, 91
Lin, M.-K., & Youdin, A. N. 2015, ApJ, 811, 17
Lindberg, J. E., Jørgensen, J. K., Brinch, C., et al. 2014, A&A, 566, A74
Liu, Y., Dipierro, G., Ragusa, E., et al. 2019, A&A, 622, A75
Long, F., Pinilla, P., Herczeg, G. J., et al. 2018, ApJ, 869, 17
Lovelace, R. V. E., Li, H., Colgate, S. A., et al. 1999, ApJ, 513, 805
Lynden-Bell, D., & Pringle, J. E. 1974, MNRAS, 168, 603
Lyra, W. 2014, ApJ, 789, 77
Lyra, W., Johansen, A., Zsom, A., et al. 2009, A&A, 497, 869
Lyra, W., & Klahr, H. 2011, A&A, 527, A138
Lyra, W., & Lin, M.-K. 2013, ApJ, 775, 17
Lyra, W., Richert, A. J. W., Boley, A., et al. 2016, ApJ, 817, 102
Lyra, W., & Umurhan, O. M. 2019, PASP, 131, 072001
MacFadyen, A. I., & Milosavljević, M. 2008, ApJ, 672, 83
Malygin, M. G., Klahr, H., Semenov, D., et al. 2017, A&A, 605, A30
Mamajek, E. E. 2009, in AIP Conf. Ser. 1158 (Melville, NY: AIP), 3
Manoj, P., Bhatt, H. C., Maheswar, G., et al. 2006, ApJ, 653, 657
Marcus, P. S., Pei, S., Jiang, C.-H., & Hassanzadeh, P. 2013, PhRvL, 111, 084501
Marcus, P. S., Pei, S., Jiang, C.-H., et al. 2015, ApJ, 808, 87
Marcus, P. S., Pei, S., Jiang, C.-H., et al. 2016, ApJ, 833, 148
Marois, C., Macintosh, B., Barman, T., et al. 2008, Sci, 322, 1348
Marois, C., Zuckerman, B., Konopacky, Q., Macintosh, B., & Barman, T. 2010, Natur, 468, 1080
Marsh, K. A., & Mahoney, M. J. 1992, ApJ, 395, L115
Martin, D. V., Triaud, A. H. M. J., Udry, S., et al. 2019, A&A, 624, A68
Martin, R. G., & Lubow, S. H. 2011, ApJ, 740, L6
Martin, R. G., & Lubow, S. H. 2019, MNRAS, 490, 1332
Martin, R. G., Lubow, S. H., Pringle, J. E., et al. 2019, ApJ, 875, 5
Martinache, F. 2010, ApJ, 724, 464
Masset, F. 2000, A&As, 141, 165
Masset, F. S., Morbidelli, A., Crida, A., et al. 2006, ApJ, 642, 478
Matsumura, S., & Pudritz, R. E. 2006, MNRAS, 365, 572
Maury, A. J., André, P., Testi, L., et al. 2019, A&A, 621, A76
McCaughrean, M. J., & O'dell, C. R. 1996, AJ, 111, 1977
McClure, M. K., Bergin, E. A., Cleeves, L. I., et al. 2016, ApJ, 831, 167
Meheut, H., Casse, F., Varniere, P., et al. 2010, A&A, 516, A31

Meheut, H., Keppens, R., Casse, F., et al. 2012, A&A, 542, A9
Meijerink, R., Glassgold, A. E., & Najita, J. R. 2008, ApJ, 676, 518
Meijerink, R., Pontoppidan, K. M., Blake, G. A., et al. 2009, ApJ, 704, 1471
Mejía, A. C., Durisen, R. H., Pickett, M. K., et al. 2005, ApJ, 619, 1098
Mendigutía, I., Mora, A., Montesinos, B., et al. 2012, A&A, 543, A59
Merín, B., Brown, J. M., Oliveira, I., et al. 2010, ApJ, 718, 1200
Meschiari, S. 2012, ApJL, 761, L7
Metchev, S., & Hillenbrand, L. 2006, ApJ, 651, 1166
Mignone, A., Bodo, G., Massaglia, S., et al. 2007, ApJS, 170, 228
Min, M., Dullemond, C. P., Dominik, C., et al. 2009, A&A, 497, 155
Miotello, A., van Dishoeck, E. F., Williams, J. P., et al. 2017, A&A, 599, A113
Mohanty, S., Jankovic, M. R., Tan, J. C., et al. 2018, ApJ, 861, 144
Monnier, J. D., & Millan-Gabet, R. 2002, ApJ, 579, 694
Montet, B. T., Crepp, J. R., Johnson, J. A., et al. 2014, ApJ, 781, 28
Mordasini, C., Klahr, H., Alibert, Y., et al. 2014, A&A, 566, A141
Morris, M. A., Desch, S. J., & Ciesla, F. J. 2009, ApJ, 691, 320
Mulders, G. D., Dominik, C., & Min, M. 2010, A&A, 512, A11
Müller, A., Keppler, M., Henning, T., et al. 2018, A&A, 617, L2
Murillo, N. M., Lai, S.-P., Bruderer, S., Harsono, D., & van Dishoeck, E. F. 2013, A&A, 560, A103
Murphy, S. J., Lawson, W. A., & Bessell, M. S. 2013, MNRAS, 435, 1325
Muto, T., Grady, C. A., Hashimoto, J., et al. 2012, ApJL, 748, L22
Muzerolle, J., Allen, L. E., Megeath, S. T., et al. 2010, ApJ, 708, 1107
Muzerolle, J., Calvet, N., & Hartmann, L. 2001, ApJ, 550, 944
Muzerolle, J., Luhman, K. L., Briceño, C., et al. 2005, ApJ, 625, 906
Myers, P. C., Adams, F. C., Chen, J., & Schaff, E. 1998, ApJ, 492, 703
Myers, P. C., & Ladd, E. F. 1993, ApJ, 413, 47
Najita, J. R., & Bergin, E. A. 2018, ApJ, 864, 168
Najita, J., Carr, J. S., & Mathieu, R. D. 2003, ApJ, 589, 931
Najita, J. R., Carr, J. S., Pontoppidan, K. M., et al. 2013, ApJ, 766, 134
Najita, J. R., Strom, S. E., & Muzerolle, J. 2007, MNRAS, 378, 369
Natta, A., Prusti, T., Neri, R., et al. 2001, A&A, 371, 186
Natta, A., Testi, L., Calvet, N., et al. 2007, in Protostars and Planets V (Tucson, AZ: Arizona Univ. Press), 767
Natta, A., Testi, L., Neri, R., et al. 2004, A&A, 416, 179
Nelson, R. P., Gressel, O., & Umurhan, O. M. 2013, MNRAS, 435, 2610
Neslušan, L. 2007, A&A, 461, 741
Nixon, C. J., King, A. R., & Pringle, J. E. 2018, MNRAS, 477, 3273
Nomura, H., & Millar, T. J. 2005, A&A, 438, 923
Öberg, K. I., Guzmán, V. V., Furuya, K., et al. 2015, Natur, 520, 198
Öberg, K. I., Guzmán, V. V., Merchantz, C. J., et al. 2017, ApJ, 839, 43
O'dell, C. R., Wen, Z., & Hu, X. 1993, ApJ, 410, 696
Ogihara, M., Kokubo, E., Suzuki, T. K., et al. 2018, A&A, 615, A63
Ogihara, M., Morbidelli, A., & Guillot, T. 2015, A&A, 584, L1
Ohashi, N., Hayashi, M., Ho, P. T. P., et al. 1997, ApJ, 488, 317
Ohashi, N., Saigo, K., Aso, Y., et al. 2014, ApJ, 796, 131

Okuzumi, S., Momose, M., Sirono, S. I., et al. 2016, ApJ, 821, 82
Okuzumi, S., & Tazaki, R. 2019, arXiv:1 904.038 69,
Owen, J. E., Clarke, C. J., & Ercolano, B. 2012, MNRAS, 422, 1880
Owen, J. E., Ercolano, B., Clarke, C. J., et al. 2010, MNRAS, 401, 1415
Paardekooper, S.-J., Lesur, G., & Papaloizou, J. C. B. 2010, ApJ, 725, 146
Paardekooper, S.-J., Thébault, P., & Mellema, G. 2008, MNRAS, 386, 973
Padgett, D. L., Brandner, W., Stapelfeldt, K. R., et al. 1999, AJ, 117, 1490
Pandey, B. P., & Wardle, M. 2012, MNRAS, 423, 222
Papaloizou, J. C. B., Nelson, R. P., & Masset, F. 2001, A&A, 366, 263
Pascucci, I., & Sterzik, M. 2009, ApJ, 702, 724
Pascucci, I., Sterzik, M., Alexander, R. D., et al. 2011, ApJ, 736, 13
Pascucci, I., Wolf, S., Steinacker, J., et al. 2004, A&A, 417, 793
Pérez, L. M., Carpenter, J. M., Andrews, S. M., et al. 2016, Sci, 353, 1519
Pérez, L. M., Isella, A., Carpenter, J. M., et al. 2014, ApJL, 783, L13
Perez-Becker, D., & Chiang, E. 2011, ApJ, 735, 8
Petersen, M. R., Stewart, G. R., & Julien, K. 2007, ApJ, 658, 1252
Pierens, A., & Lin, M.-K. 2018, MNRAS, 479, 4878
Piétu, V., Dutrey, A., & Guilloteau, S. 2007, A&A, 467, 163
Piétu, V., Dutrey, A., Guilloteau, S., Chapillon, E., & Pety, J. 2006, A&A, 460, L43
Pinilla, P., Pohl, A., Stammler, S. M., et al. 2017, ApJ, 845, 68
Pinilla, P., Tazzari, M., Pascucci, I., et al. 2018, ApJ, 859, 32
Pinte, C., Dent, W. R. F., Ménard, F., et al. 2016, ApJ, 816, 25
Pinte, C., Ménard, F., Duchêne, G., et al. 2006, A&A, 459, 797
Pinte, C., Ménard, F., Duchêne, G., et al. 2018, A&A, 609, A47
Pinto, R. F., Brun, A. S., Jouve, L., & Grappin, R. 2011, ApJ, 737, 72
Pontoppidan, K. M., Blake, G. A., & Smette, A. 2011, ApJ, 733, 84
Pontoppidan, K. M., Blake, G. A., van Dishoeck, E. F., et al. 2008, ApJ, 684, 1323
Pontoppidan, K. M., Salyk, C., Blake, G. A., et al. 2010, ApJ, 720, 887
Price, D. J., Cuello, N., Pinte, C., et al. 2018, MNRAS, 477, 1270
Powell, D., Murray-Clay, R., Pérez, L. M., Schlichting, H. E., & Rosenthal, M. 2019, ApJ, 878, 116
Powell, D., Murray-Clay, R., & Schlichting, H. E. 2017, ApJ, 840, 93
Prandtl, L. 1925, ZaMM, 5.2, 136
Pringle, J. E. 1981, ARA&A, 19, 137
Pudritz, R. E., Ouyed, R., Fendt, C., et al. 2007, in Protostars and Planets V (Tucson, AZ: Arizona Univ. Press), 277
Qi, C., Öberg, K. I., Wilner, D. J., et al. 2013, Sci, 341, 630
Raghavan, D., McAlister, H. A., Henry, T. J., et al. 2010, ApJS, 190, 1
Rafikov, R. R. 2009, ApJ, 704, 281
Reffert, S., Bergmann, C., Quirrenbach, A., Trifonov, T., & Künstler, A. 2015, A&A, 574, 115
Ribas, Á., Merín, B., Buoy, H., & Maud, L. T. 2014, A&A, 561, A54
Ricci, L., Testi, L., Natta, A., et al. 2010, A&A, 512, A15
Rice, W. K. M., Armitage, P. J., Bate, M. R., et al. 2003, MNRAS, 339, 1025
Rice, W. K. M., Lodato, G., & Armitage, P. J. 2005, MNRAS, 364, L56
Rice, W. K. M., Wood, K., Armitage, P. J., et al. 2003, MNRAS, 342, 79
Richard, S., Nelson, R. P., & Umurhan, O. M. 2016, MNRAS, 456, 3571

Robitaille, T. P., Whitney, B. A., Indebetouw, R., et al. 2006, ApJS, 167, 256
Romanova, M. M., Ustyugova, G. V., Koldoba, A. V., & Lovelace, R. V. E. 2012, MNRAS, 421, 63
Ros, K., & Johansen, A. 2013, A&A, 552, A137
Rosenfeld, K. A., Qi, C., Andrews, S. M., et al. 2012, ApJ, 757, 129
Rosotti, G. P., Ercolano, B., Owen, J. E., et al. 2013, MNRAS, 430, 1392
Ruge, J. P., Flock, M., Wolf, S., et al. 2016, A&A, 590, A17
Sadavoy, S. I., Myers, P. C., Stephens, I. W., et al. 2018, ApJ, 869, 115
Salinas, V. N., Hogerheijde, M. R., Bergin, E. A., et al. 2016, A&A, 591, A122
Sallum, S., Follette, K. B., Eisner, J. A., et al. 2015, Natur, 527, 342
Salyk, C., Herczeg, G. J., Brown, J. M., et al. 2013, ApJ, 769, 21
Salyk, C., Lacy, J., Richter, M., et al. 2019, ApJ, 874, 24
Sano, T., Miyama, S. M., Umebayashi, T., et al. 2000, ApJ, 543, 486
Schwarz, K. R., & Bergin, E. A. 2014, ApJ, 797, 113
Schwarz, K. R., Bergin, E. A., Cleeves, L. I., et al. 2016, ApJ, 823, 91
Schwarz, K. R., Teague, R., & Bergin, E. A. 2019, ApJL, 876, L13
Segura-Cox, D. M., Looney, L. W., Tobin, J. J., et al. 2018, ApJ, 866, 161
Semenov, D., Wiebe, D., & Henning, T. 2004, A&A, 417, 93
Shakura, N. I., & Sunyaev, R. A. 1973, A&A, 500, 33
Sheehan, P. D., & Eisner, J. A. 2017, ApJ, 840, L12
Simon, J. B., Armitage, P. J., & Beckwith, K. 2011, ApJ, 743, 17
Simon, J. B., Bai, X.-N., Flaherty, K. M., et al. 2018, ApJ, 865, 10
Simon, J. B., Bai, X.-N., Stone, J. M., et al. 2013, ApJ, 764, 66
Simon, J. B., Lesur, G., Kunz, M. W., et al. 2015, MNRAS, 454, 1117
Simon, M., Dutrey, A., & Guilloteau, S. 2000, ApJ, 545, 1034
Spiegel, E. A. 1971, ARA&A, 9, 323
Stapelfeldt, K. R., Krist, J. E., Ménard, F., et al. 1998, ApJL, 502, L65
Stoll, M. H. R., & Kley, W. 2014, A&A, 572, A77
Stone, J. M., Gardiner, T. A., Teuben, P., et al. 2008, ApJS, 178, 137
Stone, J. M., & Norman, M. L. 1992a, ApJS, 80, 753
Stone, J. M., & Norman, M. L. 1992b, ApJS, 80, 791
Strom, R. G., Malhotra, R., Ito, T., et al. 2005, Sci, 309, 1847
Suriano, S. S., Li, Z.-Y., Krasnopolsky, R., et al. 2019, MNRAS, 484, 107
Suzuki, T. K., Muto, T., & Inutsuka, S. I. 2010, ApJ, 718, 1289
Takahashi, S. Z., & Muto, T. 2018, ApJ, 865, 102
Takakuwa, S., Saito, M., Saigo, K., et al. 2014, ApJ, 796, 1
Tamayo, D., Triaud, A. H. M. J., Menou, K., et al. 2015, ApJ, 805, 100
Tanaka, K. E. I., Tan, J. C., & Zhang, Y. 2017, ApJ, 835, 32
Tatematsu, K., Ohashi, S., Sanhueza, P., et al. 2016, PASJ, 68, 24
Tazzari, M., Testi, L., Ercolano, B., et al. 2016, A&A, 588, A53
Teague, R., Guilloteau, S., Semenov, D., et al. 2016, A&A, 592, A49
Testi, L., Natta, A., Shepherd, D. S., et al. 2003, A&A, 403, 323
Thalmann, C., Janson, M., Garufi, A., et al. 2016, ApJ, 828, L17
Thun, D., & Kley, W. 2018, A&A, 616, A47
Tobin, J. J., Looney, L. W., Wilner, D. J., et al. 2015, ApJ, 805, 125

Tobin, J. J., Bos, S. P., Dunham, M. M., Bourke, T. L., & van der Marel, N. 2018, ApJ, 856, 164
Tokovinin, A. 2014, AJ, 147, 87
Toomre, A. 1964, ApJ, 139, 1217
Trapman, L., Miotello, A., Kama, M., et al. 2017, A&A, 605, A69
Turner, N. J., Carballido, A., & Sano, T. 2010, ApJ, 708, 188
Turner, N. J., Fromang, S., Gammie, C., et al. 2014, in Protostars and Planets VI (Tucson, AZ: Arizona Univ. Press), 411
Turner, N. J., & Sano, T. 2008, ApJL, 679, L131
Turner, N. J., Willacy, K., Bryden, G., et al. 2006, ApJ, 639, 1218
Ueda, T., Okuzumi, S., & Flock, M. 2017, ApJ, 843, 49
Umebayashi, T., & Nakano, T. 1988, PThPS, 96, 151
Umebayashi, T., & Nakano, T. 2009, ApJ, 690, 69
Umurhan, O. M., Shariff, K., & Cuzzi, J. N. 2016, ApJ, 830, 95
van Boekel, R., Min, M., Waters, L. B. F. M., et al. 2005, A&A, 437, 189
van der Marel, N., Dong, R., di Francesco, J., et al. 2019, ApJ, 872, 112
van der Marel, N., van Dishoeck, E. F., Bruderer, S., et al. 2013, Sci, 340, 1199
van der Marel, N., Williams, J. P., & Bruderer, S. 2018, ApJ, 867, L14
Varnière, P., & Tagger, M. 2006, A&A, 446, L13
Vlemmings, W. H. T., Lankhaar, B., Cazzoletti, P., et al. 2019, A&A, 624, L7
Vorobyov, E. I., & Basu, S. 2006, ApJ, 650, 956
Wagner, K., Dong, R., Sheehan, P., et al. 2018, ApJ, 854, 130
Walsh, C., Nomura, H., Millar, T. J., et al. 2012, ApJ, 747, 114
Wang, L., Bai, X.-N., & Goodman, J. 2019, ApJ, 874, 90
Wardle, M. 2007, Ap&SS, 311, 35
Ward-Thompson, D., André, P., Crutcher, R., et al. 2007, in Protostars and Planets V (Tucson, AZ: Arizona Univ. Press), 33
Weidenschilling, S. J. 1977a, MNRAS, 180, 57
Weidenschilling, S. J. 1977b, Ap&SS, 51, 153
Weissman, P. R. 1990, Natur, 344, 825
White, R. J., & Basri, G. 2003, ApJ, 582, 1109
Whitney, B. A., Robitaille, T. P., Bjorkman, J. E., et al. 2013, ApJS, 207, 30
Whitney, B. A., Wood, K., Bjorkman, J. E., et al. 2003, ApJ, 598, 1079
Willacy, K., & Millar, T. J. 1998, MNRAS, 298, 562
Williams, J. P., & McPartland, C. 2016, ApJ, 830, 32
Winn, J. N., & Fabrycky, D. C. 2015, ARA&A, 53, 409
Wise, A. W., & Dodson-Robinson, S. E. 2018, ApJ, 855, 145
Woitke, P., Kamp, I., Antonellini, S., et al. 2019, PASP, 131, 64301
Woitke, P., Kamp, I., & Thi, W.-F. 2009, A&A, 501, 383
Woitke, P., Min, M., Pinte, C., et al. 2016, A&A, 586, A103
Xu, R., Bai, X.-N., Öberg, K., et al. 2019, ApJ, 872, 107
Yang, C.-C., Mac Low, M.-M., & Johansen, A. 2018, ApJ, 868, 27
Yang, H., Herczeg, G. J., Linsky, J. L., et al. 2012, ApJ, 744, 121
Yen, H.-W., Koch, P. M., Takakuwa, S., et al. 2017, ApJ, 834, 178
Young, C. H., & Evans, N. J. II 2005, ApJ, 627, 293
Yu, M., Evans, N. J., Dodson-Robinson, S. E., et al. 2017a, ApJ, 841, 39

Yu, M., Evans, N. J., Dodson-Robinson, S. E., et al. 2017b, ApJ, 850, 169
Yu, M., Willacy, K., Dodson-Robinson, S. E., et al. 2016, ApJ, 822, 53
Zanazzi, J. J., & Lai, D. 2018, MNRAS, 473, 603
Zhang, K., Blake, G. A., & Bergin, E. A. 2015, ApJ, 806, L7
Zhu, Z., Nelson, R. P., Hartmann, L., et al. 2011, ApJ, 729, 47

Chapter 3

Microscopic to Macroscopic: Grain Growth and Pebble Formation

3.1 Introduction

The building blocks of most giant planets are submicron-size dust grains injected into the interstellar medium by supernovae and winds from AGB stars (Todini & Ferrara 2001; Valiante et al. 2009). The same process that forms dust bunnies under your bed provides the bottom rung of a size ladder that ultimately leads to planetesimals and planet cores. But astrophysicists haven't always been convinced that giant planets could have such commonplace origins. James Jeans, whose work on star formation we encountered in Section 2.4.7, suggested that the solar system planets condensed out of a filament of gas that was tidally stripped from the Sun by a close encounter with another star, an event so rare as to render exoplanet searches useless (Jeans 1919).

While the 1950s and 1960s saw theorists beginning to envision a "bottom-up" planet-formation pathway built on inelastic collisions between dust grains (e.g., Schmidt 1959; Lyttleton 1961; Woolfson 1969), observational evidence that stars could host dusty disks didn't emerge until the launch of the Infrared Astronomical Satellite (IRAS; Neugebauer et al. 1984), which detected infrared excess emission from over 40 stars (Aumann et al. 1984; Smith & Terrile 1984; Aumann 1985a, 1985b).[1] Then, as the discovery of the first exoplanet orbiting a Sunlike star (Mayor & Queloz 1995) and the subsequent scramble to find exoplanets was unraveling the "planets are rare" paradigm, the 2003 launch of the Spitzer Space Telescope (Werner et al. 2004) allowed observers to see grain growth in action in hundreds

[1] Morrison & Simon (1973) discovered Vega's 20 μm excess a full decade before IRAS was launched, but tentatively attributed it to chromospheric emission. T. Simon was, however, instrumental in advancing the theory that dust shells could cause infrared excess (e.g., Finn & Simon 1977; Dyck & Simon 1975). Now that ALMA has the sensitivity to detect submillimeter stellar emission (e.g., MacGregor et al. 2018), distinguishing chromospheric activity from warm asteroidal debris may again become a challenging problem.

doi:10.1088/2514-3433/ac1db7ch3

of protoplanetary disks (e.g., Furlan et al. 2006; Kessler-Silacci et al. 2006; Sicilia-Aguilar et al. 2006; Hernández et al. 2007; Bouwman et al. 2008). Today we can directly connect the dots between grain growth and giant-planet formation by examining ALMA, Submillimeter Array, and Very Large Array observations of macroscopic grains in disks that are being sculpted by giant planets (e.g., Andrews et al. 2011; ALMA Partnership et al. 2015; Pérez et al. 2015; Dong et al. 2018; Keppler et al. 2018).

This chapter describes the earliest stages of giant-planet formation, as microscopic grains collide to form loose aggregates, millimeter-size dust, and finally the centimeter-size pebbles that provide the raw material for rubble-pile comets and asteroids (Blum et al. 2017; Shinbrot et al. 2017). Here we will explore the bottom rungs of a size ladder that runs from submicron to giant planet—a dynamic range of over 10^{13} in radius. In Section 3.2 we analyze the ways gas drag and turbulence affect particle motion, as dust–gas interaction largely determines the outcomes of particle collisions. In Section 3.3 we summarize collision physics and describe analytical and numerical models of grain growth. In Section 3.4, we conclude by assessing which pebble-growth pathways predicted by theorists are most physically realistic. We will then be ready to explore planetesimal formation in Chapter 4.

This chapter includes several insets that describe concepts that may be new to nonexpert readers: the size scales of planet formation (inset box 3.1), the shearing-box approximation (inset box 3.2), and the ice line (inset box 3.3). Table 3.1 defines new variables introduced in this chapter; previously introduced variables will retain their definitions from Table 2.1. All variables that describe gas–solid interaction, collisions, and grain growth are functions of $r_{\rm eff}$.

Table 3.1. Definitions of Variables

Variable	Units	Definition
Star and disk properties		
g	cm s^{-2}	radial component of stellar gravity
Δg	cm s^{-2}	pressure-gradient contribution to radial effective gravity
v_K	cm s^{-1}	Keplerian orbital speed
Dust properties		
m	g	mass of dust particle
m_m	g	mass of a single monomer
$m_{\rm min}$	g	minimum particle mass present in mass distribution
ρ_m	g cm^{-3}	material density of particle (includes correction for porosity)
$r_{\rm eff}$	cm	effective radius of particle (includes correction for nonspherical structure)
$r_{\rm eff,max}$	cm	effective radius of largest particle in size distribution
$r_{\rm eff,fmax}$	cm	largest effective radius in size distribution controlled by fragmentation
$r_{\rm eff,dmax}$	cm	largest effective radius in size distribution controlled by radial drift

$r_{\mathrm{eff,min}}$	cm	effective radius of the smallest particle in size distribution
r_{large}	cm	characteristic radius of large particles in the two population model of (Birnstiel et al. 2012; Section 3.3.3.1)
σ_r	cm^2	particle cross section
n_p, $n_{p,0}$	cm^{-4}	number density of dust particles, midplane number density
n_m	$\mathrm{cm}^{-3}\ \mathrm{g}^{-1}$	number density of dust particles per mass bin
N		total number of particles
$N_{>m}$		number of particles more massive than m
ρ_p	$\mathrm{g\ cm}^{-3}$	solid mass per unit volume of *mixed* gas and dust
Σ_p	$\mathrm{g\ cm}^{-2}$	vertically integrated surface density of particles ($\Sigma_p = \int \rho_p dz$)
$\dot{M}_p$	$M_\oplus\ \mathrm{yr}^{-1}$ or $\mathrm{g\ s}^{-1}$	steady-state pebble accretion rate
ϕ_c		volume filling factor of dust particles
Gas properties		
$\overline{m}$	g	mean molecular mass
d_m	Å	effective molecular diameter
$\overline{v}$	$\mathrm{cm\ s}^{-1}$	average thermal speed of gas molecules
λ_g	cm	mean free path of gas molecules
ν_m	$\mathrm{cm}^2\ \mathrm{s}^{-1}$	molecular viscosity
P_v	$\mathrm{dyn\ cm}^{-2}$	partial pressure of water vapor
Properties of turbulence		
Re_t		ratio of turbulent viscosity to molecular viscosity
D_g	$\mathrm{cm}^2\ \mathrm{s}^{-1}$	turbulent diffusion coefficient of massless tracer particles
Sc		ratio of viscosity to diffusion coefficient of tracer particles
ℓ	au or cm	eddy size
L	au or cm	size of largest eddy
η	au or cm	size of smallest eddy
k	cm^{-1}	eddy wavenumber
k_L	cm^{-1}	wavenumber of largest eddy
k_η	cm^{-1}	wavenumber of smallest eddy
τ_e	yr or s	characteristic eddy turnover time
τ_k	yr or s	turnover time of eddy with wavenumber k
$E(k)$	$\mathrm{erg\ cm\ g}^{-1}$	eddy energy per unit mass per wavenumber
ϵ	$\mathrm{erg\ g}^{-1}\ \mathrm{s}^{-1}$	rate of energy injection into the turbulent cascade
*Variables related to gas–solid interactions**		
h	$\mathrm{cm}^2\ \mathrm{s}^{-1}$	angular momentum of solid body
F_d	dyn	drag force on solid body
C_d		drag coefficient
Re_p		ratio of inertial forcing from solid body to viscous dissipation in gas
Δv	$\mathrm{cm\ s}^{-1}$	relative speed of gas and solids of a given radius
u	$\mathrm{cm\ s}^{-1}$	radial drift speed/wind speed in the rest frame of solid body
$u_{\max}$	$\mathrm{cm\ s}^{-1}$	maximum possible radial drift speed/wind speed

(*Continued*)

Table 3.1. (*Continued*)

Variable	Units	Definition
v_{adv}	cm s^{-1}	speed of radial motion induced by advection
$\bar{v}_p$	cm s^{-1}	mass-weighted average radial speed from both drift and advection
w	cm s^{-1}	azimuthal wind speed in the rest frame of solid body
v_w	cm s^{-1}	magnitude of wind speed in the rest frame of solid body
τ_f	s or yr	friction time
τ_z, τ_{zs}, τ_{zl}	s or yr	settling time for (any particle, small particles, large particles)
τ_{dr}	s or yr	radial drift time
τ_d	s or yr	particle diffusion time
St		Stokes number: ratio of friction time to orbital time
St_m		Stokes number in the midplane
St_{turb}		turbulent Stokes number: ratio of friction time to eddy turnover time
D_p	cm^2 s^{-1}	turbulent diffusion coefficient of solid particles in the disk
v_{pt}	cm s^{-1}	rms speed of particles moving through turbulence
h_p	au	scale height of Gaussian $n_p(z)$
ℓ^*	au or cm	size of the smallest eddy that can hold a particle for its full friction time
k^*	cm^{-1}	wavenumber of eddy with size ℓ^*
S/G		dust/gas mass ratio
*Variables related to collisions and grain growth**		
m_1	g	mass of target particle
m_2	g	mass of the projectile particle ($m_2 < m_1$)
Δv_{12}	cm s^{-1}	collision speed
$\langle \Delta v_{12}^2 \rangle^{1/2}$	cm s^{-1}	rms collision speed
$\langle \Delta v_{12}^2 \rangle_B^{1/2}$	cm s^{-1}	rms speed of collisions driven by Brownian motion
$\langle \Delta v_{12}^2 \rangle_t^{1/2}$	cm s^{-1}	rms speed of collisions driven by turbulence
$\Delta v_{12,\, sys}$	cm s^{-1}	systematic (i.e., nonrandom) component of collision speed contributed by drift and settling
v_{com}	cm s^{-1}	particle speed in the center-of-mass frame
v_f	cm s^{-1}	fragmentation threshold speed in the center-of-mass frame
$v_{12,s}$	cm s^{-1}	relative speed at sticking threshold
$v_{12,b}$	cm s^{-1}	relative speed at the bouncing threshold
$\dot{r}_{\mathrm{eff}}$	cm s^{-1}	growth rate of effective radius
τ_c	s	collision timescale
τ_g	s or yr	grain-growth timescale
τ_{max}	s or yr	time it takes a particle to reach its maximum possible radius
St_f		Stokes number of the largest particle that survives fragmentation
St_{dr}		Stokes number of the largest particle that survives radial drift
$\dot{M}_p(R)$	$M_\oplus$ yr^{-1} or g s^{-1}	pebble flux drifting past disk radius R

P_s		sticking probability
m_{lr}	g	mass of the largest remnant
m_{lf}	g	mass of the largest fragment
m_{er}	g	mass eroded in collision
m_0	g	particle's precollision mass
ϵ_{ac}		fragment reaccretion efficiency
ζ		index of the power-law fragment mass distribution
$\mathcal{M}$	$cm^3\ g^{-1}\ s^{-1}$	Smoluchowski equation collision kernel
$\mathcal{C}_{ij}$	$s^{-1}\ cm^{-3}$	collision rate per unit volume for particles in mass bins i, j
$\mathcal{S}_{ij}$		complete set of particles produced in the collision of particles in mass bins i, j
Shearing-box setup		
x, y, z	au	Cartesian representation of (azimuthal, radial, vertical) coordinates
R_0	au	distance from star to box center
Ω_0	yr^{-1}	Keplerian angular speed at R_0
$\vec{v}$, v_x, v_y	$cm\ s^{-1}$	velocity vector, x and y components
Vortex parameters		
a, b	au	semimajor axis, semiminor axis
ζ		aspect ratio a/b
u_x, u_y	$cm\ s^{-1}$	x and y components of gas velocity within a vortex in shearing-sheet coordinates
$\vec{\omega}$	s^{-1}	vorticity

Notes:
*All variables that describe gas–solid interaction, collisions, and grain growth—are functions of $r_{\rm eff}$

3.1 Planet Formation: The Size Scale

A complete model of planet formation must stretch across an enormous size range: the ratio of Jupiter's radius (7×10^4 km) to the size of the largest dust grains inherited from the ISM ($r_{\rm eff} \sim 1\ \mu$m) is 7×10^{13}. Each step up the size ladder is governed by different physics, and both theory and experiments have revealed impediments to planet growth that frustrate efforts to produce a rock-solid model. Yet the fact that the Milky Way has more planets than stars (Cassan et al. 2012) means planet formation must be an incredibly robust process. Here we outline the "best-guess" model for how solids ascend the stairway from dust grains to giant planets, with Figure 3.1 as a guide to the properties of objects at each size scale. Each step is open to more investigation, and the rest of this chapter is peppered with descriptions of the analytical and numerical methods that will allow you to start your own planet-formation research.

Hit and stick (Section 3.3): Dust grains move up the lowest rungs of the size ladder using physics that is familiar to us all—electrostatic sticking, which forms dust bunnies under the bed. Experiments show that micron-size SiO_2 spheres that collide at speeds well below 1 m s^{-1} tend to stick, forming fractal aggregates (right panel of Figure 3.2; see also Figure 2 of Blum & Wurm 2008). When the aggregates collide with each other, they stick and compact, creating porous (but not fractal) particles like the interplanetary dust particle in Figure 3.1 (Blum & Wurm 2008; Weidling et al. 2009). Once silicate grains reach millimeter sizes, however, increasing collision

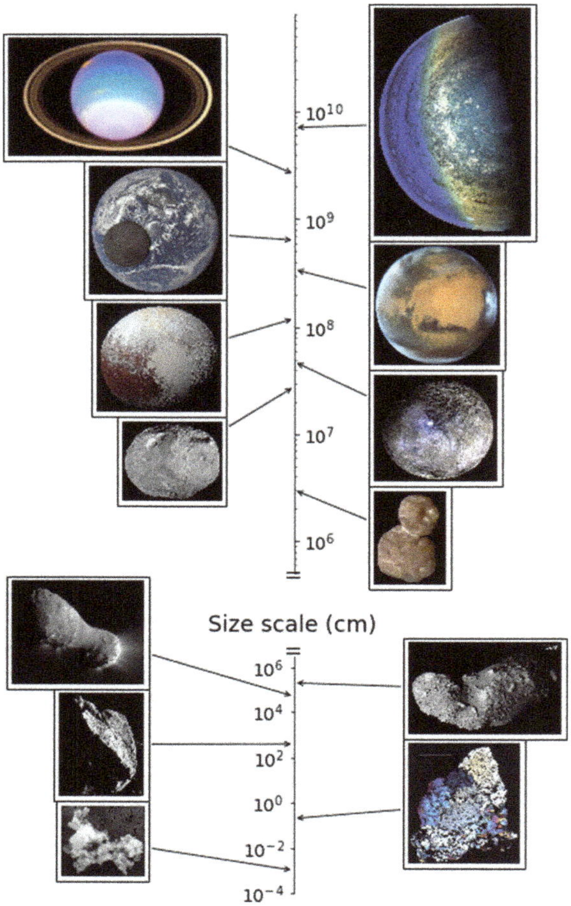

Figure 3.1. The planet-forming size "ladder" from dust grains to fully formed giant planets. From bottom to top, the images are (1) a 10 μm interplanetary dust particle likely ejected by a passing comet that was collected by a high-flying aircraft;[2] (2) a thin slice of a meteorite that was originally part of the asteroid Vesta (scale bar: 2 mm);[3] (3) a 3.8 m boulder located near the south pole of asteroid Bennu, photographed by the PolyCam camera on the OSIRIS-REx spacecraft;[4] (4) asteroid Itokawa (0.5 km) photographed by Japan's asteroid explorer Hayabusa;[5] (5) the nucleus of comet Hartley 2 (2 km on long axis) from the EPOXI mission;[6] (6) Kuiper Belt object Arrokoth (30 km on the long axis) photographed by the New Horizons spacecraft;[7] (7) asteroid Vesta (effective radius $r_{\rm eff}$ = 250 km) photographed by the Dawn spacecraft;[8] (8) false-color image of Ceres (volumetric mean radius r = 476 km) from the Dawn spacecraft;[9] (9) color image of Pluto (volumetric mean radius r = 1118 km) from blue, red, and infrared images taken by the Ralph/Multispectral Visual Imaging Camera (MVIC) on the New Horizons spacecraft;[10] (10) Hubble Space Telescope image of Mars (volumetric mean radius r = 3 389.5 km);[11] (11) Deep Space Climate observatory photograph of the full Moon eclipsing Earth (volumetric mean radius r = 6 371.0 km)—note that the far side of the Moon faces the spacecraft, which is located between Earth and the Sun;[12] (12) Hubble Space Telescope image of the ice giant Uranus (volumetric mean radius r = 25, 362 km);[13] (13) twilight at the south pole of Jupiter (volumetric mean radius r = 69, 911 km) viewed by the Juno spacecraft on its 11th close flyby—image processing by citizen scientist Gerald Eichstädt.[14]

speeds and decreasing surface-area-to-volume ratios force the particles to bounce or break (Blum & Münch 1993; Langkowski et al. 2008; Kelling et al. 2014). Grains with H_2O-ice mantles have high porosity and "premelted" surfaces that suppress bouncing (Shimaki & Arakawa 2012; Gundlach & Blum 2015; Gärtner et al. 2017), possibly allowing growth to larger sizes. In fact, as one crosses the ice line (box 3.3) at which the planet-forming disk becomes cold enough for water vapor to freeze, the increasing particle porosity, decreasing planetesimal collision speed, and high abundance of solid mass all boost the efficiency of giant-planet formation.

Though there is no firm consensus on how grains reach the centimeter-/decimeter-size range, promising possibilities include collisions between particles of very different sizes (Teiser & Wurm 2009; Güttler et al. 2010), sweep-up of microscopic grains by seed particles (Windmark et al. 2012a; though one must still grapple with the question of how to form the seeds), direct condensation of decimeter-size icy pebbles at the H_2O-ice line, where the water vapor pressure is high (Ros & Johansen 2013; Zhang et al. 2015), low-speed collisions (e.g., Windmark et al. 2012b; Booth et al. 2018), and inefficient growth in which collisions result in some breakage, but the larger of the colliders still gains a little bit of mass (Beitz et al. 2011; Deckers & Teiser 2016).

Planetesimal formation (Chapter 4): A simple experiment in a public park is enough to convince us that new physics is required to progress upward from the centimeter-/decimeter-size regime: take two pebbles from a gravel path, push them together, and note their utter inability to stick. Even pebble-size iceballs won't stick if they are compact, sintered (Okuzumi et al. 2016; Section 2.4.2), or colliding too fast (Hill et al. 2015; Deckers & Teiser 2016). It seems we must depend on gravity to move us further up the size ladder, but a solitary pebble is far too small to gravitationally attract nearby solids. Fortunately, the streaming instability (Youdin & Goodman 2005; Johansen et al. 2007; Chiang & Youdin 2010; Chapter 4) can organize clumps of pebbles that collapse directly into planetesimals under their mutual gravity. The process is similar to the Tour de France bicycle race, where riders cluster tightly behind a leader who bears the brunt of the headwind. The streaming instability requires local gas/dust mass ratios near unity, much lower than the naïve value of 100 we expect from microscopic grains well mixed with gas (Youdin & Goodman 2005; Johansen et al. 2007). Such gas/dust ratios are achievable at the midplane given efficient growth and settling, especially in disks that are enriched in heavy elements (Johansen et al. 2009). Though the streaming instability proceeds fastest for centimeter/decimeter pebbles in our model disk from Chapter 2 (Figure 1 of Youdin & Goodman 2005), it can also work for smaller particles (Lin 2019). Given that streaming instability produces planetesimals of sizes 100–400 km (Simon et al. 2016; Abod et al. 2019), the entire size distribution from 1 m to 100 km—which includes the boulder on

---

[2] NASA.

[3] NASA/JPL-Caltech/Hap McSween (Univ. Tennessee), A. Beck & T. McCoy (Smithsonian Inst.).

[4] NASA/Goddard/Univ. Arizona.

[5] NASA/JPL.

[6] NASA, JPL-Caltech, UMD, EPOXI Mission.

[7] NASA, JHU's APL, SwRI; Color Processing: Thomas Appéré.

[8] NASA, JPL-Caltech, UCLA, MPS, DLR, IDA.

[9] NASA/JPL-CalTech/UCLA/MPS/DLR/IDA.

[10] NASA/Johns Hopkins University Applied Physics Laboratory/Southwest Research Institute.

[11] NASA, ESA, and STScI.

[12] NASA, NOAA/DSCOVR.

[13] NASA, Eric Karkoschka (Univ. of Arizona), Heidi Hammell (MIT), and STScI.

[14] NASA/JPL-Caltech/SwRI/MSSS/Gerald Eichstädt

asteroid Bennu, asteroid Itokawa, comet Hartley 2, and Arrokoth in Figure 3.1—might be "backfilled" by destructive planetesimal collisions, pieces from which could later coalesce into irregularly shaped objects (Stern et al. 2019). However, we note that turbulent concentration of dust grains and hit-and-stick collisions of extremely porous aggregates may also be viable methods of forming boulders or planetesimals up to 100 m in size (Cuzzi et al. 2008; Garaud et al. 2013; Wada et al. 2011; Okuzumi et al. 2012).

Planetesimal collisions (Chapter 5): Once our disk has Ceres-size planetesimals, there are two possible pathways up the size distribution: runaway → oligarchic growth, which involves planetesimal collisions (Wetherill & Stewart 1989; Kokubo & Ida 1996; Barnes et al. 2009), and pebble accretion (Lambrechts & Johansen 2012; Levison et al. 2015; see below). Runaway growth derives its name from the fact that gravitational focusing makes the largest bodies grow fastest, accreting smaller planetesimals until they "run away" from the rest of the size distribution. Growth slows when the runaways reach the Pluto–Moon mass range (0.002–0.012$M_\oplus$) and start to dominate the dynamics of smaller planetesimals, scattering them away from easily "accretable" orbits. We are now in the slow oligarchic growth phase (Kokubo & Ida 1998, 2000), where we may be lucky if an Earth-mass object forms at 20 au (near Uranus' present orbit) within 10 Myr (Thommes et al. 2003). Accounting for gas drag on small planetesimals shortens the oligarchic growth timescale (Inaba & Ikoma 2003; Rafikov 2004; Tanigawa & Ohtsuki 2010; Fortier et al. 2013), as does increasing the total disk mass (Thommes et al. 2008; Coleman & Nelson 2014). According to Morbidelli et al. (2012), the solar system's terrestrial protoplanets were still in the oligarchic phase when the nebula gas dissipated; Earth continued to grow for the next 30–100 Myr (Touboul et al. 2007), but Mars ($M = 0.11M_\oplus$) is probably a stranded oligarch that formed within 2 Myr (Hansen 2009; Dauphas & Pourmand 2011). The Kuiper Belt never even made it as far as oligarchic growth: every known Kuiper Belt object (KBO) except Pluto and Eris fits into a continuous size distribution, indicating that only Pluto and Eris experienced runaway growth (Adams et al. 2014; Fraser et al. 2014).

and/or

Pebble accretion (Chapter 5): The pebble accretion theory was largely motivated by a need to reconcile the oligarchic growth timescale with $\lesssim$5 Myr gas disk lifetimes (Section 2.3.1). In pebble accretion, opportunistic planetesimals rapidly accrete pebbles that drift inward under the influence of gas drag (Section 3.2.2). In some models, pebble accretion continues unabated until the largest objects beyond the H_2O-ice line become gas giant cores with $M \gtrsim 10M_\oplus$ (Morbidelli et al. 2015). Pebble accretion is not a bulletproof solution to the solid-accretion timescale problem, however: it produces oligarchs in the Moon–Mars mass range (0.012–0.1 $M_\oplus$) instead of giant-planet cores if the bulk of the centimeter-size pebbles form early in the disk lifetime (Kretke & Levison 2014; Levison et al. 2015). Pebble accretion and planetesimal collisions may occur in different regions of the same disk or even concurrently.

Gas accretion (Chapter 6): The final step up the size ladder—the one that separates the giant planets from the terrestrial planets—is gas accretion. Even a Mars-size protoplanet has a tiny protoatmosphere of mass 5×10^{-9}–$10^{-5}M_\oplus$, depending on disk conditions (Erkaev et al. 2014), but this primordial atmosphere lasts less than 100 Myr. To retain a H_2/He atmosphere for several Gyr, a solid protoplanet must reach at least $2M_\oplus$ (Bodenheimer & Lissauer 2014), and possibly $6M_\oplus$ (Rogers 2015). Squeaking the largest few protoplanets over the $2M_\oplus$ threshold may only require placing them outside the H_2O-ice line, as pebble accretion favors the centimeter–decimeter drifting bodies that naturally form in ice-rich regions (Ros & Johansen 2013; Gundlach & Blum 2015) and planetesimal accretion benefits from the extra solid material provided by ice (Morbidelli et al. 2015). (Place the protoplanets too far from the star, however, and their growth will stall due to low collision rates.) In the early stages, gas accretion is roughly three dimensional, with the atmosphere and surrounding disk melding seamlessly. The gas accretion rate is opacity limited, as the protoatmosphere must cool and contract in order to allow new gas to enter the gravitationally bound region (Pollack et al. 1996; Podolak 2003; Hubickyj et al. 2005). The gas inflow will later assume more complex patterns (Kuwahara et al. 2019) and form a circumplanetary disk (Bate et al. 2003). Finally, Jupiter-type planets separate themselves from Uranus–Neptune types by nucleating an instability: the Hill sphere—the region in which gas is gravitationally bound to the planet—expands rapidly, capturing gas in the process, further expanding the Hill sphere and creating a positive feedback loop (Pollack et al. 1996).

3.2 Gas–Dust Interaction

We begin our story of giant-planet growth with the submicron dust particles that account for 1% of the mass of the interstellar medium at the Sun's location in the Galaxy (Dwek 1998). Initial radii of particles produced by direct condensation of silicates and carbon from supernova ejecta are $<0.1\ \mu m$ (Nozawa et al. 2003). By the time the grains are mixed into the densest phase of the interstellar medium, to which star-forming clouds belong (McKee & Ostriker 1977), most have already accreted mantles of thickness ~200 Å (Jenniskens et al. 1993). The grain-growth mechanism then switches from condensation and mantle accretion to hit-and-stick collisions. Protoplanetary disks inherit the collision products, some of which may become macroscopic as particles move through the infalling protostellar envelope. For example, dust grains that will eventually provide the raw material for new planets have grown beyond millimeter size in the L1157-mm and IRAS 16293 envelopes (Chiang et al. 2012; Sadavoy et al. 2018) and to centimeter size in the IRAS 4A envelope (Cox et al. 2015). All three Class 0 sources have expected ages of <50, 000 years (Kristensen & Dunham 2018) and evidence of nascent disks.

Grains in molecular clouds collide gently, at speeds of order 1 cm s^{-1} set by Brownian motion in gas with temperature 10 K (Mathis et al. 1983). Once in the disk, the collisions become both more frequent and more energetic. The gas density contrast between the inner 5 au of our model disk from Chapter 2 and the starless molecular cloud core L1554 (Ward-Thompson et al. 1999) is $10^{-10}/5 \times 10^{-18} \sim 2 \times 10^7$, which reduces the collision timescale for $0.1\ \mu m$ particles from ~60 Myr to 1 year. The rapidly growing particles become too large to flow with the gas and instead develop systematic velocities relative to both the gas and the smaller particles. Grain collision outcomes are determined by the collision speed, which is regulated by the coupling between the grains and the gas. In this section, we discuss gas–solid interaction, which sets the distribution of collision speeds for microscopic to meter-size bodies. We move on to collision outcomes in Section 3.3.

3.2.1 Foundations: Drag Laws, Friction Time, and Stokes Number

Consider a dust particle in a gas-free environment, such as a debris disk. If the particle is large enough and far enough away from the star to be impervious to radiative drag—as are centimeter-size pebbles in the Kuiper Belt—it will follow a Keplerian orbit. However, giant planet-forming environments are necessarily gas rich. As our particle interacts with the gas disk, its orbital speed deviates from the Keplerian value. Interactions with gas deliver a drag force of

$$\frac{F_d(\Delta v)}{m} = \frac{\Delta v}{\tau_f}, \tag{3.1}$$

where m is the particle mass and τ_f is the friction time over which relative motion with speed Δv between the dust particle and the gas damps out. In the time interval τ_f, a dust particle and its nearby gas molecules exchange momentum such that

$$\Delta p \sim p_0, \tag{3.2}$$

where p_0 is the particle's initial momentum. We will first consider gas–grain interactions in the Epstein regime, where the mean free path of gas molecules λ_g is similar to or larger than the grain size:

$$\lambda_g = \frac{\overline{m}}{\sqrt{2}\,\pi\rho d_m^2} \tag{3.3}$$

$$r_{\rm eff} < \frac{9}{4}\lambda_g, \tag{3.4}$$

where $\overline{m}$ is the mean molecular mass, $d_m \sim 3$ Å is the effective molecular diameter, and $r_{\rm eff}$ is the effective grain radius, which incorporates a correction for porous and nonspherical grains. Figure 3.2 illustrates $r_{\rm eff}$ for compact (left) and open (right) particle structures.

Figure 3.3 (top panel, blue) shows λ_g in the inner 20 au of our model disk midplane from Chapter 2. In all but the inner $\sim$ 1 au, grains grown by electrostatic sticking ($r_{\rm eff} \lesssim 1$ cm) are solidly in the Epstein regime. A scaled-up model of Epstein gas–grain collisions would be billiard balls pelting a planetarium dome. From Equation (2.35), we can calculate the difference between the pressure-supported azimuthal speed v_ϕ and the Keplerian speed $v_K = R\Omega_K$, shown in Figure 3.3 (top panel, red) for our model disk midplane. Because $|\Delta v| < |v_\phi - v_K|$ (i.e., dragged particles can never quite reach the Keplerian speed), we see that even for the coldest reasonable disk temperature $T \sim 10$ K $\Rightarrow c_s \sim 0.2$ km s^{-1}, given by thermal equilibrium with the surrounding molecular cloud, Δv is highly subsonic. The Epstein drag law is

$$F_d \simeq \frac{4}{3}\pi\,\rho\,r_{\rm eff}^2\,|\Delta v|\,\overline{v}, \tag{3.5}$$

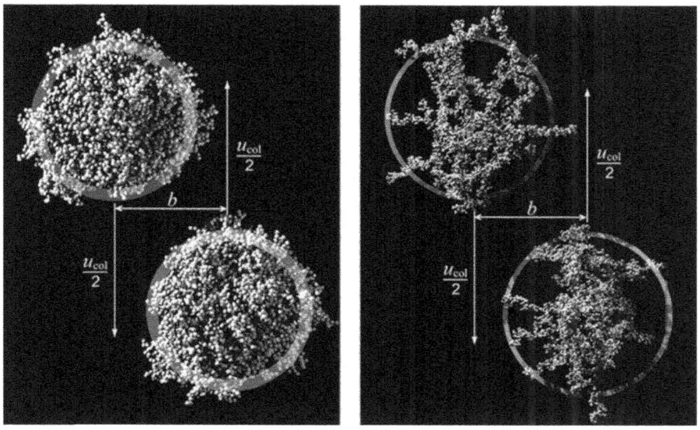

Figure 3.2. Electrostatic sticking can create particle aggregates with a compact internal structure (left) or an open fractal structure (right). A sphere with radius $r_{\rm eff}$ encloses each aggregate. In each panel, two similarly structured particles are colliding with speed $u_{\rm col}$ and impact parameter b. Figure from Wada et al. (2009), reproduced with permission.

where

$$\bar{v} = \sqrt{\frac{8}{\pi}}\, c_s \tag{3.6}$$

is the mean thermal speed of gas molecules.[15] Given $m = (4/3)\pi r_{\rm eff}^3 \rho_m$, where ρ_m is the material density of the dust, Equations (3.1) and (3.5) yield

$$\tau_f = \frac{r_{\rm eff}\rho_m}{\bar{v}\rho}. \tag{3.7}$$

The bottom panel of Figure 3.3 shows friction time τ_f as a function of $r_{\rm eff}$ and R for particles in the Epstein drag regime, given material density $\rho_m = 2$ g cm^{-3} (appropriate for silicate aggregates with some porosity). The range of friction times shown encompasses $\tau_f = 0.01$ s for ($R = 1$ au, $r_{\rm eff} = 0.1$ μm) to $\tau_f = 0.6$ yr for ($R = 20$ au, $r_{\rm eff} = 1$ cm).

To quantify the coupling between gas and dust, we examine the dimensionless Stokes number, which is the ratio of the friction time to the orbital time $1/\Omega_{\rm K}$:

$$St = \tau_f \Omega_{\rm K}. \tag{3.8}$$

Bodies with $St \gg 1$, which are much larger than the size achievable by electrostatic sticking, are almost impervious to gas drag: their dynamics are governed primarily by N-body interactions. At $St \ll 1$, a particle's momentum is almost completely determined by gas–grain collisions; such particles are entrained in the gas. The particles that are most vulnerable to gas drag (or gas "push" for positive $\partial P/\partial R$) have $St \sim 1$. To inherit the random motion of the gas, a particle must collide with

[15] In Equation (3.5), Whipple (1972) suggests replacing $4\pi/3$ with a more experimentally motivated $5.6\pi/3$.

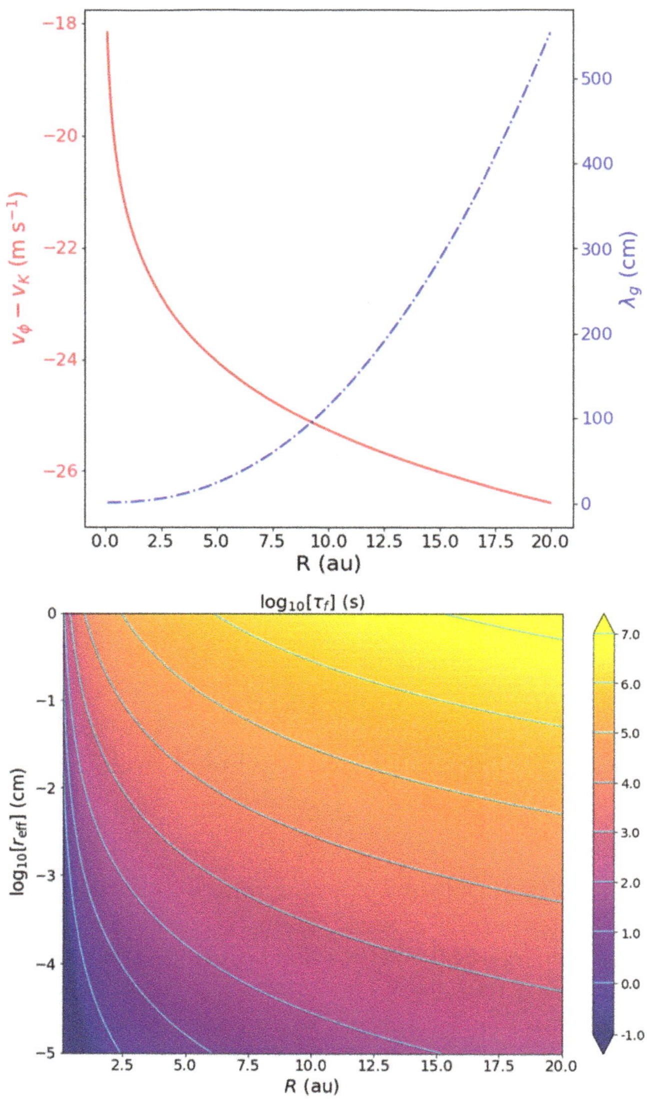

Figure 3.3. Top, red: The negative pressure gradient $\partial P/\partial R$ of our model disk from Chapter 2 gives the sub-Keplerian gas orbital speed v_ϕ; here we plot $v_\phi - v_{\rm K}$ at the midplane. Top, blue: The midplane mean free path of gas molecules in our model disk is larger than grains can grow by electrostatic sticking everywhere except the inner ~1 au. Bottom: The midplane stopping time as a function of effective radius $r_{\rm eff}$ and distance from the star R for particles in the Epstein drag regime.

enough molecules over the course of an orbit to completely average over all 4π sr of possible collision angles. At $St \sim 1$, the particle only covers 4π sr once per orbit—not enough to erase the preference for collisions on the particle's leading edge set by negative $v_\phi - v_{\rm K}$ (or on the trailing edge for positive $\partial P/\partial R$ and $v_\phi - v_{\rm K}$). In our model disk, a centimeter-size particle at 100 au has $St = 0.22$. Such a particle can neither remain still in the rest frame of the rotating gas nor orbit at the Keplerian speed.

Drifting particles may not be small enough to be in the Epstein drag regime. In fact, the particles most vulnerable to gas drag are in the decimeter to meter-size range (Whipple 1972; Weidenschilling 1977; Brauer et al. 2008). Luckily, Equation (3.1) is always valid. When the grains become larger than the gas mean free path (i.e., Equation (3.4) is no longer true), we simply replace Equation (3.5) with a fluid-based expression:

$$F_d(\Delta v) = \frac{\pi}{2} C_d \, r_{\text{eff}}^2 \, \rho \, (\Delta v)^2, \tag{3.9}$$

where C_d is the drag coefficient. Equation (3.9) is the Stokes drag law. C_d is determined by the extent to which the gas flow around the particle breaks up into turbulence. For molecular viscosity,

$$\nu_m = \frac{1}{2}\bar{v}\lambda_g \tag{3.10}$$

(important note: ν_m is different from the turbulent viscosity ν defined in Chapter 2, though it does use the same (velocity scale) × (length scale) formalism), the level of breakup turbulence is parameterized by the particle Reynolds number:

$$Re_p = \frac{2r_{\text{eff}}|\Delta v|}{\nu_m} = \frac{4r_{\text{eff}}|\Delta v|}{\lambda_g \bar{v}}. \tag{3.11}$$

Re_p is the ratio of inertial forcing from the solid body to viscous dissipation in the gas. The drag coefficient is a function of Re_p:

$$C_d = \frac{24}{Re_p} \qquad Re_p < 1 \tag{3.12}$$

$$C_d = \frac{24}{Re_p^{0.6}} \qquad 1 \leqslant Re_p < 800 \tag{3.13}$$

$$C_d = 0.44 \qquad Re_p \geqslant 800 \tag{3.14}$$

(Whipple 1972). Equations (3.9) and (3.1) and their associated values of Re_p and C_d must be solved iteratively to find consistent values of F_{d}, Δv, and τ_f.

3.2.2 Vertical Settling and Radial Drift

Armed with our understanding of drag laws and friction time, we now examine large-scale particle motion driven by gravity and gas drag. The vertical component of stellar gravity g_z forces grains to settle toward the midplane. Because particles with $St \ll 1$ are well coupled to the gas, they fall at terminal speed v_z:

$$\frac{F_d(v_z)}{m} = g_z. \tag{3.15}$$

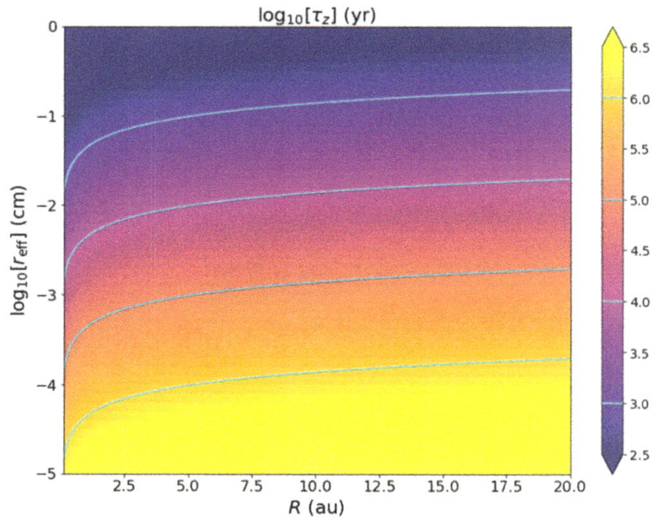

Figure 3.4. Settling time τ_z as a function of effective grain size $r_{\rm eff}$ and distance from the star R for particles initially at height $z = H$ in our analytical model disk, assuming the flow is perfectly laminar. This calculation does not include particle growth during settling.

If the disk is fully laminar, with no random gas motion except heat, the Epstein drag law gives the vertical speed v_z and settling time for small particles τ_{zs} of

$$v_z = -\Omega_{\rm K}^2 z \tau_f \tag{3.16}$$

$$\tau_{zs} = \frac{z}{|v_z|} = \frac{1}{\Omega_{\rm K}^2 \tau_f}. \tag{3.17}$$

We investigate the settling of large particles with $St \gtrsim 1$ by noting that in the absence of gas drag, the vertical component of stellar gravity (Equation (2.8)) would force all particles to oscillate about the midplane such that $\ddot{z} = -\Omega_{\rm K}^2 z$. Gas drag damps the oscillations on the friction timescale τ_f; large particles with $St \gtrsim 1$ therefore have settling time $\tau_{zl} = \tau_f$. Following Youdin & Lithwick (2007), we combine τ_{zs} and τ_{zl} to find the settling timescale for particles of any size:

$$\tau_z = \frac{1}{\Omega_{\rm K}^2 \tau_f} + \tau_f. \tag{3.18}$$

Figure 3.4 shows the settling time from height H as a function of $r_{\rm eff}$ and R, again for porous silicate grains with $\rho_m = 2$ g cm^{-2} in our analytical disk model. While centimeter-size grains will fall to the midplane within 10^3 years of forming, microscopic grains can stay lofted at height H for longer than the average disk lifetime (Section 2.2). Such large values of τ_z seem to be at odds with the requirement that giant planets form within a few million years, given that rapid planet growth requires solids to be concentrated at the midplane. Fortunately, Figure 3.4 hints at a solution to the problem: if grains can reach macroscopic sizes over a growth timescale $\tau_g < \tau_z$, they will settle and participate in planet formation. In fact, growth and settling form a positive feedback loop: rapid settling increases collision rates and allows large particles to grow by "sweeping up" small grains, while rapid growth increases the settling speed $|v_z|$.

We now turn to radial motion in the midplane of a laminar disk. For the rest of this section, all variables assume their midplane values. If dP/dR is negative, the gas orbital speed is slightly sub-Keplerian because the outward pressure support lowers the effective gravity that the orbiting gas feels: $g_{\mathrm{r,eff}} < GM_*/R^2$ (Section 2.4.1). In this case, F_d is a drag force that saps angular momentum from the particle, which, lacking pressure support, feels a headwind as it tries to orbit at the Keplerian speed. If unchecked, the particle will eventually spiral into the star; the natural structure of a protoplanetary disk, where flaring $H(R)$ leads to negative density and pressure gradients (Equation (2.14)) would suppress planet formation (Weidenschilling 1977) if not for particle-concentrating gas pileups, vortices, or instabilities (Section 3.2.3; Chapter 4). However, for positive dP/dR, the gas orbital speed is super-Keplerian: $\Omega > \Omega_\mathrm{K}$ (Equation (2.37)). Now the particle feels a tailwind, which accelerates it and increases its angular momentum. The positive pressure gradient has long captured the imagination of those who hope Earth is not an oddity (e.g., Haghighipour & Boss 2003; Kretke & Lin 2007). Both the plenitude of exoplanets (Cassan et al. 2012; Fressin et al. 2013) and the structures observed in resolved disks (Section 2.4.2) suggest that positive pressure gradients may play a strong role in halting pebble drift, though there are other ways to frustrate the rapid inspiral of dust grains (Section 3.2.3).

Here we will derive the radial drift speed and timescale using the formalism of Weidenschilling (1977), who first put forward the notion of a "meter-size barrier" to planet formation when describing the fate of particles with $St \sim 1$. For more compact notation, we will rewrite Equations (2.33) and (2.34) as

$$\frac{v_\phi^2}{R} = g + \Delta g \tag{3.19}$$

$$v_\phi^2 = v_\mathrm{K}^2\left(1 + \frac{\Delta g}{g}\right), \tag{3.20}$$

where v_K is the Keplerian orbital speed, $g = v_\mathrm{K}^2/R$, and $\Delta g = (1/\rho_0)\, dP/dR$. In general Δg is negative, but local surface density maxima can yield positive Δg. Because $|\Delta g/g| \ll 1$, we can use a binomial expansion to approximate v_ϕ:

$$v_\phi \simeq v_\mathrm{K}\left(1 + \frac{\Delta g}{2g}\right). \tag{3.21}$$

In the gas rest frame, the dragged particle spirals inward with velocity $\Delta\vec{v} = -u\hat{R} + w\hat{\phi}$ (Figure 3.5). The particle feels a headwind with velocity $-\Delta\vec{v}$, which has radial component u and azimuthal component $-w$ (if the particle is pushed

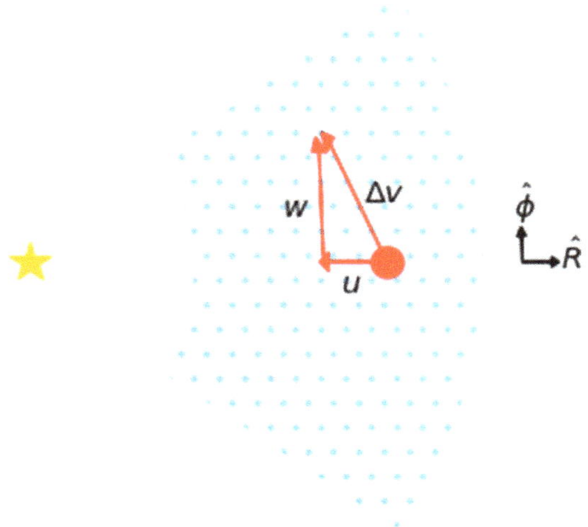

Figure 3.5. Components of inspiral speed measured in the rest frame of gas. Arrows indicate the direction of each vector component, while u, w, and Δv are the component magnitudes. In the rest frame of the particle, the wind speed is $u\hat{R} - w\hat{\phi}$.

outward by a positive pressure gradient, simply reverse the signs of all wind vectors). If the particle moves at terminal speed, the forces balance in the radial direction: in the inertial frame,

$$\frac{(v_\phi + w)^2}{R} + \frac{F_d}{m}\frac{u}{\Delta v} - g = 0. \tag{3.22}$$

Substituting Equation (3.21) into Equation (3.22) and noting that $(v_\phi + w)^2 \simeq v_K^2(1 + \Delta g/g + 2w/v_K)$ for $w/v_K \ll 1$, we find that

$$\frac{F_d}{m}\frac{u}{\Delta v} + \Delta g + \frac{2wv_K}{R} \simeq 0. \tag{3.23}$$

Next we examine the effect of azimuthal drag on the particle's specific angular momentum h. The drag torque on the particle comes from the azimuthal component of F_d:

$$\frac{dh}{dt} = -R\frac{F_d}{m}\frac{w}{\Delta v} \tag{3.24}$$

$$h = R(v_\phi + w). \tag{3.25}$$

Differentiating Equation (3.25), noting that $dR/dt = -u$ (where dR/dt describes the particle's radial motion due to the headwind *only* and does not include advection along with the accreting gas, an effect that we will discuss below) and rewriting d/dt as $(d/dR)(dR/dt)$, we have

$$\frac{dh}{dt} = -Ru\frac{d}{dR}[v_\phi + w] - u(v_\phi + w). \tag{3.26}$$

Given $dv_\phi/dR = (dv_\phi/dv_{\rm K})(dv_{\rm K}/dR)$,

$$\frac{dv_\phi}{dR} = -\frac{1}{2}\frac{v_{\rm K}}{R}\left(1 + \frac{\Delta g}{2g}\right). \tag{3.27}$$

Because $w(R)$ is smoothly varying, $dw/dR \simeq w/R$. Equation (3.26) becomes

$$\frac{dh}{dt} = -uv_{\rm K}\left(1 + \frac{\Delta g}{2g}\right) + \frac{uv_{\rm K}}{2}\left(1 + \frac{\Delta g}{2g}\right) - 2uw. \tag{3.28}$$

When we divide Equation (3.28) by $v_{\rm K}^2$ and substitute Equation (3.24) in for dh/dt, we find

$$-\frac{F_d}{m}\frac{w}{\Delta v} + \frac{uv_{\rm K}}{2R} \simeq 0 \tag{3.29}$$

for $u/v_{\rm K} \ll 1$ and $w/v_{\rm K} \ll 1$. The maximum radial drift speed $(dR/dt)_{\rm max} = -u_{\rm max}$ is given by

$$u_{\rm max} = -\frac{\Delta g}{2g}v_{\rm K} = v_{\rm K} - v_\phi. \tag{3.30}$$

For negative $\Delta g/g$, $u_{\rm max}$ is positive, indicating that the radial component of the headwind felt by the particle points outward. In the limit $u \to u_{\rm max}$, $w \to (v_{\rm K} - v_\phi)/2$.

Equations (3.23) and (3.29), plus either the Epstein or Stokes drag law (Equation (3.5) or (3.9)), form a system that can be solved iteratively to find self-consistent values of u, w, and $F_{\rm d}$. A correct solution requires the drag regime to be consistent with the gas mean free path and, for Stokes drag, the particle Reynolds number. A simplifying assumption is that the Stokes number and particle/gas relative velocity are independent; in that case, our system of equations reduces to

$$w = \frac{v_{\rm K}}{2}\frac{\Delta g}{g}\left(\frac{1}{1 + St^2} - 1\right) \tag{3.31}$$

$$u = \frac{2u_{\rm max}}{St + St^{-1}}, \tag{3.32}$$

where u and dP/dR have opposite signs (recall that our sign convention has $u = -dR/dt$; Weidenschilling et al. 1977). From Equations (3.7) and (3.8), we see that St and Δv are always independent in the Epstein drag regime; the same is true in the Stokes drag regime when $Re_p < 1$. For intermediate-size particles with $1 < Re_p < 800$, the relationship between St and Δv is weak at $St \propto (\Delta v)^{-2/5}$ and Equation (3.32) is an adequate approximation. For particles that are large relative to the mean free path, with $Re_p \geqslant 800$, $St \propto (\Delta v)^{-1}$ (Equations (3.1), (3.9), (3.11), and (3.14)). Use of Equation (3.32) is therefore often restricted to analyses of the behavior of drifting pebbles (e.g., Birnstiel et al. 2010; Pinilla et al. 2012).

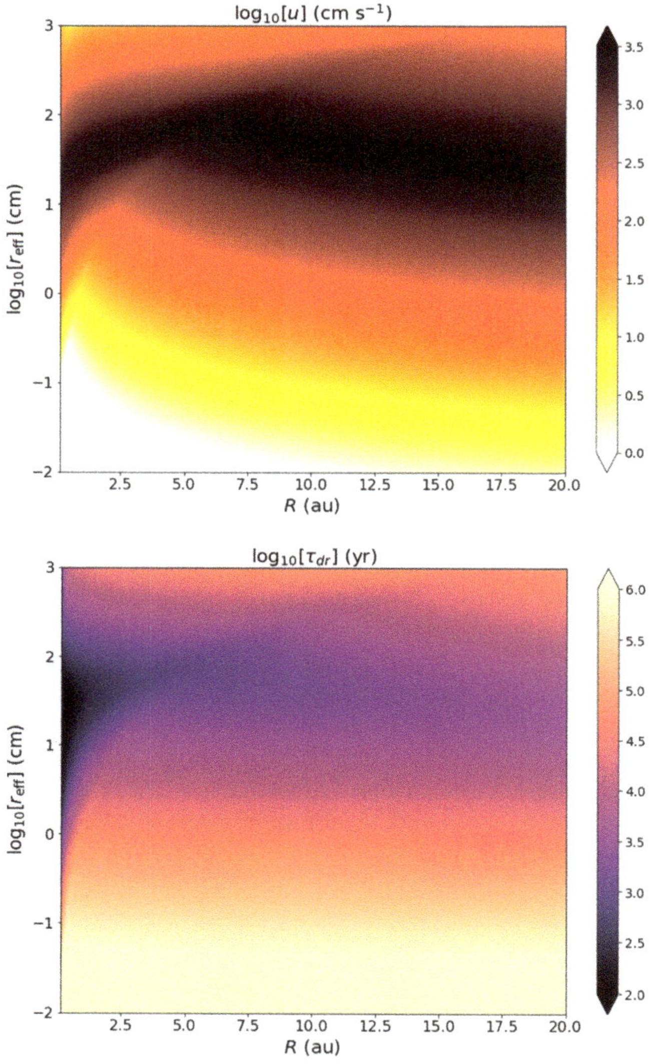

Figure 3.6. Top: midplane radial drift speed $|u(R, r_{\rm eff})|$ for the disk model defined in Chapter 2. Drift speeds reach > 30 m s^{-1}. Bottom: radial drift timescale for solids in the midplane. In our maximum-mass solar nebula, the most vulnerable bodies with $\rho_m = 2$ g cm^{-3} have $r_{\rm eff} \sim 20$ cm.

When we calculate the radial drift timescale,

$$\tau_{dr} = \frac{R}{|u|}, \tag{3.33}$$

we see how severely gas drag can affect the inventory of planet-forming material. Figure 3.6 shows the radial wind speeds $u = -dR/dt$ (top) and drift timescales τ_{dr} (bottom) for grains with 0.1 mm $< r_{\rm eff} <$ 10 m. For decimeter particles in the inner 2 au,

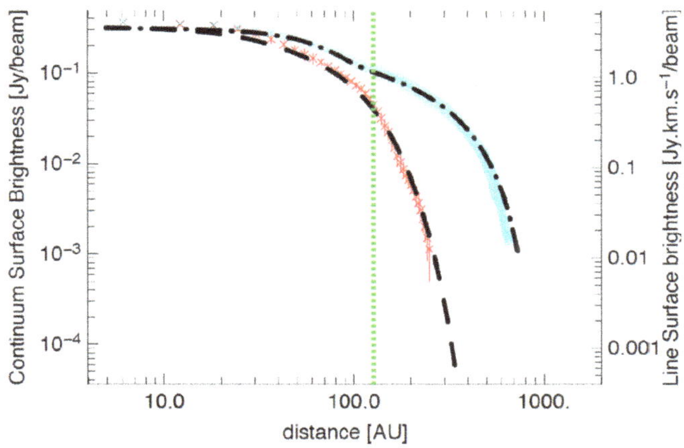

Figure 3.7. Surface brightness profiles of the disk surrounding HD 163296 inferred from ALMA images of dust continuum emission at 850 μm (red crosses, dashed line) and CO $J = 3 \rightarrow 2$ (blue crosses, dashed–dotted line). The CO gas disk appears to be more extended than the disk of millimeter-size dust grains; a possible explanation is that the grains are succumbing to radial drift. Image credit: de Gregorio-Monsalvo et al. (2013), their Figure 3.2, used with permission © ESO.

the drift timescale is ~100–1000 years. Figure 3.6 (bottom) also shows that even for particles with $r_{\rm eff} \lesssim 1$ cm, drag-induced radial drift is nonnegligible over the lifetime of the protoplanetary disk. To make matters worse, there is another contribution to radial drift: solids are advected inward along with the accreting gas at speed

$$v_{\rm adv} = \frac{v_R}{1 + St^2} \tag{3.34}$$

(Takeuchi & Lin 2002), where v_R is the gas radial velocity introduced in Section 2.3.3 (Equation (2.34)). Equation (3.34) allows for outward motion when the disk gas has positive v_R, as we expect at the outer edges of viscously spreading disks.

While we cannot detect decimeter particles in astrophysical disks, observational evidence for the radial drift of millimeter and centimeter pebbles comes from comparing the outer radii of dust continuum emission and CO gas. $R_{\rm out,gas} > 4R_{\rm out,dust}$ indicates that drift must be ongoing and that different radial extents for the gas and dust cannot be explained by optical depth effects (Trapman et al. 2019). Pebble drift is clearly happening in the disk surrounding CX Tau, which has $R_{\rm out,gas}/R_{\rm out,dust} > 5$ (Facchini et al. 2019), while Andrews et al. (2012) argue that the TW Hya disk, with $R_{\rm out,gas}/R_{\rm out,dust} > 3.6$, also has drifting millimeter-size grains. Ansdell et al. (2018) found median $R_{\rm out,gas}/R_{\rm out,dust} \sim 2$ for disks in the Lupus star-forming region and suggested that dust/gas opacity differences and radial drift are both in play. Figure 3.7 shows azimuthally averaged surface brightness profiles of 850 μm continuum and ^{12}CO (3–2) emission in the HD 163296 disk, one of many systems with $R_{\rm out,gas} > R_{\rm out,dust}$ (de Gregorio-Monsalvo et al. 2013). Although it is theoretically possible to avoid radial drift if the dust is highly porous, with $\rho_m \lesssim 0.1$ g cm^{-3} (Wada et al. 2011; Okuzumi et al. 2012), collisions tend to compact millimeter-size aggregate particles, increasing ρ_m and speeding up radial drift (Weidling et al. 2009; see Section 3.3.2).

3.2.3 Particle Trapping

If we wish to keep planet-building solids from falling into the star, we need to either (1) bypass the (deci)meter-size barrier by allowing particles to jump rapidly from centimeter to >10m sizes, or (2) trap some of the drifting solids. We note that the maximum-mass solar nebula model (MAXSN; Nixon et al. 2018) that we adopt in this book is built on the assumption that planet formation is wasteful: most solids are not locked into planets that survive well into the star's main-sequence lifetime. Radial drift almost certainly reduces the mass of solids available for planet formation, but we don't yet know its importance relative to other mass sinks such as planet(esimal) scattering and disk winds. Still, the fact that the Milky Way has as many planets as stars (Cassan et al. 2012) means we must develop a model in which planet formation is the default, not the exception. We will explore particle-growth scenarios that move directly from pebbles to planetesimals in Chapter 4. Here we model the trapping effects of (1) a local maximum in the surface density distribution and (2) a vortex.

If gas piles up anywhere in the disk, a positive pressure gradient that halts inward-drifting solids may develop. The positive pressure gradient reverses the signs of Δg and u, pushing particles outward instead of inward and trapping solids in the pressure maximum associated with the gas pileup. Possible locations of gas pileups are the centrifugal radius when there is still significant infall from the protostellar envelope (Stage I disk; Section 2.2) or the edge of a magnetically dead zone where a sharp change in turbulent viscosity traps inward-flowing gas (e.g., Kretke & Lin 2007; Lyra et al. 2009a; Dzyurkevich et al. 2010; Ruge et al. 2016). There is also a positive pressure gradient at the outer edge of a gap opened by a planet (Section 2.4.2; though relying on planets to form pressure maxima that enable planet formation recalls the paradox of which came first, the chicken or the egg; Paardekooper & Mellema (2004); Zhu et al. 2014) or a gap cleared by sublimation of metals (Dzyurkevich et al. 2013). ALMA surveys of star-forming regions show that almost every disk has a ring structure in dust continuum emission (Figure 2.14; Andrews et al. 2018). Such observations suggest that dust trapping is a robust process. Yet it is not clear that dead-zone edges reliably form density/pressure maxima: Dzyurkevich et al. (2013) find that $\rho_0(R)$ decreases smoothly across the outer edge of the dead zone for a wide range of disk parameters. Furthermore, if young stars typically form a "T-Tauriosphere" that excludes incoming cosmic rays (Cleeves et al. 2013), the entire planet-forming region may be magnetically dead, with no MRI turn-on anywhere. Instead, Riols & Lesur (2018) argue that magnetohydrodynamic (MHD) winds and nonideal MHD effects such as ambipolar diffusion create dust-trapping zonal flows and density bumps. Because we need a mechanism for halting particle drift that is firmly grounded in reality rather than just the product of theorists' hopeful imaginations, the origin of dust rings and pressure maxima in observed disks is fertile ground for future research.

To build our physical intuition about how pressure maxima trap particles, we construct a toy-model disk with two components: a smooth power law with $\Sigma \propto R^{-1}$ and a Gaussian bump centered at 10 au. Figure 3.8 shows the surface density distribution in the inner 20 au of our "bumpy" disk (red, left y-axis). The total disk mass is $0.1 M_{\odot}$ within 100 au, as in Chapter 2, and the power-law component contains

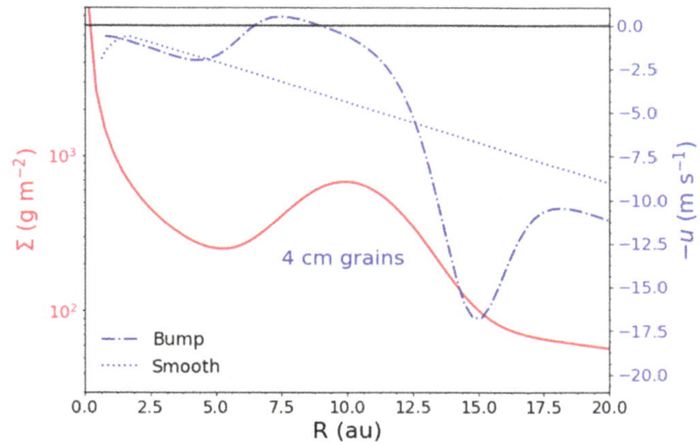

Figure 3.8. Red solid line, left y-axis: surface density profile of the inner 20 au of a disk with a Gaussian surface density bump (mean 10 au, standard deviation 2 au) overlaid on a standard disk model with $\Sigma \propto R^{-1}$. The bump has one-fourth the mass of the underlying power-law surface density distribution. Blue lines, right y-axis: drift speed $-u$ for the disk with the surface density bump (dashed–dotted) and underlying power-law disk (dotted).

four times as much mass as the Gaussian component. The flaring H/R and the radially decreasing temperature profile (Section 2.3.2) combine to drive the pressure maximum to 9.1 au, inside the local surface density maximum. The blue curves and right-hand y-axis of Figure 3.8 show the drift speed $dR/dt = -u$ for 4 cm particles, as calculated from Equations (3.23) and (3.29) and the drag laws. dR/dt is positive between 6.2 au and the pressure maximum at 9.1 au. Making the assumption that the dust/gas mass ratio and the grain-size distribution are initially uniform throughout the disk, we allow the 4 cm pebbles to drift for 4500 years. Figure 3.9 shows snapshots of the distribution of 4 cm solids at $t = 0$, 1500 yr, 3000 yr, and 4500 yr. Pebbles quickly concentrate in two annuli—one centered at the pressure maximum and one in the inner 2 au, which is in the process of draining onto the star. Our toy model shows the power of pressure maxima in halting pebble drift. Even a pressure maximum with a much shorter lifetime than the disk can salvage a significant fraction of the disk's drifting pebbles, as the planetesimal formation timescale for highly concentrated pebbles is only a few orbits (Chapter 4; Johansen et al. 2007).

To study dust trapping in vortices, we adapt the model first proposed by Barge & Sommeria (1995) and expanded by Tanga et al. (1996), Klahr & Henning (1997), Provenzale (1999), and Inaba & Barge (2006). Our analytical vortex model does not include any pressure gradients; instead the dust is trapped by the Coriolis deflection. We note that real anticyclonic vortices (those with spin opposite to the Keplerian rotation) are high-pressure regions, which enhances their particle-trapping efficiency. Within the shearing-box coordinate system presented in the inset on box 3.2, we place a 2D vortex with concentric elliptical streamlines:

$$u_x = \frac{3\Omega_0 \zeta y}{2(\zeta - 1)} \tag{3.35}$$

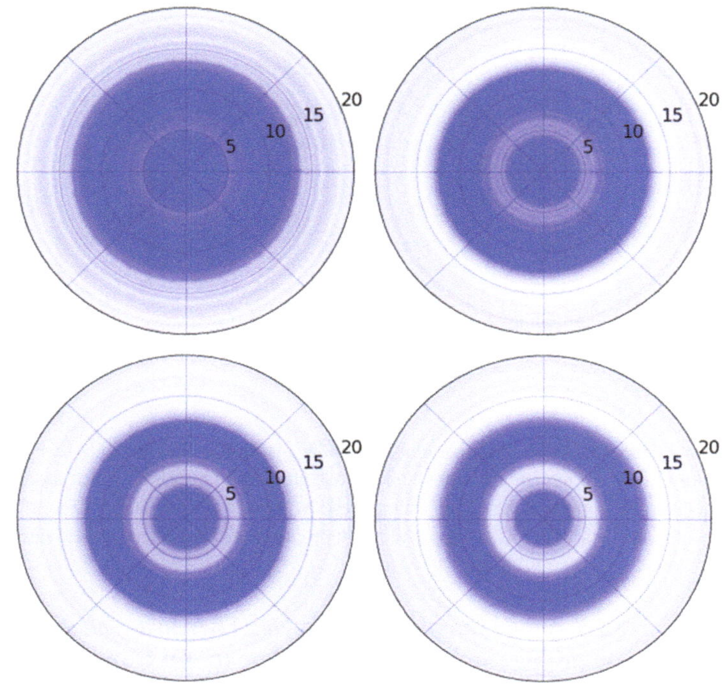

Figure 3.9. Evolution of the spatial distribution of 4 cm pebbles in the inner 20 au of a disk with surface density profile and drift speed shown in Figure 3.8. The surviving pebbles become concentrated near the pressure maximum centered at 9.2 au. Top left: $t = 0$; top right: $t = 1500$ yr; bottom left: $t = 3000$ yr; bottom right: $t = 4500$ yr.

$$u_y = -\frac{3\Omega_0 x}{2\zeta(\zeta - 1)}. \tag{3.36}$$

This is the Kida (1981) vortex model, where Ω_0 is the Keplerian angular speed at the shearing-box center, u_x is the gas azimuthal speed, u_y is the gas radial speed, and $\zeta = a/b$ is the ratio of the vortex semimajor axis a to the semiminor axis b. Outside the vortex, the gas follows Keplerian orbits with $u_x = (3/2)\Omega_0 y$ and $u_y = 0$. The Kida approximation treats the vortex as a self-contained unit with uniform vorticity $\vec{\varpi}$, where

$$\vec{\varpi} = \vec{\nabla} \times \vec{u}. \tag{3.37}$$

In Equation (3.37), $\vec{u}$ is the gas vector velocity within the vortex. Nauta (2000) found that a single vortex inserted into a 2D (R, ϕ) simulation of a Keplerian disk showed Kida-like evolution. The Kida model also works well for large vortices formed by mergers of smaller vortices in an inverse turbulent cascade (Lin & Papaloizou 2011). We note that the inverse cascade may also form zonal flows where the azimuthal velocity differs from the Keplerian value across a wide annulus; the zonal flows in turn create particle-trapping pressure maxima (Uribe et al. 2011). Here we use the Kida model for a simple demonstration of how particles behave when they encounter Coriolis deflection in vortices, but the literature on planet formation includes numerical simulations of dust dynamics in vortices that include pressure gradients and arise self-consistently from disk turbulence (e.g., Lyra et al. 2009b). Other studies of the behavior of dust in vortices come from Chavanis (2000), Godon & Livio (2000), Johansen et al. (2004), Zhu et al. (2014), and Lovascio & Paardekooper (2019). A clump of dust trapped in a vortex may also find itself stretched by the shear flow into a narrow ring, a structure that appears in many observed disks (Figure 2.14) and creates favorable conditions for planetesimal formation (Surville & Mayer 2019).

The drag force, Coriolis deflection, and centripetal acceleration together give the equations of motion

$$\ddot{x} = 2\Omega_0\dot{y} - \frac{F_d(\Delta v_x)}{m} \tag{3.38}$$

$$\ddot{y} = 3\Omega_0^2 y - 2\Omega_0\dot{x} - \frac{F_d(\Delta v_y)}{m}, \tag{3.39}$$

where x and y are the azimuthal and radial coordinates of the particle in the shearing-box coordinate system and Δv_x and Δv_y are the corresponding vector components of particle/gas relative velocity. The top panel of Figure 3.10 shows the results of a toy model with a vortex centered at 10 au in the model disk from Section 2.3.2. The vortex has a semimajor axis $a = 0.5H$ and $\zeta = 4$. We follow the motion of a vortex--approaching gas parcel plus three solid particles for 10 orbital periods (300 years). In the Kida approximation, no gas crosses the vortex boundary (black dotted line), so our parcel of gas (purple dashed–dotted line) skirts the vortex edge before departing on a circular orbit. Particles, however, can permeate the vortex boundary because they move relative to the gas. All three of our test particles get trapped by a combination of drag and Coriolis deflection, with the particle most vulnerable to gas drag ($r_{\rm eff} = 40$ cm, $St \sim 1$, red dashed line) diving straight into the vortex center while the particle that is best coupled to the gas ($r_{\rm eff} = 0.4$ cm, $St \ll 1$, green solid line) spirals gently in from the vortex edge. Raettig et al. (2015) suggest that pebbles trapped in vortices may go on to form planetesimals via an in-vortex streaming instability.

Dubrulle et al. (1995) and Youdin & Lithwick (2007) warn that vortices can only trap particles if they persist longer than all of the following timescales: the friction time τ_f, the orbital time $1/\Omega_{\rm K}$, and the vortex circulation time (i.e., gas must swirl around the vortex many times before the vortex dissipates). Still, for the centimeter-size pebbles that form remnant planetesimals such as comet 67P (Fulle et al. 2016), τ_f is never more than 1 yr in the inner 20 au of our model disk, and planetesimal formation by the streaming instability takes less than ~10 orbits once particles are highly concentrated in the midplane (Youdin & Goodman 2005). Thus, over the disk evolution timescale, any given particle-trapping vortex may be considered transient: it only needs to exist long enough to turn concentrated dust into planetesimals (Heng & Kenyon 2010). Vortices naturally arise in baroclinic regions (Section 2.4.6) with temperature gradients steeper than $T \propto R^{-1/4}$ (Petersen et al. 2007).

There is observational evidence for long-lived vortices in disks. In addition to Figure 2.24, which shows a giant dust trap likely formed by the Rossby wave instability (van der Marel et al. 2013), the disk MWC 758 has two vortices that shine in 0.87-mm emission (bottom of Figure 3.10; labeled "North Clump" and "South Clump") that are the result of dynamical influence from planets (Dong et al. 2018; Baruteau et al. 2019).

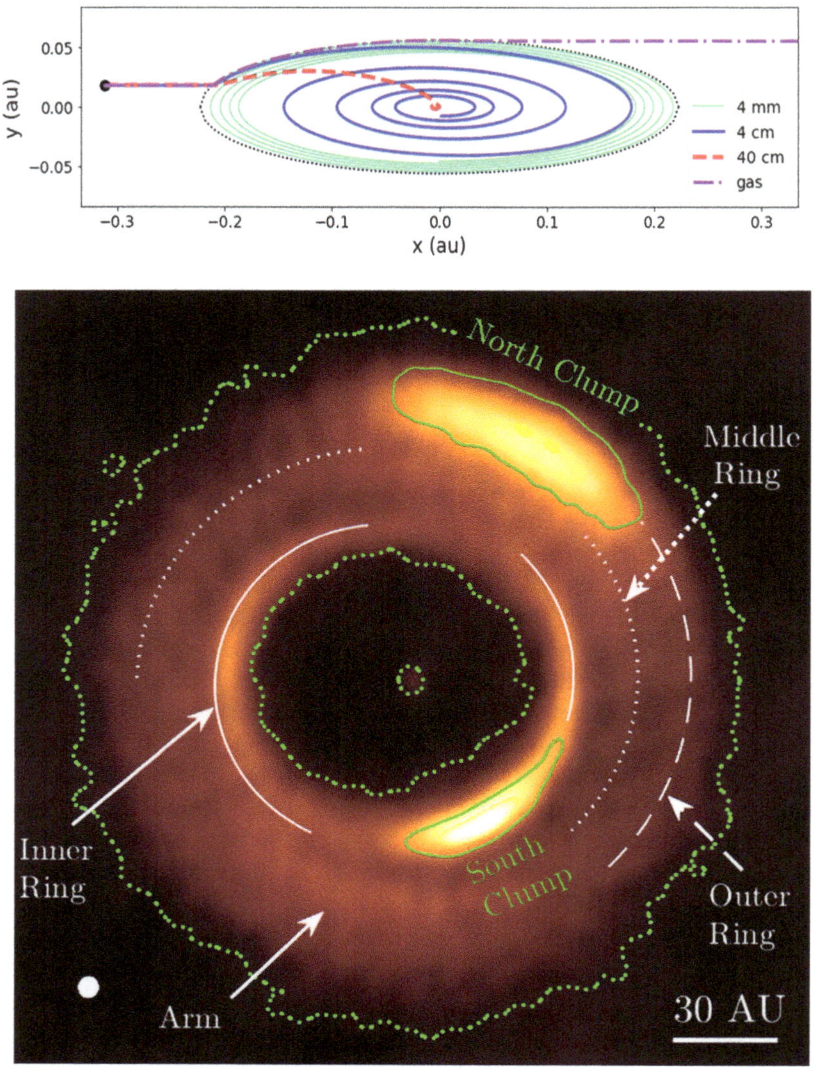

Figure 3.10. Top: trajectories of a gas parcel and three particles with sizes 0.4 cm, 4 cm, and 40 cm that all approach an elliptical Kida vortex from the same starting point (black dot on left). The vortex is centered at 10 au, has semimajor axis $0.5H$, and is azimuthally stretched so that $\zeta = 4$. The $+y$ direction points toward the star. Bottom: large vortices in the MWC 758 disk, age 3.2 Myr (Meeus et al. 2012). Although these vortices are probably a dynamical response to one or more giant planets (Dong et al. 2018; Baruteau et al. 2019), the image —which traces dust grains of size ~0.1–1 mm—illustrates the vortices' particle-trapping ability. Image credit: ALMA (ESO/NAOJ/NRAO)/Dong et al.

Of course, if our hypothesis is that vortices salvage planet-forming pebbles from their death spiral into the star, we need some vortices to appear before any giant planets grow. Given that the first solar system planetesimals formed less than 50,000 years after the first solids condensed out of the solar nebula (Jacobsen et al. 2008;

MacPherson et al. 2010),[16] another disk forming its first planetesimals could still be in the Stage I phase (Dauphas & Chaussidon 2011) and any vortices would be undetectable, obscured by the infalling envelope. Furthermore, a small vortex that naturally arises from α-model turbulence (Sections 2.3.3 and 2.4.6) would be too small to be resolved even by a high-powered interferometer. While the vortices in Figure 3.10 might not be analogous to planetesimal-forming vortices (if vortices drive planetesimal formation at all), the image showcases their particle-trapping power.

3.2 The shearing box

The shearing sheet, shown in Figure 3.11, is a 2D rectangle in the (R, ϕ) plane that is small enough for us to ignore the curvature of circular orbits (Goldreich & Lynden-Bell 1965). The 3D extension, which treats the vertical dimension instead of using the thin-disk approximation, is called a shearing box. The box is centered at distance from the star R_0; we map azimuthal flow onto the x-coordinate and radial flow onto the y-coordinate. Looking down on the midplane from high on the $+z$ axis, we measure box dimensions $L_y \ll R_0$ and $L_x \ll 2\pi R_0$. In 3D, the vertical dimension is related to the disk scale height H; box heights of $L_z = 6H$–$10H$ are common. During time Δt, the box rotates through angle $\Omega_0 \Delta t$ in the (R, ϕ) inertial frame. If the fluid element is on a circular orbit, $v_y = 0$ and $\Delta\phi_0 = \Delta\phi$. To find $v_x(y)$ for a circular orbit without pressure support, we write a first-order Taylor expansion of $\Omega_K(R)$ centered around R_0:

$$\Omega_K(R - R_0) = \Omega_0 + \frac{3}{2}\frac{\Omega_0 y}{R_0}, \tag{3.40}$$

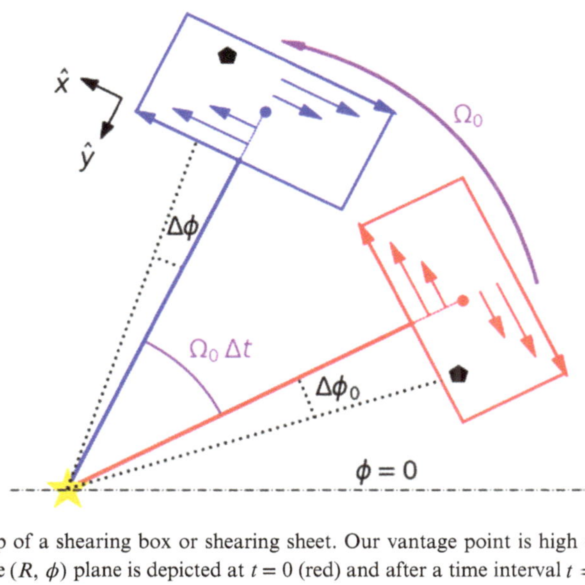

Figure 3.11. Setup of a shearing box or shearing sheet. Our vantage point is high on the $+z$ axis; the box position in the (R, ϕ) plane is depicted at $t = 0$ (red) and after a time interval $t = \Delta t$ (blue). Round points indicate the box center at a distance R_0 from the star, and the entire box rotates around the star

[16] The time at which the first nebular condensates appear does not coincide with the onset of star formation; see Section 4.2 on the meteoritic clock.

at Keplerian angular speed $\Omega_0 = \Omega_K(R_0)$. The $(\hat{x}, \hat{y})$ unit vectors of the box coordinate system at a time Δt are shown in black. The arrows within the boxes represent the speed of fluid elements on circular orbits, as measured in the box reference frame. The black pentagons show the positions at $t = 0$ and $t = \Delta t$ of an arbitrary fluid element. Because $+y$ points to the star, the fluid element's radial coordinate in the (R, ϕ) inertial frame is $R_0 - y$. The box center, the star, and the fluid element define angles $\Delta\phi_0$ at $t = 0$ and $\Delta\phi$ at $t = \Delta t$. In the inertial frame, our fluid element rotates through an angle $\Delta\phi_0 + \Omega_0\Delta t + \Delta\phi$ in time Δt and moves radial distance $\Delta R = -\Delta y$.

where $R = R_0 - y$ and $\Omega_0 = \Omega_K(R_0)$. Defining $v_x(R_0) = 0$, and recalling that $v_\phi = R\Omega_K$ in the inertial frame, we find

$$v_x = \frac{3}{2}\Omega_0 y. \tag{3.41}$$

Shearing-box simulations are usually run for hundreds to thousands of dynamical timescales $1/\Omega_0$, much less than the lifetime of the disk but far longer than the time it takes gas on circular orbits to move the length of the box L_x. Shearing boxes are often used to study turbulence and planetesimal formation; in order to let the instabilities responsible for each phenomenon fully develop and saturate, we must recycle material that leaves the simulation box. The boundary conditions are therefore periodic. For example, a particle leaving the box through the azimuthal boundary at $x = -L_x/2$ will reenter through the opposite azimuthal boundary at $x = L_x/2$ with $y_{\rm entry} = y_{\rm exit} + v_y\Delta t$, where $y_{\rm exit}$ and $y_{\rm entry}$ are the y-coordinate upon exiting and reentering the box, respectively. In 3D, a gas parcel moving out the top of the box $(+z)$ will reenter through the bottom of the box with a negative z-coordinate. Gas leaving the radial boundaries at $-L_y/2$ and $L_y/2$ has its reentry position modified by the Keplerian shear as well as its exit speed. A very small sample of the many disk and planet-related papers that have used the shearing box are Hawley et al. (1995; MRI turbulence), Fromang & Papaloizou (2006; dust settling), Johansen et al. (2007; planetesimal formation), Dzyurkevich et al. (2010; particle trapping at dead-zone edges), Lyra & Klahr (2011; baroclinic vortex formation), Paardekooper (2012; gravitoturbulence and fragmentation), and Bai & Stone (2013; magnetocentrifugal wind launching). For more information about how to set up shearing-box simulations, see Hawley et al. (1995) and Stone & Gardiner (2010).

We caution that not all shearing-box simulations use the same coordinate system. Sometimes $\hat{x}$ points toward the star (e.g., Paardekooper 2012) and occasionally the Keplerian flow is set up to be clockwise rather than counterclockwise (Barge & Sommeria 1995). When comparing the results of shearing-box simulations in the literature, a good first step is to sketch all the different coordinate systems and Keplerian flow directions.

3.2.4 Turbulent Stirring

In Chapter 2 we discussed how hydrodynamic turbulence may be present even in regions of the disk where the ionization fraction is too low for the magnetorotational instability. Even if accretion is primarily driven by large-scale winds (e.g., Bai & Stone 2013; Klaassen et al. 2013; Chambers 2019), hydrodynamic instabilities create minimum values of α and v_t (e.g., Marcus et al. 2015). We must therefore consider the effect of turbulence on the dust dynamics. As in Chapter 2, we will assume that the turbulent viscosity is given by the α prescription (Equation (2.30)) and that energy cascades into smaller and smaller eddies until it is dissipated as heat. (Because some (magneto)hydrodynamic instabilities either lead to a "truncated" turbulent cascade that does not reach all the way to the dissipation scale or to an inverse cascade that forms large vortices (Lyra & Umurhan 2019), one must verify that the assumptions used in this section are valid for the type of turbulence under consideration.) The turbulent eddies randomize particle motions on the orbital timescale or less, playing a role analogous to pressure support in gas: they give the particles random "kicks" so that rather than settling into the midplane, the particles have some steady-state vertical number density distribution. Here we use the scaling arguments of Youdin & Lithwick (2007) to develop physical intuition about dust dynamics in turbulent gas. We will calculate (1) the rms particle speed due to turbulence $v_{pt}(r_{\rm eff})$, (2) the particle diffusion coefficient $D_p(r_{\rm eff})$, and (3) the scale height $h_p(r_{\rm eff})$ of the dust distribution. In the next section, we will use our values of v_{pt} to determine particle collision speeds.

We start by considering the gas diffusion coefficient D_g, which encodes the efficiency with which turbulence redistributes massless tracer particles. We expect D_g to be similar to the turbulent viscosity ν (Equation (2.30)). In fact, the ratio $Sc = \nu/D_g$—called the Schmidt number in hydrodynamics[17]—varies from $Sc < 1$ to $Sc > 10$ in simulations, depending on whether net vertical magnetic fields are included in MRI turbulence (Carballido et al. 2005; Johansen & Klahr 2005; Turner et al. 2006; Fromang & Papaloizou 2006). Here we will use $S_c = 1$ and $D_g = \nu$ for simplicity, but we note that Okuzumi & Hirose (2011) recommend $S_c \approx 3$ in a dead zone (their equation 35). To measure particle coupling to the turbulence rather than the orbital motion (as in Equation (3.8)), we define a turbulent Stokes number:

$$St_{\rm turb} = \frac{\tau_f}{\tau_e}, \tag{3.42}$$

where τ_e is the characteristic eddy overturn timescale. The ratios St and $St_{\rm turb}$ determine how particles respond to turbulence. For turbulence in which the slowest eddies turn over in time $t \lesssim 1/\Omega_{\rm K}$, grain–eddy interaction can be broken down into three regimes:

- $St < 1$, $St_{\rm turb} < 1$: The smallest particles are well coupled to both the Keplerian orbit ($St < 1$) and the turbulence ($St_{\rm turb} < 1$). They pick up the gas velocity field and inherit an rms turbulent speed of $v_{pt} = v_t$, where v_t is the characteristic turbulent speed of the gas (Equation (2.31)). The particle diffusion coefficient matches the gas diffusion coefficient: $D_p = D_g$.
- $St < 1$, $St_{\rm turb} > 1$: For slightly larger particles or faster eddy overturn, particles remain coupled to the gas on the orbital timescale ($St < 1$), but not on the eddy turnover time ($St_{\rm turb} > 1$). Instead of getting trapped in eddies, the particles get a "kick" in a random direction from each eddy they cross. During one friction time, a particle crosses $N = \tau_f/\tau_e$ eddies, each of which delivers a kick of velocity amplitude $v_{\rm kick} = v_t\tau_e/\tau_f$. v_{pt} is then the result of the random walk through velocity space:

$$v_{pt} \simeq \sqrt{N} v_{\rm kick} \simeq \frac{v_t}{\sqrt{St_{\rm turb}}}. \tag{3.43}$$

[17] In some simulations, the diffusion coefficient is calculated for small dust grains that are well coupled to the gas instead of massless tracer particles; here the Schmidt number is defined as $Sc = \nu/D_p$ with the understanding that the measured diffusion coefficient is for particles with $St \ll 1$. One could wish for more consistent notation, but because $D_p = D_g$ for small particles (Equation (3.44)), both types of simulations are measuring essentially the same thing.

From the (length scale) × (velocity scale) formalism, the dust diffusion coefficient is

$$D_p = (v_{pt}\tau_f) \times v_{pt} = v_t^2 \tau_e = D_g. \tag{3.44}$$

- $St > 1 \rightarrow St_{\rm turb} > 1$: When $St > 1$, particle orbital motion starts to differ from gas orbital motion: now the particle undergoes epicyclic (radial) and vertical oscillations on timescale $\sim 1/\Omega_{\rm K}$. The eddies give the particle many random velocity kicks during each oscillation, yielding $v_{pt} \sim v_t / \sqrt{St_{\rm turb}}$ as above. Over the course of an orbit, turbulence changes the oscillation epicenter by a distance of

$$d_{\rm epi} \simeq v_{pt}/\Omega_{\rm K}. \tag{3.45}$$

The diffusion timescale is τ_f; writing the dust diffusion coefficient as $D_p \sim$ (length scale)^2/ (timescale), we find

$$D_p \simeq \frac{d_{\rm epi}^2}{\tau_f} \simeq \frac{v_t^2 \tau_e}{\Omega_{\rm K}^2 \tau_f^2} \simeq \frac{D_g}{St^2}. \tag{3.46}$$

Instead of falling inexorably toward the midplane, particles in a turbulent disk assume a vertical Gaussian distribution as random kicks from turbulent eddies counterbalance settling (Section 3.2.2). The number density of particles with an effective radius between $r_{\rm eff}$ and $r_{\rm eff} + dr_{\rm eff}$ is

$$n_p(r_{\rm eff}, z) = n_{p,0}(r_{\rm eff}) \exp\left[-\frac{1}{2}\frac{z^2}{h_p(r_{\rm eff})^2}\right], \tag{3.47}$$

where $n_{p,0}$ is the midplane number density. We find $h_p(r_{\rm eff})$ by equating the settling time (Equation (3.18)) with the particle diffusion time τ_d:

$$\tau_d = \frac{h_p^2}{D_p}, \tag{3.48}$$

$$h_p^2 \simeq \frac{D_p}{\Omega_{\rm K}}\left(St + \frac{1}{St}\right). \tag{3.49}$$

Substituting max[St, $1/St$] for $\left(St + \frac{1}{St}\right)$ on the right-hand side of Equation (3.49) and using Equation (3.44) or (3.46) for D_p, depending on St, we find

$$h_p \simeq \sqrt{\frac{D_g}{\Omega_{\rm K} St}}. \tag{3.50}$$

In Equation (3.50), D_g and St are calculated at the midplane. Equation (3.50) is not always well behaved for small St, as it can yield $h_p > H$. Setting $D_g = \nu$ and adding a correction factor to enforce $h_p < H$ gives

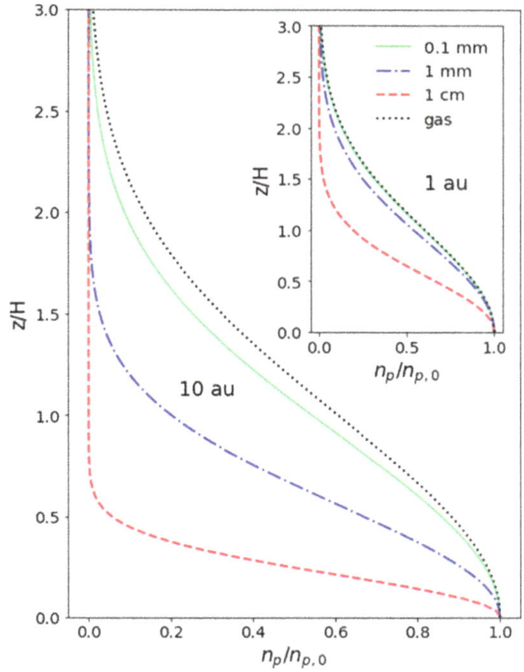

Figure 3.12. The dust number density distribution as a function of height at 1 au (inset) and 10 au (main plot) for particles with $r_{\rm eff} = 0.1$ mm, 1 mm, and 1 cm. Here we use the disk model from Chapter 2 with $\alpha = 10^{-3}$, though purely hydrodynamic turbulence in the interior of a magnetically dead zone would yield $\alpha \sim 10^{-4}$. Number densities $n_p(z)$ are scaled to the midplane value $n_{p,0}$ for each particle size. In a realistic size distribution, $n_{p,0}$ decreases as $r_{\rm eff}$ increases.

$$h_p^2 = \frac{H^2}{1 + St/\alpha}. \tag{3.51}$$

(Remember that h_p, n_p, τ_d, and D_p are all functions of $r_{\rm eff}$.) Figure 3.12 shows the vertical number density distribution of particles with $r_{\rm eff} = 0.1$ mm, 1 mm, and 1 cm in our model disk with $\alpha = 10^{-3}$. Macroscopic particles occupy vertical distributions with much lower scale heights than the gas, which increases the midplane dust/gas mass ratio and favors planetesimal formation (Chapter 4).

Finally, because the α model of turbulence assumes the largest eddies overturn on the orbital timescale, we will set $St_{\rm turb} = St$ for the remainder of this chapter. Youdin & Lithwick (2007) present a more complete model featuring characteristic eddies with overturn timescales both smaller and larger than $1/\Omega_{\rm K}$.

3.3 Grain Growth

Armed with our understanding of gas–solid interactions, we now turn to grain growth. Hit-and-stick collisions with speeds determined by gas dynamics move particles up the size ladder from submicron size to at least millimeter size, and probably to centimeter size beyond the water-ice line (see the size scale inset box 3.1). Our model of collisional grain growth is based on an extensive set of laboratory experiments, with further guidance from numerical simulations. When solids reach centimeter sizes, however, they are highly vulnerable to gas drag and their surface area/mass ratios are simply too low to allow growth by electrostatic sticking. We must then turn to a combination of gas–solid interactions and gravitational instability to preserve the pebbles by turning them into planetesimals. Although it's not possible to reproduce planetesimal-forming conditions in the lab, meteorites give clues about when the first macroscopic particles and planetesimals formed and how long the gas-assisted planetesimal formation epoch lasted (see Section 4.2). For the rest of this chapter, we will enjoy the wealth of experimental and observational data while highlighting the processes that are not fully understood. After discussing collision speeds and postcollision outcomes, we will put the pieces together to build models of grain growth.

3.3.1 Collision Speeds

The outcome of a collision is largely determined by its speed: slow collisions allow electrostatic growth to progress to sizes larger than 1 mm, while fast collisions force particles to either bounce or break (e.g., Güttler et al. 2010; Blum 2018). The relative speeds of macroscopic particles have systematic components from settling (v_z), radial drift ($-u$ and $v_{\rm adv}$), and azimuthal drift (w; Section 3.2.2), plus a random component from turbulence (v_{pt}; Section 3.2.4). For microscopic particles that are well coupled to the gas, Brownian motion forces collisions (e.g., Testi et al. 2014, see their Figure 3.3). The expected collision speed for any two particles can be found by adding all five components in quadrature:

$$\langle \Delta v_{12}^2 \rangle^{1/2} = \sqrt{\sum_j \Delta v_{12,j}^2}. \tag{3.52}$$

The settling and drift contributions to Equation (3.52) give a mean-square expected collision speed of

$$\Delta v_{12,sys}^2 = ([-u_1 + v_{{\rm adv},1}] - [-u_2 + v_{{\rm adv},2}])^2 + (w_1 - w_2)^2 + (v_{z1} - v_{z2})^2, \tag{3.53}$$

where the "sys" subscript indicates systematic contributions to relative speed as opposed to random motion. The Brownian contribution to collision speed, which dominates for particles with $St \ll 1$, is

$$\langle \Delta v_{12}^2 \rangle_B^{1/2} = \sqrt{\frac{8k_B T(m_1 + m_2)}{\pi m_1 m_2}}, \tag{3.54}$$

where m_1 and m_2 are the masses of particles 1 and 2 and k_B is the Boltzmann constant. The quantity $(m_1 + m_2)/(m_1 m_2) = 1/m_{\rm red}$ is the inverse reduced mass of the two-body system; Equation (3.54) states that two bodies collide at the mean thermal speed of a single particle with their reduced mass. Because Brownian motion is

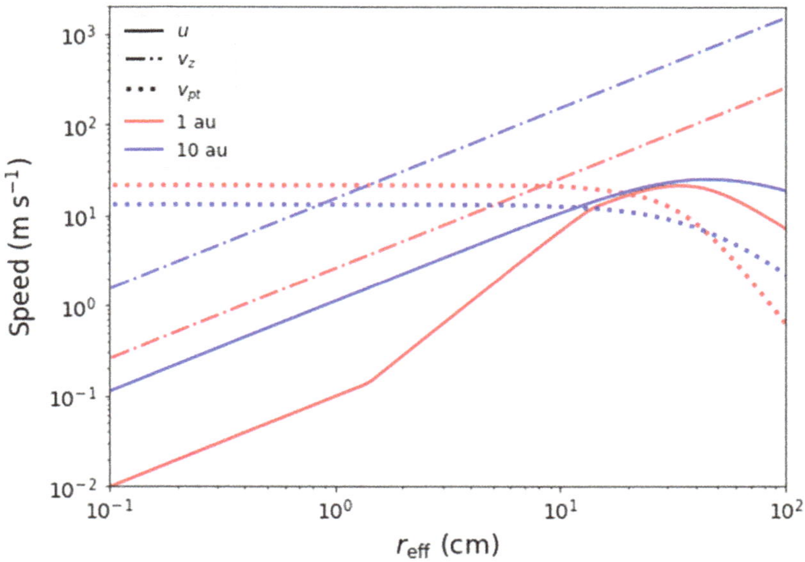

Figure 3.13. Components of particle speed in the size range that bridges electrostatic sticking and gas-driven particle concentration. Dashed–dotted lines show settling speed v_z at height H, while solid lines show radial drift speed $|-u|$. The dotted lines show v_{pt}. In the midplane, turbulence provides the dominant contribution to speed for particles smaller than ~10–30 cm.

composed of random walks, Equation (3.54) gives the expected value from a distribution of possible collision speeds.

For guidance on how to treat larger particles, we turn to Figure 3.13, which shows the turbulent, radial drift, and settling contributions to particle speed (v_{pt}, $-u$, v_z) as a function of $r_{\rm eff}$ for macroscopic solids at 1 au and 10 au in our model disk with $\alpha = 10^{-3}$. Turbulent stirring is the biggest speed contributor for $r_{\rm eff} \lesssim 1$ cm. Above the midplane at $z = H$, the vertical settling speed dominates for particles in the 1–10 cm size range, but $v_z = 0$ in the midplane. As particles reach decimeter sizes, radial and azimuthal drift take over as the most important speed contributors in the midplane (w is not shown in Figure 3.13, but $w \sim u$ for the fastest-drifting particles). Increasing α would increase v_{pt} along with the particle size at which settling and drift start to dominate collision speeds. Even at $\alpha = 10^{-4}$, $-u + v_{\rm adv}$ and w don't reach the same order of magnitude as v_{pt} until particles cross the centimeter-size threshold. Given that $\alpha = 10^{-4}$ may be a minimum realistic value, ignoring v_z, w, and $-u + v_{\rm adv}$ and calculating collision speeds based only on turbulent stirring is often justifiable for millimeter-size grains, which radiate the bulk of the light captured in ALMA observations, or centimeter-size pebbles, which provide the raw material for planetesimals (see size scale inset box 3.1).

To find $\Delta v_{12,t}$, we use the methods of Ormel & Cuzzi (2007). The expected mean-square turbulence-driven collision speed for particles of size $r_{\rm eff,1}$ and $r_{\rm eff,2}$ is

$$\langle \Delta v_{12}^2 \rangle_t = v_1^2 + v_2^2 - 2\overline{v_1 v_2}, \tag{3.55}$$

where $v_1 = v_{pt}(r_{\mathrm{eff},1})$ and $v_2 = v_{pt}(r_{\mathrm{eff},2})$ (Markiewicz et al. 1991). The cross-term $2\overline{v_1 v_2}$ corrects for correlated motion between particles 1 and 2. For now, we develop a simple model that assumes all turbulence-driven collisions between particles of sizes $r_{\mathrm{eff},1}$ and $r_{\mathrm{eff},2}$ take place at rms speed $\langle \Delta v_{12}^2 \rangle_t^{1/2}$ (e.g., Windmark et al. 2012a). Following Markiewicz et al. (1991) and Ormel & Cuzzi (2007), we rewrite the cross-term as

$$2\overline{v_1 v_2} = v_{c1}^2 + v_{c2}^2, \tag{3.56}$$

where v_{c1} and v_{c2} trace each particle's interaction with the size spectrum of eddies that have overturn time $\tau_e > \tau_f$ (that is, eddies that can "catch" the particle and force it to forget its previous motion before leaving the eddy). From the Kolmogorov (1941) turbulence model,

$$\frac{1}{2} v_t^2 = \int_{k_L}^{k_\eta} dk E(k) \tag{3.57}$$

$$E(k) = \frac{v_t^2}{3k_L} \left(\frac{k}{k_L} \right)^{-5/3}, \tag{3.58}$$

where $k = 1/\ell$ is the wavenumber of an eddy with size ℓ, $dkE(k)$ is the energy per unit mass in eddies with wavenumbers in the range k to $k + dk$, L and k_L are the size and wavenumber of the largest eddy (the α model of turbulence gives $L = \sqrt{\alpha} H$; see Equations (2.30) and (2.32)), and η and k_η are the size and wavenumber of the smallest eddy. For $i = [1, 2]$,

$$v_{ci}^2 = \frac{2\tau_{f,i}}{\tau_{f,1} + \tau_{f,2}} \left(\int_{k_L}^{k^*} dk\, E(k) - \int_{k_L}^{k^*} dkE(k) \left[\frac{1}{1 + \tau_k/\tau_{f,i}} \right]^2 \right). \tag{3.59}$$

In Equation (3.59), $r_{\mathrm{eff},1} > r_{\mathrm{eff},2}$, $St_1 > St_2$, τ_k is the turnover time of eddies with wavenumber k, and k^* is the wavenumber of the smallest eddy that has $\tau_k > \tau_{f,1}$. In the second integral on the right-hand side of Equation (3.59), the contribution of large eddies with long overturn timescales relative to $\tau_{f,i}$ is downweighted so that eddies with wavenumbers near k^* contribute the most. Subtracting $v_{c1}^2 + v_{c2}^2$ from the mean-square collision speed corrects for the fact that the largest eddies tend to capture both particles, aligning their trajectories; the eddies with $\tau_k \sim \tau_f$ are most effective at randomizing particle motion and driving collisions.

From Cuzzi & Hogan (2003) and Ormel & Cuzzi (2007), the solution to Equation (3.59) is

$$v_{ci}^2 = v_t^2 \left(\frac{\tau_{f,i}}{\tau_{f,1} + \tau_{\tau,2}} \right) \left[(1 - x^{*-2/3}) - \left(\frac{St_i}{1 + St_i} - \frac{St_i}{1 + St_i x^{*2/3}} \right) \right], \tag{3.60}$$

where $x^* = k^*/k_L$.[18] To find k^*, we turn back to the Kolmogorov (1941) turbulence model, which states

[18] In Equation (3.60), Ormel & Cuzzi (2007) use $St = \tau_f/\tau_L$, where τ_L is the overturn time of the largest eddy. They then take $\tau_L \sim 1/\Omega_K$ so that their definition of St is the same as in Equation (3.8). If you are studying dust behavior in turbulence with $\tau_L \neq 1/\Omega_K$, be sure to calculate the Stokes number using the appropriate eddy overturn time.

$$\tau_e \simeq \epsilon^{-1/3}\ell^{2/3}. \tag{3.61}$$

In Equation (3.61), ϵ is the rate at which the Keplerian shear flow funnels energy into the turbulent cascade (units erg g^{-1} s^{-1} = cm^2 s^{-3}). Because the largest eddies have $\tau_e \simeq 1/\Omega_K$ and $\ell \simeq \sqrt{\alpha}\, H$,

$$\epsilon \simeq \alpha H^2 \Omega_K^3. \tag{3.62}$$

Substituting Equation (3.62) into Equation (3.61) and setting $\tau_\kappa = \tau_{f,1}$ for wave-number k^* yield

$$\frac{\ell^*}{H} = \alpha^{1/2}\left(\tau_{f,1}\Omega_K\right)^{3/2}, \tag{3.63}$$

$$k^* = 1/\ell^*, \tag{3.64}$$

where ℓ^* is the size of the smallest eddy that can hold the larger of the colliding particles for one full friction time.

All that remains is to treat the cases where (a) $\ell^* < \eta$, in which both particles have such short stopping times that any eddies that could randomize their motion would have been smaller than the dissipation scale, and (b) $\ell^* > L$, where even the largest eddy turns over too slowly to catch particles with long stopping times. When taking into account both possibilities, the Ormel & Cuzzi (2007) turbulence-driven collision speed solution is approximated as

$$\langle \Delta v_{12}^2 \rangle_t^{1/2} = v_t \times \begin{cases} Re_t^{1/4}\Omega_K(\tau_{f,1} - \tau_{f,2}) & \tau_{f,1} \ll \tau_\eta \\ 1.6\left(\Omega_K \tau_{f,1}\right)^{1/2} & \tau_\eta \ll \tau_{f,1} \ll 1/\Omega_K \\ \left(\dfrac{1}{1 + \Omega_K \tau_{f,1}} + \dfrac{1}{1 + \Omega_K \tau_{f,2}}\right)^{1/2} & \tau_{f,1} \gg 1/\Omega_K. \end{cases} \tag{3.65}$$

In Equation (3.65), the Reynolds number of the turbulence is $Re_t = \nu/\nu_{\rm mol}$, which is the ratio of the turbulent viscosity to the molecular viscosity; τ_η is the overturn time of the smallest eddy, which we find from Equation (3.61) with $\eta = \nu_{\rm mol}^{3/4}\epsilon^{-1/4}$ (Kolmogorov 1941). Figure 3.14 shows the expected collision speeds given by Equation (3.65) for particles with ρ_m = 2 g cm^{-3} in the midplane of our model disk at 1 au (top) and 10 au (bottom). While turbulence is the main collision speed contributor for the particle sizes shown here, radial and azimuthal drift begin to drive collisions for particles that reach 10 cm at 10 au, or 1 m at 1 au.

In reality, not all collisions have $\Delta v_{12,B} = \langle \Delta v_{12}^2 \rangle_B^{1/2}$ and $\Delta v_{12,t} = \langle \Delta v_{12}^2 \rangle_t^{1/2}$. A more complete collision model has the random components of Δv_{12} drawn from a probability distribution function. Experiments by Windmark et al. (2012b) show that using a speed distribution instead of the rms Brownian and turbulent speed components $\langle \Delta v_{12}^2 \rangle_B^{1/2}$ and $\langle \Delta v_{12}^2 \rangle_t^{1/2}$ materially alters the grain-growth pathway, allowing "lucky" particles that experience low-speed collisions to reach centimeter sizes, then sweep up smaller particles. A sampling of other grain-growth simulations

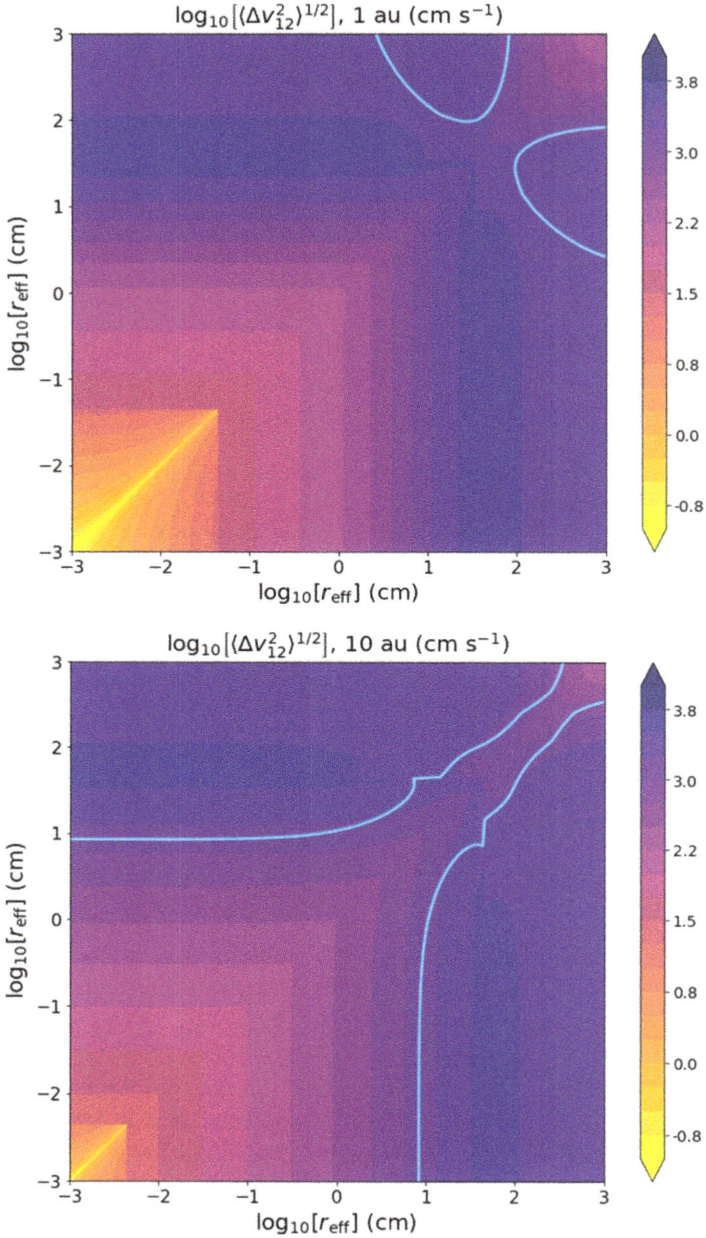

Figure 3.14. Expected collision speeds (Equation (3.52)) in the midplane of our model disk with $\alpha = 10^{-3}$. Top: 1 au. Bottom: 10 au. Because particles with $v_z \sim \langle\Delta v_{12}^2\rangle^{1/2}$ make only small excursions from the midplane (Figures 3.12 and 3.13), we assume $v_{z1} = v_{z2} = 0$. Discontinuities correspond to transitions between the three different regimes in Equation (3.65). Following Ormel & Cuzzi (2007), we use $Re{=}10^8$ to determine the size of the smallest eddies. The light blue lines show the transition between turbulence-driven collisions (bottom and left) and drift-driven collisions (top and right).

that include a distribution of collision speeds is Okuzumi & Hirose (2011), Garaud et al. (2013), Drażkowska et al. (2014), Estrada et al. (2016), and Booth et al. (2018). The Maxwellian speed distribution model of Windmark et al. (2012b) assumes collisions are driven only by turbulence and has

$$P(\Delta v_{12}) = \sqrt{\frac{54}{\pi}} \frac{\Delta v_{12}^2}{\langle \Delta v_{12}^2 \rangle_t^{3/2}} \exp\left[-\frac{3}{2} \frac{\Delta v_{12}^2}{\langle \Delta v_{12}^2 \rangle_t}\right], \tag{3.66}$$

where P is the probability density. The Gaussian model of Garaud et al. (2013) (supported by Pan & Padoan 2013) includes all possible contributors to collision speed:

$$\begin{aligned} P(\Delta v_{12}) = \frac{1}{\sqrt{2\pi}\,\langle \Delta v_{12}^2 \rangle^{1/2}} \frac{\Delta v_{12}}{\Delta v_{12,sys}} \Bigg(& \exp\left[-\frac{(\Delta v_{12} - \Delta v_{12,sys})^2}{2\langle \Delta v_{12}^2 \rangle}\right] \\ & - \exp\left[-\frac{(\Delta v_{12} + \Delta v_{12,sys})^2}{2\langle \Delta v_{12}^2 \rangle}\right]\Bigg). \end{aligned} \tag{3.67}$$

Equation (3.67) works for either Brownian or turbulent random motion. In the small-particle limit where $\Delta v_{12,sys}^2 \to 0$, Equation (3.67) reduces to a Maxwellian distribution.

3.3.2 Collision Outcomes

From the overview of the planet-forming size ladder in box 3.1, we already know that not all collisions between dust grains lead to perfect sticking. Comprehensive guides to the "zoo" of possible collision outcomes—which include bouncing, erosion, fragmentation, and cratering—come from Güttler et al. (2010) and Blum (2018). Although there may be a two-body collision pathway that goes all the way to Ceres size (Garaud et al. 2013), the prevalence of millimeter- and centimeter-size particles in asteroids and comets (Bizzarro et al. 2004; Cuzzi et al. 2008; Fulle et al. 2016, 2019; Herique et al. 2019, e.g.) suggests that gravitational collapse of a pebble cloud is the more likely planetesimal formation pathway. We will therefore focus on collisions that affect the abundance of planetesimal-forming pebbles with $r_{\rm eff} \lesssim 10$cm. Possible collision outcomes are:

- **Sticking**. In hit-and-stick growth, van der Waals forces attach a small dust grain with $St \ll 1$ to either another small grain or to a larger aggregate (a collection of individual grains, or monomers, bound by surface attraction). Hit-and-stick growth may also bind two fluffy aggregates with high surface area-to-mass ratios if they collide at low speed (Dominik & Tielens 1997). A second type of sticking happens when the collision deforms the particle surfaces, increasing the contact area at the collision site (Chokshi et al. 1993; Weidling et al. 2009); in the language of Güttler et al. (2010), this type of growth is "sticking with surface effects." Both types of sticking happen when the surface attraction is strong enough to overcome the rebound energy, which is the leftover collision energy that did not go into rearranging aggregates or making their monomers roll. A third type of sticking occurs when a compact projectile embeds itself in a larger, porous target (Blum & Wurm 2008; Langkowski et al. 2008). Although most simulations and lab experiments find that sticking cannot form aggregates beyond ~1 cm in size (e.g., Beitz et al. 2011; Birnstiel et al. 2012), Wada et al. (2011) and Okuzumi et al. (2012) suggest that icy particles that remain extremely porous, with $\rho_m \sim 0.1$ g cm^{-3}, may be able to reach planetesimal sizes by sticking alone.

- **Bouncing**. Collisions that are too energetic for sticking, but not strong enough to break either of the colliders, lead to bouncing. Bouncing does not move solids up the size ladder (though Blum & Wurm 2008 found that such collisions sometimes involve a small amount of mass transfer) and can be a significant barrier to planet formation (Zsom et al. 2010; Kruss et al. 2016). Aggregates get slowly compacted as they experience repeated bouncing collisions; they keep the same mass but decrease in $r_{\rm eff}$ and porosity, or fraction of volume filled by vacuum (Blum & Münch 1993; Weidling et al. 2009). The accompanying loss of surface area makes it less likely that an aggregate can grow by sticking in a "lucky" low-speed collision (e.g., Estrada et al. 2016). Because icy particles have stronger surface attraction than silicate grains (see the inset box 3.1), bouncing may not be a common collision outcome beyond the H_2O-ice line (Shimaki & Arakawa 2012; Gundlach & Blum 2015; Gärtner et al. 2017). Xiang et al. (2019) and Steinpilz et al. (2019) propose that electrical charging during collisions may increase the efficiency of electrostatic sticking, while Kruss & Wurm (2018) suggest that large-scale disk magnetic fields might turn on ferromagnetism in iron-rich dust, again suppressing bouncing. Furthermore, simulations that draw the random collision speed components from a distribution rather than using only rms values show that low-speed collisions create enough large aggregates to sweep up the smaller victims of bouncing (Windmark et al. 2012b; Drażkowska et al. 2014; Estrada et al. 2016). Finally, Seizinger & Kley (2013) argue that particles prepared in the lab for drop-tower collision experiments are more likely to bounce than aggregates that grow naturally in protoplanetary disks due to their ordered internal structure. Given the abundance of planets in the Galaxy, it is clear that the bouncing barrier does not prevent planet growth. Exactly how the barrier is overcome is an ongoing area of research.
- **Erosion, cratering, and abrasion**. While porous particles can grow to large sizes in low-speed collisions (~cm s^{-1}), they are vulnerable at higher speeds. Erosive collisions scrape a small amount of material off of one or both colliders. After many rounds of collisions, an aggregate's surface may be smooth enough to halt erosion (Schräpler & Blum 2011; Schräpler et al. 2018). In cratering, a projectile excavates a cavity in a target aggregate (Wurm et al. 2005; Paraskov et al. 2007). The transition from erosion to cratering is smooth, with larger targets and higher collision speeds tending toward cratering (Blum 2018). Abrasion, which is similar to erosion, involves gradual mass loss from centimeter-size aggregates undergoing slow collisions, $\lesssim$1 m s^{-1} (Kothe 2016). The aggregates may survive ~1000 such collisions before being finally being destroyed by abrasion.

Figure 3.15. Two dust aggregates just after release in the drop tower at Technische Universität Braunschweig (top) and their postcollision fragments (bottom). Image reprinted from Schräpler et al. (2012), used with permission.

- **Catastrophic fragmentation**. The most destructive collision outcome involves the breakup of both the projectile (the smaller of the two colliding particles) and the target. Unlike in bouncing, erosion, and cratering, the collision leftovers are not similar in size to the input particles. Instead, the fragments follow a power-law size distribution where both the exponent and the size of the largest remnant are functions of collision speed (Bukhari Syed et al. 2017). (Simulations often use simplified fragmentation prescriptions where the power-law index and the mass ratio of the largest remnant and target are fixed, e.g., Sengupta et al. 2019). See Figure 3.15 for before and after pictures of a fragmenting collision. While catastrophic fragmentation of rocky grains is usually associated with speeds $\Delta v_{12} \gtrsim 1$ m s^{-1} (Blum & Münch 1993), more recent experiments have yielded fragmentation of aggregates of microscopic silicate grains at only 10–16 cm s^{-1} (Schräpler et al. 2012; Deckers & Teiser 2013). Icy aggregates may be more robust to fragmentation than silicate aggregates (Okuzumi et al. 2012; Hill et al. 2015; Pinilla et al. 2016; Cridland et al. 2017), though experimental results are limited. If collisions between icy particles of similar size cannot reach the threshold speed for fragmentation (Figure 3.14), their growth is instead limited by erosion (Krijt et al. 2015). Because drift speeds reach 1 m s^{-1} for particles with $r_{\rm eff} \sim 3$ cm (Figure 3.6), the combination of radial and azimuthal drift could certainly drive destructive collisions between (e.g.) millimeter- and centimeter-size silicate particles. So could vertical settling, though centimeter-size particles usually don't stray very far beyond the midplane (Figure 3.12) and so typically have low vertical speeds. Although destructive collision outcomes such as fragmentation, erosion, and cratering impede planet growth, they are robustly documented in the lab and necessary for maintaining the population micron-size grains that contribute to the infrared part of star/disk SEDs (Dullemond & Dominik 2005; Section 2.2). The combination of astrophysical, meteoritic, and laboratory results suggests that there is a continuous cycle of grain growth, planetesimal formation, and catastrophic collisions between both planetesimals and pebbles throughout the planet-growth epoch (Dauphas & Chaussidon 2011).

- **Fragmentation with mass transfer**. If there is hope for a planetesimal formation pathway via two-body collisions, it comes from fragmentation with mass transfer (e.g., Garaud et al. 2013; Deckers & Teiser 2014). Here the collision speed is above the fragmentation limit for the projectile, but not the target. Although the projectile jettisons some fragments, the target gains between a few percent and 50% of the original projectile mass (Teiser et al. 2011a, 2011b; Meisner et al. 2013). According to the new collision model presented by Blum (2018), meter-size boulders make ideal projectiles: if they collide with any target that has $r_{\rm eff} > 1$ m, they will donate some of their mass to the target. However, the bridge from boulders to planetesimals is narrow, as the projectile size requirement is extremely exact. Some mass transfer is possible at centimeter to meter sizes, but Booth et al. (2018) suggest that breakthrough growth from centimeter → meter → kilometer is unlikely because the required meter-size projectiles are vulnerable to radial drift.

In the next section, we will use our knowledge of collision outcomes to construct models that predict the grain-size distribution and dust surface density available for planetesimal formation.

3.3.3 Grain Growth Models

A grain-growth model uses a prescription for collision outcome as a function of speed and particle properties to simulate the evolution of the grain-size distribution. Numerical simulations that include most or all of the collision outcomes described in Section 3.3.2 are useful for predicting the dominant particle-growth pathways (e.g., Brauer et al. 2008; Zsom et al. 2010; Drazkowska et al. 2019) and the grain-size distribution (e.g., Carballido et al. 2016; Sengupta et al. 2019) and for evaluating the importance of rare, "lucky" events (e.g., low-speed collisions that lead to the growth of centimeter-size aggregates; Estrada et al. 2016; Booth et al. 2018). However, because such computationally expensive simulations cannot always cover an entire disk lifetime or explore a wide variety of disk masses and turbulent speeds, one may wish to use a simpler model that includes a subset of collision outcomes to develop physical intuition about grain growth (e.g., Kornet et al. 2001; Rózyczka et al. 2004; Klahr & Bodenheimer 2006; Garaud 2007; Hughes & Armitage 2010; Krijt et al. 2016). We will first study the Birnstiel et al. 2012 semianalytical model, which has been used to trace the growth of planet-forming pebbles in numerous ALMA-observed disks (e.g., Ricci et al. 2017; Tripathi et al. 2018; Stammler et al. 2019). We will then explore the more complex grain-growth model of Windmark et al. (2012a) and discuss techniques for using it in numerical simulations.

3.3.3.1 Semianalytical Model

The Birnstiel et al. (2012) model, which is benchmarked against numerical simulations of dust evolution by Birnstiel et al. (2010), predicts the particle surface density $\Sigma_p(R)$ available for planetesimal formation and the pebble flux $\dot{M}_p(R)$ that can contribute to pebble accretion (see the inset box 3.1). We divide the continuous grain-size distribution into two populations: small particles with $|v_{\rm adv}| > |u|$ whose radial motion is dominated by advection, and large particles with $|v_{\rm adv}| < |u|$ whose radial motion is governed by drag-induced drift. Macroscopic particles of sizes shown in Figure 3.13 usually satisfy the large-particle condition. The simplest grain-size distribution is a power law:

$$\frac{dn_p}{dr_{\rm eff}} = N_r\left(\frac{r_{\rm eff}}{r_r}\right)^{-q}, \tag{3.68}$$

where n_p is the grain number density (previously seen in Equation (3.47)), N_r is the normalization (units of cm^{-4}), and r_r is the reference size. Equation (3.68) describes grains in the interstellar medium ($q = 3.5$; Mathis et al. 1977) and in debris disks ($q = 3.36$; MacGregor et al. 2016). For both q values, small grains provide almost all the optical cross section while large grains hold almost all the mass (see Table 3.1 of Dohnanyi 1969). Even in a more complex size distribution, it is still almost always true that the mass is concentrated in the largest bodies (e.g., Brauer et al. 2008; Schlichting et al. 2013). We can therefore keep an approximate inventory of solids available for planetesimal formation—which is seeded by pebbles and/or boulders instead of small grains—by tracking only the maximum grain size $r_{\rm eff,max}(R, t)$ and ignoring the detailed size evolution of smaller particles.

In the Birnstiel et al. (2012) model, the only two collision outcomes included are sticking and fragmentation: particles at a given location R grow until either their characteristic collision speeds reach the fragmentation threshold or they start to decouple from the gas and drift inward. The chosen fragmentation threshold speed is $v_f = 10$ m s^{-1}, higher than the ~1 m s^{-1} usually quoted for silicates (Blum & Wurm 2008) but more conservative than the largest estimates of fragmentation speed for icy grains, which can reach 50 m s^{-1} (Wada et al. 2009; Krijt et al. 2015). The particles are assumed to be icy because the vast majority of the solid mass in any disk is located outside the water-ice line (see the inset below). We ignore the dependence of v_f on particle size (Güttler et al. 2010). Because microscopic grains almost always stick when they collide, we only need to consider fragmentation as a possible collision outcome for pebbles, which we define as macroscopic particles in the millimeter to centimeter-size range. Catastrophic collisions therefore occur in the friction time regime $\tau_\eta \ll \tau_{f,1} \ll 1/\Omega_{\rm K}$ and are driven primarily by turbulence in disks with uniform $\alpha(R, z) = 10^{-3}$–10^{-2}, though radial and azimuthal drift can drive fragmentation in disks with weaker turbulence ($\alpha \sim 10^{-4}$).

3.3 Where is the H_2O-ice line?

Since H_2O-ice mantles speed up pebble growth and alter grain-growth pathways (for example, by switching the dominant growth mode for millimeter-size particles from fragmentation with mass transfer to direct sticking; Wada et al. 2011), we can only construct realistic planet-formation models if we know where a disk's H_2O-ice line is at any given time. The simplest approach to finding the water-ice line is to construct a disk model using your favorite heating prescription (passive, viscous, or both; see Sections 2.3.2, 2.4.4, and 2.4.5) and find the locations where $T = 160$ K. From Figures 2.18 and 2.19, we see that the water-ice "line" is really 2D surface: vertical motion, as well as radial motion, can cause ice to sublimate or freeze (Davis 2005; Isella & Natta 2005; Oka et al. 2011). One can find the location of the ice line more accurately by taking into account the pressure dependence of the sublimation temperature (Podolak & Zucker 2004). From Min et al. (2016), sublimation occurs when

$$P_v(T) = \frac{R_g}{\mu_{H_2O}} \cdot 10^{(B-A/T)} \cdot (1\text{K}), \rho \tag{3.69}$$

where P_v is the partial pressure of H_2O vapor, μ_{H_2O} is the molar mass of H_2O, and A and B are experimentally measured constants ($A = 2\,827.7$ K, $B = 7.720\,5$; Pollack et al. 1994; Kama et al. 2009). To find the partial pressure of water vapor, simply multiply the total gas pressure by the mole fraction of water vapor. For solar composition, water makes up $\sim 10^{-4}$ of the total number of molecules in the disk, and there are equal numbers of ice molecules and gas molecules at the sublimation front. The water vapor partial pressure is then $\sim 5 \times 10^{-5} P$.

Added complexity comes from the kinetic nature of freezeout. To freeze, water molecules attach to grain surfaces, which have sticking efficiencies that vary with composition (Ros et al. 2019). The most exact computations of the ice-line location come from fully kinetic chemical reaction networks that include nonequilibrium processes. For example, Equation (3.69) requires modification when UV photodesorption is important (photodesorption happens when a photon hits a grain surface and gives a water molecule enough kinetic energy to escape into the gas phase). Chemical reactions and photodissociation can also change the water abundance so that the approximation $P_v \sim 5 \times 10^{-5} P$ is invalid. Another complication is that the H_2O-ice-line location changes as a function of time: star evolution along the Hayashi track plus the grain-settling-induced decrease in angle φ at which starlight intercepts the disk surface both lead to cooling (e.g., Yu et al. 2016), while viscous spreading enables cosmic-ray penetration, which strengthens MRI turbulence and leads to heating (e.g., Landry et al. 2013). A partial reference list on the behavior of H_2O in molecular clouds and protoplanetary disks is Aikawa et al. (1997, 1999), Lecar et al. (2006), Dodson-Robinson et al. (2009), Visser et al. (2009), Semenov & Wiebe (2011), Martin & Livio (2012), Henning & Semenov (2013), Furuya et al. (2013), van Dishoeck et al. (2013), Du & Bergin (2014), van Dishoeck et al. (2014), Baillié et al. (2015), and Mulders et al. (2015).

Until recently, most chemical evolution codes were proprietary, though the gas-phase reaction rates were based on publicly available databases such as UMIST and KIDA (Woodall et al. 2007; McElroy et al. 2013; Wakelam et al. 2012). Now the ProDiMo disk-modeling code (Woitke et al. 2009), which includes chemistry and freezeout, is available by registration.[19] Kamp et al. (2017) compare model results for

[19] http://www.astro.rug.nl/~prodimo/

different "under-the-hood" reaction networks in ProDiMo, focusing on molecular emission-line fluxes. Not surprisingly, changes in the reaction network strongly affect the locations of the ice reservoirs and line-emitting areas. There are still significant uncertainties relating to water chemistry and freezeout in disks (and, indeed, the chemistry and freezeout of any molecule).

Observations provide some guidance on where the H_2O-ice line falls in Stage II and transitional evolutionary phases. Some astrophysical disks have H_2O ice-line locations inferred from infrared spectroscopy. Meijerink et al. (2009) constructed non-LTE models that qualitatively reproduced the Spitzer mid-infrared spectra of Stage II disks with water vapor detections by Salyk et al. (2008) and Carr & Najita (2008), showing that the water vapor must be localized inside $\lesssim$1 au. From high-resolution spectra, Salyk et al. (2019) reconstructed the velocity fields of the water vapor-rich regions in six Stage II disks, finding that the emission came from the inner few astronomical units. Their results were similar to previous studies by Pontoppidan et al. (2010), Banzatti et al. (2014), and Salyk et al. (2015). In disks without inner holes, at least, water ice resides outside ~3 au. In the disk surrounding the protostar V883 Ori, which is experiencing a short-term episode of strong accretion, Cieza et al. (2016) inferred an ice line at 42 au from the change in submillimeter intensity and spectral index (which measures grain size) as a function of distance from the star.

At only 47 pc away, the transitional disk surrounding TW Hya is bright enough for high signal-to-noise spectroscopy and subtends a large-enough angle on the sky for spatially resolved imaging. The combination of velocity-resolved spectral lines plus a well-determined temperature profile constructed from the SED and spatially resolved images allowed Zhang et al. (2013) to locate the surface H_2O-ice line, where ice near the disk surface sublimates (where water vapor is abundant, the optical depth in its infrared emission lines is too high to see into the midplane). Figure 3.16 shows the best-fit water vapor column density (number of molecules per cm^{-2}; similar to the surface density introduced in Chapter 2) as a function of radius in the TW Hya disk. The black dashed line shows the total gas column density; the deficit inside 4 au is the transitional disk's inner hole. Water vapor is not detected inside the hole. The water vapor column density has a local peak from 4 to 4.2 au, indicating that ice is sublimating right at the inner edge of the gas disk—as expected, given that the inner edge is directly irradiated by the central star. As expected for solar abundances, the water vapor mole fraction at the peak is 2.5×10^{-4}. Beyond 4.2 au, the water vapor abundance drops precipitously, consistent with freezeout. A similar analysis of four Stage II and transitional disks by Blevins et al. (2016) found surface snow lines at 3–11 au.

What about constraints from the solar system? Asteroid composition indicates that the snow line during the final epoch of planetesimal formation was at ~3 au (Abe et al. 2000; Morbidelli et al. 2000; Rivkin et al. 2002). Given that both models and observations show that the snow line can move inside 1 au during Stage II (e.g., Lecar et al. 2006; Garaud & Lin 2007; Pontoppidan et al. 2010; Oka et al. 2011; Landry et al. 2013; Salyk et al. 2019), the low water abundance in Earth and Mars and the dryness of the inner asteroid belt are puzzling. Morbidelli et al. (2016) argue that Jupiter's formation "fossilized" the H_2O- ice-line location. Because gas accretes inward at a higher velocity than any ice line's rate of inward motion, the gas in the terrestrial planet-forming region was sourced from beyond 1 au. If that gas came from beyond the H_2O-ice line, it would be water-poor, as the water in its source region would be frozen. By the time the solar nebula cooled to the point that the midplane temperature was 160 K at 1 au, the gap opened by Jupiter prevented icy pebbles from drifting into the inner disk. With water-poor gas as an accretion relic and no supply of inward-drifting icy pebbles, there was no way to form water ice at 1 au, even if

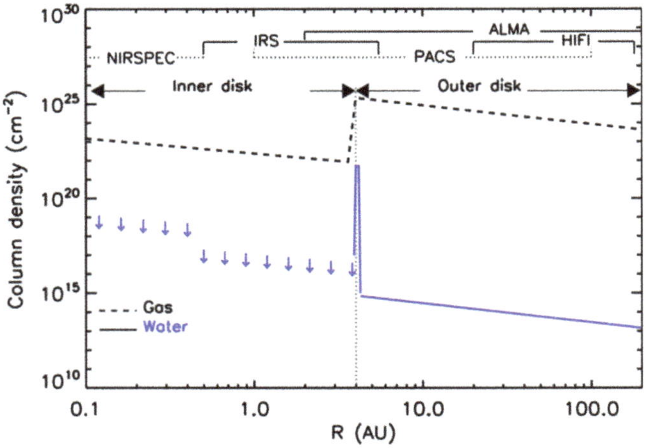

Figure 3.16. The location of water vapor in the disk surrounding TW Hya. The blue solid line shows the water vapor column density as a function of distance from the star, while the blue arrows show upper limits at locations where water vapor doesn't contribute to the infrared spectrum. The black dashed line shows the gas column density profile. The discontinuity at 4 au shows the transition radius: TW Hya is a transitional disk with an inner hole 4 au wide. The best-fit model water distribution shows ice sublimating right at the transition radius. Brackets at the top of the plot show the regions in the disk traced by different observatories and instruments. Image reprinted from Zhang et al. (2003).

the disk was cold enough. The results from Morbidelli et al. (2016) highlight the importance of dust traps and gas–solid interaction in determining the chemical inventory of the disk. Reconstructions of the solar nebula ice abundance profile are also frustrated by the fact that dynamical chaos in the planetesimal disk "scrambles" the chemical information (e.g., Raymond & Izidoro 2017).

It would be nice if we had more concrete information about the location of the midplane ice line and the water abundance in either the solar nebula or astrophysical disks, but broadly speaking, icy pebbles and planetesimals form beyond 1–3 au.

We begin by using Equations (2.14), (3.6), and (3.8) to rewrite the midplane Stokes number as

$$St_m = \frac{\pi}{2}\frac{r_{\rm eff}\rho_m}{\Sigma}. \tag{3.70}$$

Assuming catastrophic collisions are driven by turbulence, we can set $\langle \Delta v_{12}^2 \rangle_t^{1/2} = v_f$. Using the approximation

$$\langle \Delta v_{12}^2 \rangle_t^{1/2} \simeq \sqrt{3\alpha St}\ c_s \tag{3.71}$$

for collisions between equal-size particles (Ormel & Cuzzi 2007), we find the midplane Stokes number for which collisions lead to catastrophic fragmentation:

$$St_f \simeq \frac{1}{3}\frac{v_f^2}{\alpha c_s^2}. \tag{3.72}$$

St_f gives the maximum possible size any particle can reach, but the largest particles that survive long-term without being fragmented have Stokes numbers slightly below St_f. Introducing f_f, an order-unity factor that enforces $St_m < St_f$, we find that the largest surviving particles have size

$$r_{\text{eff,fmax}} \simeq f_f \frac{2}{3\pi} \frac{\Sigma}{\rho_m} \frac{v_f^2}{\alpha c_s^2}. \tag{3.73}$$

Fragmentation is not the only process that removes pebbles from a given location in the disk—we also have to consider radial drift. We calculate the maximum size a growing pebble in the midplane at location R can reach before it drifts inward by setting the drift timescale τ_{dr} (Equation (3.33)) equal to the grain-growth timescale $\tau_g = r_{\text{eff}}/\dot{r}_{\text{eff}}$. We approximate $\dot{r}_{\text{eff}}$ as r_{eff}/τ_c, where τ_c is the collision timescale

$$\tau_c = \frac{1}{n_p \sigma_r \left\langle \Delta v_{12}^2 \right\rangle_t^{1/2}} \tag{3.74}$$

and $\sigma_r = \pi r_{\text{eff}}^2$ is the particle cross section. With order-unity factors neglected, the grain-growth rate is

$$\dot{r}_{\text{eff}} \simeq \frac{\rho_p}{\rho_m} \langle \Delta v_{12}^2 \rangle_t^{1/2}, \tag{3.75}$$

where $\rho_p \times 1\ \text{cm}^3$ is the total mass of particles contained in $1\ \text{cm}^3$ of mixed gas and dust and $\rho_p = m n_p$. To find the grain-growth timescale, we combine Equations (3.50), (3.71), and (3.75) with Equation (2.14). We can eliminate the material density ρ_m from Equation (3.75) using

$$\frac{\rho_m}{\rho_p} = \frac{3}{4\pi r_{\text{eff}}^3 n_p} \tag{3.76}$$

$$n_p \simeq \frac{\Sigma_p}{m h_p}, \tag{3.77}$$

where $\Sigma_p = \int \rho_p dz$ is the surface density of the particles. After leaving out factors of order unity, the result is

$$\tau_g = \frac{\Sigma}{\Sigma_p \Omega_K}. \tag{3.78}$$

Rewriting Equation (3.33) as $\tau_{dr} = v_K/(St \Delta g)$ and setting $\tau_{dr} = \tau_g$ gives the Stokes number of the largest particle that can survive radial drift long-term:

$$St_{dr} = f_d \frac{\Sigma_p}{\Sigma} \frac{\Omega_K v_K}{|\Delta g|}, \tag{3.79}$$

corresponding to the drift-limited maximum size

$$r_{\text{eff, dmax}} = f_d \frac{\Sigma_p}{\Sigma} \frac{\rho}{\rho_m} \frac{v_{\text{K}} \overline{v}}{|\Delta g|}. \tag{3.80}$$

In Equations (3.79) and (3.80), f_{d} is another order-unity fudge factor that enforces $St < St_{dr}$.

Next we derive a mathematical relationship between Σ_p and the inward flux of pebbles through each disk annulus $\dot{M}_p$. We first consider the case where radial drift dominates pebble velocity and with it the local pebble surface density. Because the size of the fastest-drifting particle decreases with increasing R beyond about 6 au (Figure 3.6, top), we expect to see drift-dominated particle dynamics in the outer disk. As in Equation (2.32), the pebble flux is

$$\dot{M}_p(R) = 2\pi R \Sigma_p \overline{v}_p, \tag{3.81}$$

where the mass-averaged particle drift speed is

$$\overline{v}_p = f_s v_{\text{adv}}(r_s) + f_l \left[v_{\text{adv}}(r_{\text{large}}) - u(r_{\text{large}}) \right]. \tag{3.82}$$

In Equation (3.82), f_l is the fraction of solid mass in the large population (size r_{large}) and $f_s = 1 - f_l$ is the mass fraction in the small population (size r_s). $u(r_s)$ is not included in $\overline{v}_p$ because $u \to 0$ for the small population. Combining Equations (3.30), (3.32), and (3.79) gives the drift-dominated solid surface density

$$\Sigma_{pd}(R) = \sqrt{\frac{\dot{M}_p \Sigma}{2\pi f_d R v_{\text{K}}}}. \tag{3.83}$$

Calibrating Equation (3.83) against the Birnstiel et al. (2010) simulations gives $f_d = 0.55$. If $\dot{M}_p(R)$ is constant so that the pebble flow is in steady state, a gas surface density profile with $\xi = 1$ yields $\Sigma_{pd} \propto R^{-3/4}$.

In the inner disk, where collisions are more frequent, fragmentation sets the maximum pebble size and controls the solid surface density profile. Equations (3.30), (3.32), and (3.72) yield the fragmentation-limited particle surface density

$$\Sigma_{pf}(R) = \frac{3\dot{M}_p}{2\pi f_f R^2} \frac{g}{|\Delta g|} \frac{\alpha c_s^2}{v_f^2 \Omega_{\text{K}}} = \frac{3\dot{M}_p}{2\pi R^2 f_f} \frac{g}{|\Delta g|} \frac{\nu}{v_f^2}. \tag{3.84}$$

Constant $\dot{M}_p(R)$ and $\Sigma \propto R^{-1}$ combine to give $\Sigma_{pf} \propto R^{-3/2}$, and the calibrated value of f_f is 0.37.

The last ingredient in the model is time evolution. The disk only achieves $\dot{M}_p(R) \simeq$ constant after grains have had time to grow to size $r_{\text{eff,fmax}}$ or $r_{\text{eff,dmax}}$. The grain-growth timescale τ_g is best interpreted as an e-folding time, which means the time it takes a particle to reach its maximum effective radius is

$$\tau_{\max} = \tau_g \cdot \ln\left(\frac{r_{\text{eff,max}}}{r_s}\right), \tag{3.85}$$

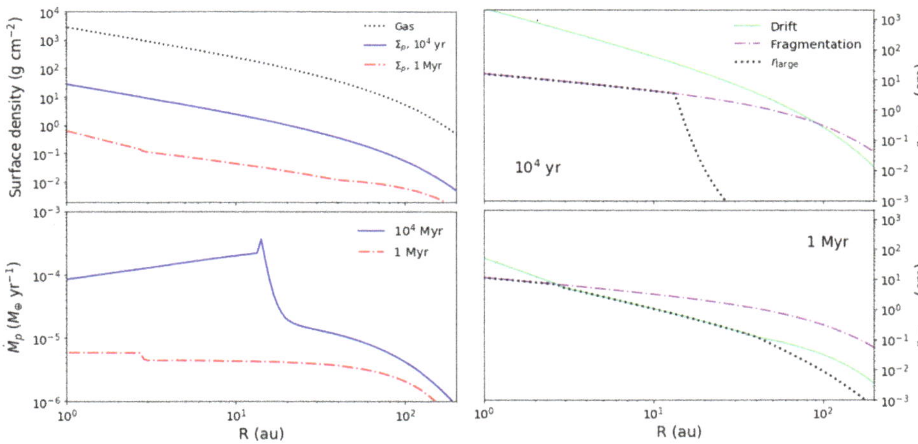

Figure 3.17. Left panels: Solid surface density (top) and pebble flux (bottom) from the Birnstiel et al. (2012) model disk with $\alpha = 10^{-3}$ at times 10^4 years and 1 Myr. Right panels: $r_{\rm eff,dmax}$ (green solid line), $r_{\rm eff,fmax}$ (purple dash-dot line), and $r_{\rm large}$ (black dotted line) for the pebbles after 10^4 years (top) and 1 Myr (bottom). The young disk has a fragmentation-dominated size distribution, while the more evolved disk becomes drift dominated outside 2.5 au due to the lower pebble flux.

where $r_{\rm eff,max} = \min\,[r_{\rm reff,fmax}, r_{\rm reff,\,dmax}]$. At any point in time, the largest particle in the midplane has a radius

$$r_{\rm large}(R,\,t) = \min\left[r_{\rm eff,max},\; r_s \exp\left(\frac{t}{\tau_g}\right)\right]. \tag{3.86}$$

We use an advection/diffusion equation to follow the solid surface density as particles drift and diffuse:

$$\frac{\partial \Sigma_p}{\partial t} + \frac{1}{R}\frac{\partial}{\partial R}\left[R\left(\Sigma_p \overline{v}_p - \Sigma\nu\frac{\partial}{\partial R}\left[\frac{\Sigma_p}{\Sigma}\right]\right)\right] = 0, \tag{3.87}$$

where Equation (3.87) can be solved using the same algorithms as Equation (2.28). Calibrated values of f_l that can be used to determine $\overline{v}_p$ are 0.97 for drift-limited growth and 0.75 for fragmentation-limited growth. If the largest particles have not yet reached either $r_{\rm eff,fmax}$ or $r_{\rm eff,dmax}$, one may use the mass ratio of particle size $r_{\rm large}$ to size $r_{\rm eff,dmax}$ to find $f_l \simeq 0.97\, r_{\rm large}^3 / r_{\rm eff,dmax}^3$.

Figure 3.17 shows the solid surface density (top left) and pebble flux (bottom left) for our model disk with $\alpha = 10^{-3}$ after 10^4 years and 1 Myr of evolution. Maximum pebble sizes $r_{\rm eff,dmax}$, $r_{\rm eff,fmax}$, and $r_{\rm large}$ after $t = 10^4$ years and $t = 1$ Myr are shown on the right. Following Birnstiel et al. (2012), we have modified the disk to use the exponential taper on the gas surface density profile (Equation (2.5)) with $R_0 = 60$ au; the gas mass in the inner 100 au is unchanged. We initialize the particle evolution with a dust/gas mass ratio of 0.01, as in the interstellar medium, and a population of small grains with $r_s = 0.1$ μm. Calculations are performed with `two-pop-py`, a lightweight and easy-to-use `python` implementation of the Birnstiel et al. (2012) model released by Birnstiel et al. (2017).[20] After only 10^4 years of evolution, not much of the disk's solid mass has had time to drift into the star and $\Sigma_p/\Sigma \simeq 0.01$ at all radii. The pebble flux is not yet in steady state (bottom left): there is a local maximum followed by a sharp dropoff at 14 au, beyond which particles have not had time to reach $r_{\text{eff,fmax}}$ (top right). After 1 Myr of evolution, the pebble flux is in steady state and the disk's solid mass budget has been dramatically depleted: the pebble mass in the inner 100 au is only $12.4 M_\oplus$, compared with $333 M_\oplus$ of solids at $t = 0$. Even with 100% efficient conversion of pebbles into planets, $12.4 M_\oplus$ barely provides enough solid mass for one giant-planet core (Pollack et al. 1996). Because 1 Myr of solids spiraling into the star has left the number density of pebbles too low for frequent collisions everywhere except in the inner 2.5 au, the outer disk is drift-dominated instead of fragmentation-dominated.

The calculations shown in Figure 3.17 underscore the need for quick planetesimal formation: a million-year time lag before we reach the "Ceres" step on the size ladder (see the inset box 3.1) leaves even the MAXSN virtually incapable of forming giant planets. Fortunately, the meteoritic record shows that the first planetesimals in the inner ~3 au of the solar nebula—the magmatic iron parent bodies—formed $\lesssim$0.1 Myr after the first calcium aluminum-rich inclusions condensed from the hot Stage I disk (Dauphas & Chaussidon 2011; see Section 4.2).

3.3.3.2 *Numerical Model*

To include other collision outcomes besides sticking and fragmentation, and to compute the time-evolving grain-size distribution $N(r_{\text{eff}}, R, z, t)dr_{\text{eff}}$,[21] we require a full-scale numerical simulation rather than a semianalytical model. Most recent dust evolution simulations are based on the work of Güttler et al. (2010), who synthesized all available experimental data into a prescription for collision outcome as a function of particle properties and speed. We will illustrate the basics of grain-growth models by following Windmark et al. (2012a), who created a simpler and faster version of the Güttler et al. (2010) prescription focusing on "compact" particles with $\phi_c > 0.4$, where ϕ_c is the fraction of the effective particle volume $4\pi r_{\text{eff}}^3/3$ filled by solid material. The material density $\rho_m = 2$ g cm^{-3} that we have used throughout this chapter corresponds to $\phi_c \simeq 0.6$ for aggregates of the olivine/pyroxene grains seen in mid-IR spectra of Stage II disks (e.g., Kessler-Silacci et al. 2006). Because repeated collisions force aggregates to become more compact, it is reasonable to assume that many macroscopic particles—even icy ones—have $\phi_c > 0.4$ (though see Okuzumi et al. 2012). However, the Windmark et al. (2012a) model does not include size-dependent porosity (ϕ_c and ρ_m are functions of r_{eff}), which may affect the drift rate and therefore the number density of large grains (Equation (3.80)).

[20] http://birnstiel.github.io/two-pop-py/

[21] Disks are usually assumed axisymmetric due to the high computational cost of simulating grain coagulation. However, Drazkowska et al. (2019) added the ϕ dependence in order to simulate grain evolution in the complex gas density profile surrounding a giant planet. One could perform a similar experiment for grain growth and motion in and around a vortex.

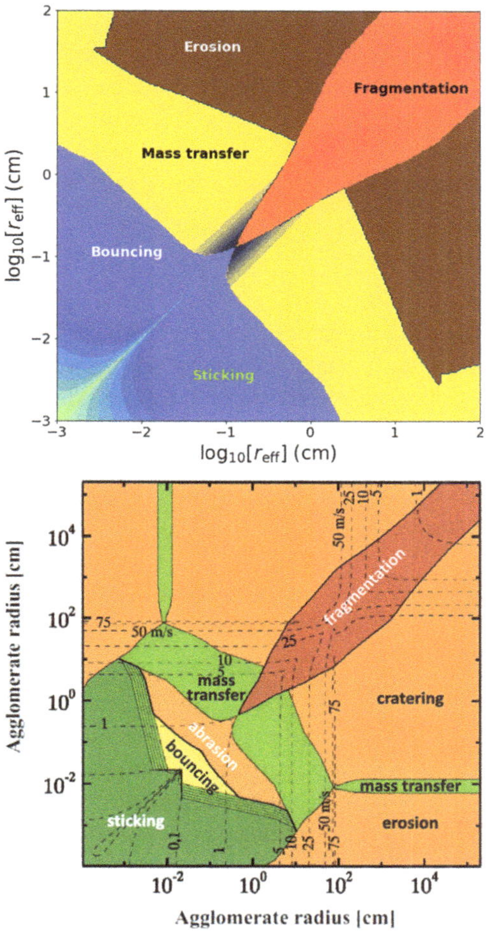

Figure 3.18. Top: Windmark et al. (2012a) collision model applied to the gas densities and velocities at 1 au in our fiducial model disk, given $\alpha = 10^{-3}$. Gray contours in the mass transfer set of collision outcomes show the mass ratio of the largest collision product to the target, $(m_1 + \epsilon_{ac}m_2 - m_{er})/m_1$, from 1 to 1.16 in steps of 0.02. Green contours show the sticking probability from 0 to 1 in steps of 0.1. Bottom: Updated collision model by Blum (2018), with permission of Springer. The new model has bouncing restricted to a much narrower size regime than in Windmark et al. (2012a) model, suggesting that collisional growth to centimeter sizes is robust. Contour lines show $\langle \Delta v_{12}^2 \rangle^{1/2}$.

The top panel of Figure 3.18 shows the Windmark et al. (2012a) collision model calculated at 1 au in our model disk from Chapter 2, with $\alpha = 10^{-3}$. The bottom panel of Figure 3.18 shows the updated model by Blum (2018), again for 1 au and $\alpha = 10^{-3}$ but with gas density taken from the minimum-mass solar nebula model.[22] While Figure 3.18 shows the collision outcome as a function of particle size, we will

[22] A preliminary version of the Blum (2018) model is available in Kothe (2016), https://nbn-resolving.org/urn:nbn:de:gbv:084-17010312131.

use mathematics describing outcomes as functions of particle mass as in Windmark et al. (2012a). For the remainder of this section we will discuss the particle mass distribution $N(m)\,dm$ rather than the size distribution $N(r_{\rm eff})\,dr_{\rm eff}$. For target mass m_1 and projectile mass m_2, where $m_1 > m_2$ as in Section 3.3.1; the collision-outcome prescription is as follows:

- **Sticking and bouncing:** All collisions with speeds below the sticking threshold

$$\Delta v_{12,s} = \left(\frac{m_2}{3.0 \times 10^{-12}\ \mathrm{g}}\right)^{-5/18} \quad \mathrm{cm\ s^{-1}} \tag{3.88}$$

  lead to perfect sticking, no matter the target mass. Equation (3.88) shows that the threshold sticking speed decreases with increasing particle mass, as expected: while colliding microscopic particles almost always stick, macroscopic aggregates with low surface area/mass ratios are not easily bonded by electrostatic forces. Sticking replaces particles 1 and 2 with a new particle of mass $m_1 + m_2$.

  All collisions with speeds above the bouncing threshold

$$\Delta v_{12,b} = \left(\frac{m_2}{3.3 \times 10^{-3}\ \mathrm{g}}\right)^{-5/18} \quad \mathrm{cm\ s^{-1}} \tag{3.89}$$

  lead to bouncing, erosion, or fragmentation (with or without mass transfer). $\Delta v_{12,b}$ decreases with increasing m_2, meaning that the probability of "lucky" low-speed collisions in which rebound energy is lost to rearranging the aggregates vanishes for large projectiles. When two particles bounce, they return to the mass distribution unchanged. The Windmark et al. (2012a) model ignores the fact that bouncing decreases porosity, which in turn increases the expected speed for the particle's next collision. Collisions with $\Delta v_{12,s} < \Delta v_{12} < \Delta v_{12,b}$ could result in sticking or bouncing, depending on the impact parameter and the internal properties of the colliding aggregates. Windmark et al. (2012a) handle such collisions probabilistically. For $\Delta v_{12,s} < \Delta v_{12} < \Delta v_{12,b}$, the sticking probability is

$$P_s = 1 - k_1 \log_{10}(\Delta v_{12}) - k_2, \tag{3.90}$$

  where $k_1 = 0.4$ and $k_2 = \log_{10}(m_2/m_s)/\log_{10}(m_b/m_s)$. Okuzumi et al. (2012) and Krijt et al. (2015) suggest that icy particles are both sticky and porous enough to eliminate any possibility of bouncing. A bounce-free collision model would prescribe that all collisions lead to sticking unless the speed is high enough to fragment at least one of the colliders (see below).
- **Erosion and/or fragmentation with mass transfer:** Whenever any collision has $\Delta v_{12} > \Delta v_{12,b}$, one must check whether one or both particles should break. Like Windmark et al. (2012a), we will make the simplifying assumption that each particle's kinetic energy is used only to rearrange or fragment that same particle and does not affect the other particle. Instead of using the relative speed Δv_{12} to find the joint collision outcome for both particles, we use each

particle's speed in the center-of-mass frame to see how the collision affects each particle individually. The center-of-mass-frame speeds are

$$v_{\rm com,2} = \frac{\Delta v_{12}}{1 + m_2/m_1} \tag{3.91}$$

$$v_{\rm com,1} = \frac{\Delta v_{12}}{1 + m_1/m_2}. \tag{3.92}$$

A particle's threshold fragmentation speed is

$$v_f = \left(\frac{m}{3.67 \times 10^7 \text{ g}}\right)^{-0.16} \text{ cm s}^{-1}. \tag{3.93}$$

Even though Equation (3.93) has v_f decreasing as m increases, Equation (3.92) shows that each particle's speed in the center-of-mass frame is inversely proportional to its mass, which guarantees that the target is less vulnerable to breakage than the projectile: the target cannot fragment or erode in a collision that leaves the projectile intact. The mass eroded from the target in a collision with $v_{\rm com,1} < v_{f,1}$ is given by

$$\frac{m_{er}}{m_2} = (9 \times 10^{-6}) \cdot \left(\frac{m_2}{m_m}\right)^{0.15} \cdot \left(\frac{\Delta v_{12}}{1 \text{ cm s}^{-1}}\right) - 0.4, \tag{3.94}$$

where m_m is the mass of the smallest possible particle or monomer. If the projectile breaks (i.e., $v_{\rm com,2} > v_{f,2}$), the target may accrete some of the fragments, an outcome we have labeled "fragmentation with mass transfer." The fragment accretion efficiency is

$$\epsilon_{ac} = \min\left[(-6.8 \times 10^{-3}) + (2.8 \times 10^{-4}) \cdot \frac{13 \text{ cm s}^{-1}}{v_{f,2}} \cdot \frac{\Delta v_{12}}{1 \text{ cm s}^{-1}}, 0.5\right], \tag{3.95}$$

where the hard upper limit of $\epsilon_{ac} = 0.5$ is observed in experiments. The target may accrete projectile fragments and get eroded within the same collision. After such a collision, the target is replaced with a new particle of mass $m_1 + \epsilon_{ac} m_2 - \max[m_{er}, 0]$. The leftover, unaccreted projectile mass $m_2(1 - \epsilon_{ac})$ and the eroded mass m_{er} (if $m_{er} > 0$) are each replaced with a set of fragments with a power-law mass distribution (see list entry below on fragmentation). Collisions with $\epsilon_{ac} m_2 > m_{er}$ help move the target up the size ladder.

- **Catastrophic fragmentation:** If $v_{\rm com,2} > v_{f,2}$ and $v_{\rm com,1} > v_{f,1}$, the outcome is catastrophic fragmentation: both particles break into pieces, which must be added to the mass distribution. Following Geretshauser et al. (2011), we assume each particle breaks into (a) a largest remnant with mass m_{lr}, and (b) a population of smaller fragments that follow a power-law mass distribution with $dN/dm = N_0(m/m_r)^{-\zeta}$, where N is the total number (not number density) of fragments, N_0 is the normalization (units g^{-1}), and m_r is the reference mass for the normalization. From experiments, the largest remnant mass is

$$\frac{m_{lr}}{m_0} = 3.27\left(\frac{m_0}{1\ \mathrm{g}}\right)^{-0.068}\left(\frac{v_{\mathrm{com,n}}}{1\ \mathrm{cm\ s^{-1}}}\right)^{-0.43}, \tag{3.96}$$

where m_0 is the precollision mass of the fragmenting particle and $n = [1, 2]$. The integrated fragment mass distribution is

$$M_{\leqslant m} = \int_{m_{\min}}^{m} m\ dm\ N_0\left(\frac{m}{m_r}\right)^{-\zeta} = \frac{N_0 m_r^{\zeta}}{2-\zeta}\left[m^{2-\zeta} - m_{\min}^{2-\zeta}\right], \tag{3.97}$$

where $M_{\leqslant m}$ is the total mass contained fragments smaller than m and $m_{\min}$ is the lower mass cutoff of the power-law population. The upper mass cutoff of the power-law population is

$$m_{\mathrm{lf}} = m_0 \cdot \min\left[\frac{m_{lr}}{m_0}, 1 - \frac{m_{lr}}{m_0}\right], \tag{3.98}$$

while setting $M_{\leqslant m_{\mathrm{lf}}} = m_0 - m_{lr}$ gives

$$N_0 = (2-\zeta)(m_0 - m_{lr})m_r^{-\zeta}\left[m_{\mathrm{lf}}^{2-\zeta} - m_{\min}^{2-\zeta}\right]. \tag{3.99}$$

When the collision outcome is erosion or fragmentation with mass transfer, rather than catastrophic fragmentation, we simply use Equations (3.96) and (3.98) to find $m_{\mathrm{lr},\,2}$ and $m_{\mathrm{lf},\,2}$ and create a power-law population of fragments with total mass $(1 - \epsilon_{ac})m_2 - m_{lr,\,2} + m_{er}$.

All that is left is to specify ζ and $m_{\min}$. Windmark et al. (2012a) recommend $\zeta = 9/8$ based on experimental data compiled by Güttler et al. (2010), but others use ISM dust as the template for their fragments and set $\zeta = 11/6$ (e.g., Drażkowska & Dullemond 2014). Some experimental studies find slightly steeper fragment mass distributions (Deckers & Teiser 2014), but $\zeta \geqslant 2$ is not usually observed in the lab. Because the fragment population is "top heavy" (i.e., the largest fragments hold most of the mass) for $\zeta < 2$, the choice of $m_{\min}$ does not strongly affect the outcome of a grain-growth simulation with fixed porosity ($\phi_c(m)$= constant) as long as grains of $m_{\min}$ are in the electrostatic sticking regime. When comparing simulation results with spectral energy distributions or near-IR observations, we recommend a minimum size cutoff of $r_{\mathrm{eff,min}} \sim 0.1$–$1\ \mu$m ($m_{\min} = 8.4 \times 10^{-15}$–$8.4 \times 10^{-12}$ g for $\rho_m = 2$ g cm^{-3}). We caution that more sophisticated simulations that treat porosity evolution find different grain-growth pathways for different values of $r_{\mathrm{eff,min}}$: dust aggregate growth is drift-limited for $r_{\mathrm{eff,min}} = 1\ \mu$m but bouncing-limited for $r_{\mathrm{eff,min}} = 0.1\ \mu$m (Lorek et al. 2018).

Figure 3.19 shows integrated fragment mass distributions given by Equation (3.97) (blue, right-hand axis) for a 1 cm particle in two destructive collisions, $v_{\mathrm{com}} = 1.5v_f$ (top) and $v_{\mathrm{com}} = 7v_f$ (bottom). Randomly selected fragments obeying the $\zeta = 9/8$ power law with $m_{\min} = 8.4 \times 10^{-15}$ g are

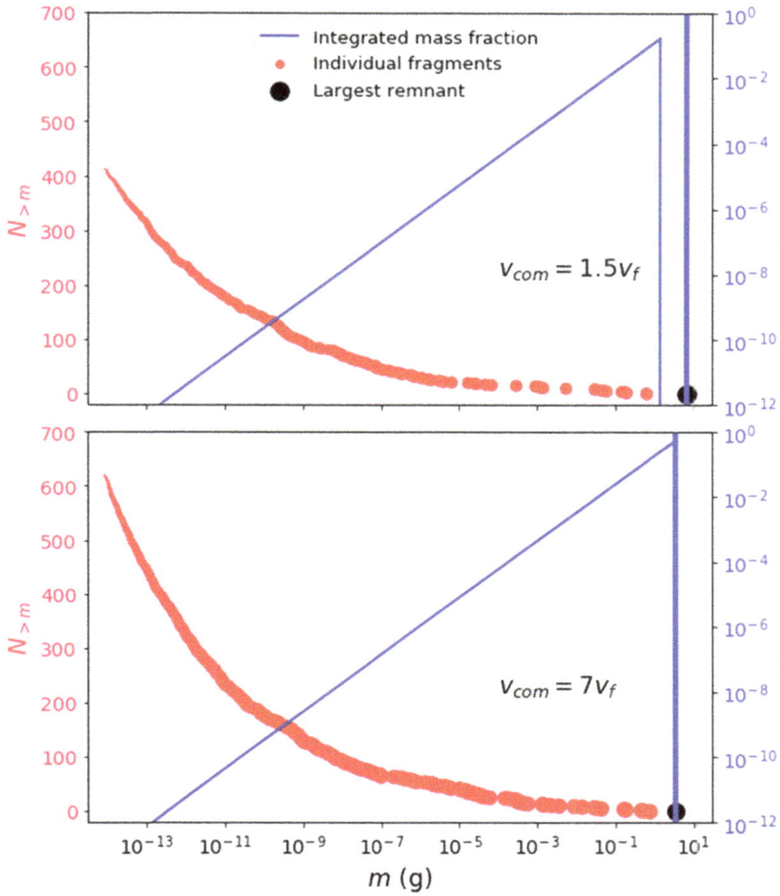

Figure 3.19. Fragments of a 1 cm particle with $\rho_m = 2\,\mathrm{g\,cm^{-3}}$ from a destructive collision with $v_{\rm com} = 1.5v_f$ (top) and a supercatastrophic collision with $v_{\rm com} = 7v_f$ (bottom). The right-hand axis and blue lines show the cumulative mass distribution $M_{\leqslant m}/m_0$ (Equation (3.97)). At $v_{\rm com} = 1.5v_f$, the largest remnant has mass $m_{lr}/m_0 = 0.83$ (Equation (3.96)) and is detached from the rest of the size distribution (Equation (3.98)), while at $v_{\rm com} = 7v_f$ the largest remnant mass of $m_{lr}/m_0 = 0.43$ is also the upper limit to the power-law fragment mass distribution. The red dots are randomly chosen fragments of mass m from the size distribution with $dN/dm = N_0(m/m_r)^{-\zeta}$, where $\zeta = 9/8$. The y value of each dot, plotted on the left-hand axis, shows $N_{>m}$, the number of fragments with mass larger than m. The largest remnant, shown in black, has $N_{>m} = 0$. This figure was adapted from Windmark et al. (2012a), Figure 5.

shown in red, with the number of fragments larger than mass m given by the left-hand axis. If $m_{lr}/m_0 \geqslant 0.5$, the largest remnant detaches from the rest of the fragment mass distribution, creating a δ function in the integrated mass distribution (top). In supercatastrophic collisions with $m_{lr}/m_0 < 0.5$, $m_{lr} = m_{\rm lf}$ and the largest remnant is attached to the power-law population (bottom).

With a complete description of collision outcomes in hand, we can now simulate the time evolution of a disk's grain-size distribution. We will use the Smoluchowski (1916) coagulation equation:

$$\frac{\partial}{\partial t}[n_m(m, R, z)] = \int_{m_i=m_{\min}}^{m_{\max}} dm_i \int_{m_j=m_{\min}}^{m_{\max}} dm_j \ \mathcal{M}(m, m_i, m_j, R, z) \ n_m(m_i, R, z) \ n_m(m_j, R, z), \tag{3.100}$$

where $n_m(m, R, z)$ (units of cm^{-3} g^{-1}) is the number density per unit mass of grains with mass m at disk coordinates (R, z) and $\mathcal{M}$ is the kernel, a collision model such as the one by Windmark et al. (2012a), which encodes how mass is redistributed over all m after a collision between particles with masses m_i and m_j. A pseudo-code for solving the Smoluchowski equation is as follows:

1. Initialize a population of small grains with either a uniform mass distribution $n_m(m)\ dm=$ constant (e.g., Brauer et al. 2008) or a power-law mass distribution to reflect grains inherited from the ISM (e.g., Sengupta et al. 2019). When Equation (3.68) is written for dn_m/dm instead of $dn_p/dr_{\rm eff}$, the power-law index for the ISM dust size distribution is 11/6. If larger grains have not yet had time to settle so that dust and gas are well mixed, n_m is related to ρ and to ρ_p, the total dust mass per volume of mixed gas and dust, by

$$\rho_p = \int_{m=m_{\min}}^{m_{\max}} n_m \ m \ dm = \rho \frac{S}{G}, \tag{3.101}$$

where S/G is the initial dust/gas mass ratio (usually ~0.01) and $m_{\min}$ and $m_{\max}$ are the minimum and maximum grain masses present at the beginning of the simulation. Grains inherited from the ISM have $r_{\rm eff,max} \sim 0.25\ \mu$m (Mathis et al. 1977), but we have seen in Section 3.2 that Stage I protostellar envelopes have larger grains, so setting $m_{\max}$ using $r_{\rm eff,max} = 10$–$100\ \mu$m may be reasonable for a power-law grain-size distribution. Accurately capturing the behavior of rare, "lucky" particles that stick instead of bouncing or breaking requires mass resolution of 30–40 bins per decade (Drażkowska et al. 2014).

2. To evolve the mass distribution in a single grid zone (R, z) for one timestep Δt:
   (a) Find C_{ij}, the collision rate per unit volume (units of collisions s^{-1} cm^{-3}), for each possible combination of m_i and m_j, where i and j are mass bins:

$$C_{ij} = \Delta v_{ij} \cdot \pi\left(r_{\rm eff,\,i}^2 + r_{\rm eff,\,j}^2\right) \cdot n_i n_j, \tag{3.102}$$

   where the number density of dust in the mass bin centered on m_i is

$$n_i = \int_{m=m_i/10^{(\Delta \log m)/2}}^{10^{(\Delta \log m)/2} \cdot m_i} n_m dm \tag{3.103}$$

   and $\Delta \log m$ is the logarithmic mass grid spacing. Use the prescription laid out in Section 3.3.1 to find $\Delta v_{ij} = \Delta v_{12}$.

   (b) Because collisions change the original particle masses m_i and m_j, remove all colliding pairs from the mass distribution:

$$n_i \leftarrow n_i - C_{ij}\Delta t, \quad i \leftrightarrow j \tag{3.104}$$

(c) Add the collision products to the mass distribution: for each (m_i, m_j) pair, use the kernel to identify the set $\mathcal{S}_{ij}$ of particles produced in a single collision. For example, sticking produces a single particle with $m = m_i + m_j$, bouncing returns two particles with unchanged masses m_i and m_j, and catastrophic fragmentation of both particles yields two groups consisting of largest remnant+fragments. Then, for all m_k in $\mathcal{S}_{ij}$,

$$n_k \leftarrow n_k + N_k C_{ij} \Delta t, \tag{3.105}$$

where N_k is the number (not number density) of particles in mass bin k produced by a single collision between particles with masses m_i and m_j.

3. For a one-zone simulation as in Windmark et al. (2012a), repeat step 2. For a 1D R or z simulation, or a 2D (R, z) simulation, apply a prescription for vertical settling, radial drift, and/or turbulent diffusion (e.g., Equations (3.17), (3.32), and (3.51)), then repeat step 2.

The pseudo-code provided here is useful for conceptualizing how a Smoluchowski solver might work, but it is missing details on how to represent a large mass range, handle rare collisions that occur fewer than once per timestep, choose appropriate timesteps, limit numerical diffusion between mass bins, and treat dust–gas momentum exchange. We also do not discuss Monte Carlo grain-growth simulations, which represent the mass distribution using tracer particles instead of binned number densities. For more information on numerical methods, see Ormel & Spaans (2008), Drażkowska et al. (2014), Paruta et al. (2016), Li et al. (2017), Booth et al. (2018), Stoyanovskaya et al. (2018), the appendix of Brauer et al. (2008), and references therein.

3.4 Conclusion: How Do Dust Grains Evolve?

Now that we are familiar with analytical and numerical tools for investigating grain growth, what conclusions can we draw about the dominant grain evolution pathways? Recent work has led to rough agreement on several key points:

- **Pebbles are the raw ingredients of the planetesimals that build giant planets.** While the original streaming instability models featured meter-size boulders as the likely planetesimal source material (Johansen et al. 2007), the European Space Agency's Rosetta mission to comet Churyumov-Gerasimenko (67P) has uncovered multiple lines of evidence that the comet formed via the collapse of a pebble cloud. The CONSERT instrument recorded heterogeneities in density, dust-to-ice ratio, and composition that are only compatible with porosity variations at size scales lower than 1 m, suggesting that the comet's constituent particles are centimeter-size pebbles (Herique et al. 2019). The pebble-cloud collapse scenario is further supported by measurements of tensile strength, thermal inertia, dust size distribution, and water vapor production rate (Poulet et al. 2016; Blum et al. 2017; Fulle & Blum 2017). Comet Hartley 2 (103P) also shows signs of a pebble-cloud origin, as EPOXI observations revealed centimeter- to decimeter-size particles floating around the comet nucleus (Kretke & Levison 2015).

It is more difficult to identify the original building blocks of asteroids and meteorites given their extensive alteration by collisions and melting (Henke et al. 2016)—for example, the boulders that populate the surfaces of asteroids Bennu and Ryugu are likely reaccreted fragments of disrupted parent bodies (Dellagiustina et al. 2019; Grott et al. 2019). However, an examination of 13 meteorite falls through Earth's atmosphere revealed that collisional fracturing of the parent body does not fully explain the low material strengths of carbonaceous objects, indicating pebble-pile origins (Popova et al. 2011). Carbonaceous asteroids may be leftovers from the population of Jupiter- and Saturn-forming planetesimals that were implanted in the asteroid belt during Jupiter's migration (Kruijer et al. 2020). Finally, the brightness evolution of debris disks is consistent with initial planetesimal size distributions predicted by pebble concentration models (Krivov et al. 2018).

- **Rescuing pebbles from radial drift requires a dust trap, a rapid planetesimal formation mechanism, or both.** For freely drifting particles, pairwise collisional growth is too slow to produce planetesimals (Brauer et al. 2008; Birnstiel et al. 2012; Birnstiel & Andrews 2014; Homma & Nakamoto 2018). In a disk with no dust traps, most pebbles would either drift into the star or get broken back down to micron to millimeter sizes in fragmenting or erosive collisions (Krijt et al. 2015; Estrada et al. 2016; Vorobyov et al. 2018; Figure 3.17). Fortunately, dust-trapping pressure maxima and vortices can form in a variety of locations; a partial list includes dead-zone edges (Kretke & Lin 2007; Ruge et al. 2016), sublimation fronts (Dzyurkevich et al. 2013), and zonal flows induced by nonideal MHD effects (Riols & Lesur 2018). It may even be possible for collisional fragmentation to create a dust trap, as highly concentrated fragments exchange enough momentum with the gas to force it to deviate from Keplerian rotation (Gonzalez et al. 2017). Finally, the simple fact that pebbles cannot move far from the midplane (Figure 3.12) can sometimes keep them concentrated enough for the streaming instability to kick in, triggering the formation of pebble-pile planetesimals (Johansen et al. 2007; Bai & Stone 2010; Drażkowska & Dullemond 2014). We discuss the streaming instability and other particle-clumping mechanisms in Chapter 4.
- **Collisions between particles with a high mass ratio m_1/m_2 may play a major role in forming pebbles** (Güttler et al. 2010; Beitz et al. 2011; Teiser et al. 2011b; Meisner et al. 2013; Deckers & Teiser 2014). Figure 3.18 shows that for $r_{\rm eff} \gtrsim 1$mm, collisions between unequal-mass particles can lead to fragmentation with mass transfer, even when the fragmentation threshold speed is independent of the mass ratio. Using a physically realistic collision model that has fragmentation threshold speed increasing with the mass ratio m_1/m_2 (Bukhari Syed et al. 2017) further improves the viability of fragmentation with mass transfer as a pebble-growth pathway. Booth et al. (2018) find that pebbles at 5 au in a disk with $\alpha = 10^{-3}$ can grow beyond 1 cm within 3×10^4 years. Pebble-growth timescales under 5×10^4 years fit the meteoritic constraints (see Section 4.2) and allow the first planetesimals to form in the protostar's Stage I or Stage II (Dauphas & Chaussidon 2011).

- **The bouncing barrier is surmountable** (Windmark et al. 2012b; Garaud et al. 2013; Seizinger & Kley 2013). Dust evolution models that use a velocity distribution instead of assuming each collision takes place at the rms speed $\langle \Delta v_{12}^2 \rangle^{1/2}$ usually find that "lucky" low-speed collisions can result in sticking even when aggregates reach millimeter sizes. The particles that reach centimeter sizes can sweep up smaller grains whose growth has been stalled by bouncing. If pebbles are prevented from drifting inward, centimeter-size seeds can "snowplow" all the way up to $r_{\rm eff} \gtrsim 10$ m (Windmark et al. 2012a; Drażkowska et al. 2013). Icy particles may be porous enough to redirect all collision energy into rearranging monomers within aggregates, allowing sticking even at velocities up to 50 m s^{-1} (Kataoka et al. 2013; Krijt et al. 2015; Kataoka 2017). Furthermore, aggregates with $T < 400$ K may have sticky organic mantles that suppress bouncing, though such aggregates are still vulnerable to radial drift (Homma et al. 2019).
- **The H_2O-ice line might be a particularly favorable location for pebble and planetesimal formation** (see the inset box 3.3 for a discussion of the ice-line location). Here, pebbles may be able to reach decimeter sizes without needing "lucky" low-speed collisions, extreme porosity, or enough time to sweep up a vast number of smaller grains. Inwardly drifting grains sublimate at the H_2O-ice line, locally enhancing the water vapor pressure by a factor of up to 10 (Stevenson & Lunine 1988; Ciesla & Cuzzi 2006). The water vapor then diffuses away from the vapor pressure maximum, recondensing to form mantles on particles located just beyond the ice line. Ros & Johansen (2013) find that ice mantle deposition can grow particles from millimeter to decimeter sizes in under 2000 orbits given turbulent efficiency $\alpha = 10^{-4}$–10^{-2}. Decimeter-size snowballs are prime candidates for planetesimal formation via the streaming instability.

There is not yet a consensus on whether grain growth is frustrated or enhanced when the grains are electrically charged (Okuzumi 2009; Steinpilz et al. 2019). Chemical evolution models usually assume that dust grains are negatively charged, with electrons on the surfaces contributing to dissociative recombination reactions (e.g., Bergin et al. 2007; Semenov et al. 2010; Eistrup et al. 2018), though the midplanes of some disks may be protected from ionization by "T-Tauriospheres" created by star winds and magnetic fields (Cleeves et al. 2013). Further research into the collisional behavior of particles with non-H_2O ice mantles is needed, given that organic and CO_2 ices may provide important contributions to the planet-forming mass inventory (Bockelée-Morvan et al. 2004; Öberg et al. 2011; Walsh et al. 2014; Yu et al. 2016). The first experimental studies of collisions between CO_2-ice particles found silicate-like behavior, with sticking speeds an order of magnitude smaller than in collisions between H_2O-ice particles (Musiolik et al. 2016a, 2016b). Yet despite the remaining uncertainties surrounding collisional growth and disk composition, the bottom rungs of the giant-planet size ladder appear to be firmly in place: every barrier to particle growth can be overcome by a physically realistic mechanism.

Armed with our new understanding of dust dynamics and collisional growth, we are ready to make the leap from pebbles to planetesimals.

References

Abe, Y., Ohtani, E., Okuchi, T., Righter, K., & Drake, M. 2000, Origin of the Earth and Moon (Tucson, AZ: Arizona Univ. Press)
Abod, C. P., Simon, J. B., Li, R., et al. 2019, ApJ, 883, 192
Adams, E. R., Gulbis, A. A. S., Elliot, J. L., et al. 2014, AJ, 148, 55
Aikawa, Y., Umebayashi, T., Nakano, T., et al. 1997, ApJL, 486, L51
Aikawa, Y., Umebayashi, T., Nakano, T., et al. 1999, ApJ, 519, 705
ALMA Partnership, Brogan C. L., Pérez, L. M., et al. 2015, ApJL, 808, L3
Andrews, S. M., Huang, J., Pérez, L. M., et al. 2018, ApJL, 869, L41
Andrews, S. M., Wilner, D. J., Espaillat, C., et al. 2011, ApJ, 732, 42
Andrews, S. M., Wilner, D. J., Hughes, A. M., et al. 2012, ApJ, 744, 162
Ansdell, M., Williams, J. P., Trapman, L., et al. 2018, ApJ, 859, 21
Aumann, H. H. 1985, PASP, 97, 885
Aumann, H. H., Gillett, F. C., Beichman, C. A., et al. 1984, ApJL, 278, L23
Bai, X.-N., & Stone, J. M. 2010, ApJ, 722, 1437
Bai, X.-N., & Stone, J. M. 2013, ApJ, 769, 76
Baillié, K., Charnoz, S., & Pantin, E. 2015, A&A, 577, A65
Banzatti, A., Meyer, M. R., Manara, C. F., et al. 2014, ApJ, 780, 26
Barge, P., & Sommeria, J. 1995, A&A, 295L, 1B
Barnes, R., Quinn, T. R., Lissauer, J. J., et al. 2009, Icar, 203, 626
Baruteau, C., Barraza, M., Pérez, S., et al. 2019, MNRAS, 486, 304
Bate, M. R., Lubow, S. H., Ogilvie, G. I., et al. 2003, MNRAS, 341, 213
Beitz, E., Güttler, C., Blum, J., et al. 2011, ApJ, 736, 34
Benz, W., Ida, S., Alibert, Y., et al. 2014, in Protostars and Planets VI, ed. H. Beuther, et al. (Tucson, AZ: Arizona Univ. Press), 691
Bergin, E. A., Aikawa, Y., Blake, G. A., et al. 2007, in Protostars and Planets V, ed. B. Reipurth, et al. (Tucson, AZ: Arizona Univ. Press), 751
Birnstiel, T., & Andrews, S. M. 2014, ApJ, 780, 153
Birnstiel, T., Dullemond, C. P., & Brauer, F. 2009, A&A, 503, L5
Birnstiel, T., Dullemond, C. P., & Brauer, F. 2010, A&A, 513, A79
Birnstiel, T., Klahr, H., & Ercolano, B. 2012, A&A, 539, A148
Birnstiel, T., Klahr, H., & Ercolano, B. 2017, TWO-POP-PY: Two-population dust evolution model, Astrophysics Source Code Library, ascl:1708.015
Bizzarro, M., Baker, J. A., & Haack, H. 2004, Natur, 431, 275
Blevins, S. M., Pontoppidan, K. M., Banzatti, A., et al. 2016, ApJ, 818, 22
Blum, J. 2018, SSRv, 214, 52
Blum, J., Gundlach, B., Krause, M., et al. 2017, MNRAS, 469, S755
Blum, J., & Münch, M. 1993, Icar, 106, 151
Blum, J., & Wurm, G. 2008, ARA&A, 46, 21
Bockelée-Morvan, D., Crovisier, J., Mumma, M. J., et al. 2004, in Comets II, 391
Bodenheimer, P., & Lissauer, J. J. 2014, ApJ, 791, 103
Boley, A. C., Morris, M. A., & Ford, E. B. 2014, ApJL, 792, L27
Booth, R. A., Meru, F., Lee, M. H., et al. 2018, MNRAS, 475, 167

Bouwman, J., Henning, T., Hillenbrand, L. A., et al. 2008, ApJ, 683, 479
Brauer, F., Dullemond, C. P., & Henning, T. 2008, A&A, 480, 859
Brauer, F., Dullemond, C. P., Johansen, A., et al. 2007, A&A, 469, 1169
Bukhari Syed, M., Blum, J., Wahlberg Jansson, K., et al. 2017, ApJ, 834, 145
Carballido, A., Matthews, L. S., & Hyde, T. W. 2016, ApJ, 823, 80
Carballido, A., Stone, J. M., & Pringle, J. E. 2005, MNRAS, 358, 1055
Carr, J. S., & Najita, J. R. 2008, Sci, 319, 1504
Cassan, A., Kubas, D., Beaulieu, J.-P., et al. 2012, Natur, 481, 167
Chambers, J. 2019, ApJ, 879, 98
Chapman, N. L., Mundy, L. G., Lai, S.-P., et al. 2009, ApJ, 690, 496
Chatterjee, S., Ford, E. B., Matsumura, S., et al. 2008, ApJ, 686, 580
Chavanis, P. H. 2000, A&A, 356, 1089
Chiang, E., & Youdin, A. N. 2010, AREPS, 38, 493
Chiang, H.-F., Looney, L. W., & Tobin, J. J. 2012, ApJ, 756, 168
Cieza, L. A., Casassus, S., Tobin, J., et al. 2016, Natur, 535, 258
Chokshi, A., Tielens, A. G. G. M., & Hollenbach, D. 1993, ApJ, 407, 806
Ciesla, F. J., & Cuzzi, J. N. 2006, Icar, 181, 178
Cleeves, L. I., Adams, F. C., & Bergin, E. A. 2013, ApJ, 772, 5
Coleman, G. A. L., & Nelson, R. P. 2014, MNRAS, 445, 479
Cox, E. G., Harris, R. J., Looney, L. W., et al. 2015, ApJ, 814, L28
Cridland, A. J., Pudritz, R. E., & Birnstiel, T. 2017, MNRAS, 465, 3865
Cuzzi, J. N., & Hogan, R. C. 2003, Icar, 164, 127
Cuzzi, J. N., Hogan, R. C., & Shariff, K. 2008, ApJ, 687, 1432
Dauphas, N., & Chaussidon, M. 2011, AREPS, 39, 351
Dauphas, N., & Pourmand, A. 2011, Natur, 473, 489
Davis, S. S. 2005, ApJ, 620, 994
de Gregorio-Monsalvo, I., Ménard, F., Dent, W., et al. 2013, A&A, 557, A133
Deckers, J., & Teiser, J. 2013, ApJ, 769, 151
Deckers, J., & Teiser, J. 2014, ApJ, 796, 99
Deckers, J., & Teiser, J. 2016, MNRAS, 456, 4328
Dellagiustina, D. N., Emery, J. P., Golish, D. R., et al. 2019, NatAs, 3, 341
Dodson-Robinson, S. E., Willacy, K., Bodenheimer, P., et al. 2009, Icar, 200, 672
Dohnanyi, J. S. 1969, JGR, 74, 2531
Dominik, C., & Tielens, A. G. G. M. 1997, ApJ, 480, 647
Dong, R., Liu, S.-y., Eisner, J., et al. 2018, ApJ, 860, 124
Drażkowska, J., & Alibert, Y. 2017, A&A, 608, A92
Drażkowska, J., & Dullemond, C. P. 2014, A&A, 572, A78
Drazkowska, J., Li, S., Birnstiel, T., Stammler, S. M., & Li, H. 2019, ApJ, 885, 91
Drażkowska, J., Windmark, F., & Dullemond, C. P. 2013, A&A, 556, A37
Drażkowska, J., Windmark, F., & Dullemond, C. P. 2014, A&A, 567, A38
Du, F., & Bergin, E. A. 2014, ApJ, 792, 2
Dubrulle, B., Morfill, G., & Sterzik, M. 1995, Icar, 114, 237
Dullemond, C. P., & Dominik, C. 2005, A&A, 434, 971
Dwek, E. 1998, ApJ, 501, 643
Dyck, H. M., & Simon, T. 1975, ApJ, 195, 689
Dzyurkevich, N., Flock, M., Turner, N. J., et al. 2010, A&A, 515, A70

Dzyurkevich, N., Turner, N. J., Henning, T., et al. 2013, ApJ, 765, 114
Eistrup, C., Walsh, C., & van Dishoeck, E. F. 2018, A&A, 613, A14
Erkaev, N. V., Lammer, H., Elkins-Tanton, L. T., et al. 2014, P&SS, 98, 106
Estrada, P. R., Cuzzi, J. N., & Morgan, D. A. 2016, ApJ, 818, 200
Facchini, S., van Dishoeck, E. F., Manara, C. F., et al. 2019, A&A, 626, L2
Finn, G. D., & Simon, T. 1977, ApJ, 212, 472
Fortier, A., Alibert, Y., Carron, F., et al. 2013, A&A, 549, A44
Fraser, W. C., Brown, M. E., Morbidelli, A., et al. 2014, ApJ, 782, 100
Fressin, F., Torres, G., Charbonneau, D., et al. 2013, ApJ, 766, 81
Fromang, S., & Papaloizou, J. 2006, A&A, 452, 751
Fulle, M., & Blum, J. 2017, MNRAS, 469, S39
Fulle, M., Blum, J., & Rotundi, A. 2019, ApJL, 879, L8
Fulle, M., Della Corte, V., Rotundi, A., et al. 2016, MNRAS, 462, S132
Furlan, E., Hartmann, L., Calvet, N., et al. 2006, ApJS, 165, 568
Furuya, K., Aikawa, Y., Nomura, H., et al. 2013, ApJ, 779, 11
Garaud, P. 2007, ApJ, 671, 2091
Garaud, P., & Lin, D. N. C. 2007, ApJ, 654, 606
Garaud, P., Meru, F., Galvagni, M., et al. 2013, ApJ, 764, 146
Gärtner, S., Gundlach, B., Headen, T. F., et al. 2017, ApJ, 848, 96
Georgakarakos, N., Eggl, S., & Dobbs-Dixon, I. 2018, ApJ, 856, 155
Geretshauser, R. J., Meru, F., Speith, R., et al. 2011, A&A, 531, A166
Godon, P., & Livio, M. 2000, ApJ, 537, 396
Goldreich, P., & Lynden-Bell, D. 1965, MNRAS, 130, 125
Gonzalez, J.-F., Laibe, G., & Maddison, S. T. 2017, MNRAS, 467, 1984
Grott, M., Knollenberg, J., Hamm, M., et al. 2019, NatAs, 3, 971
Gundlach, B., & Blum, J. 2015, ApJ, 798, 34
Güttler, C., Blum, J., Zsom, A., et al. 2010, A&A, 513, A56
Hansen, B. M. S. 2009, ApJ, 703, 1131
Hawley, J. F., Gammie, C. F., & Balbus, S. A. 1995, ApJ, 440, 742
Haghighipour, N., & Boss, A. P. 2003, ApJ, 583, 996
Heng, K., & Kenyon, S. J. 2010, MNRAS, 408, 1476
Henke, S., Gail, H.-P., & Trieloff, M. 2016, A&A, 589, A41
Henning, T., & Semenov, D. 2013, ChRv, 113, 9016
Herique, A., Kofman, W., Zine, S., et al. 2019, A&A, 630, A6
Hernández, J., Hartmann, L., Megeath, T., et al. 2007, ApJ, 662, 1067
Hill, C. R., Heißelmann, D., Blum, J., et al. 2015, A&A, 573, A49
Homma, K., & Nakamoto, T. 2018, ApJ, 868, 118
Homma, K. A., Okuzumi, S., Nakamoto, T., et al. 2019, ApJ, 877, 128
Hubickyj, O., Bodenheimer, P., & Lissauer, J. J. 2005, Icar, 179, 415
Hughes, A. L. H., & Armitage, P. J. 2010, ApJ, 719, 1633
Ida, S., Lin, D. N. C., & Nagasawa, M. 2013, ApJ, 775, 42
Inaba, S., & Barge, P. 2006, ApJ, 649, 415
Inaba, S., & Ikoma, M. 2003, A&A, 410, 711
Isella, A., & Natta, A. 2005, A&A, 438, 899
Jacobsen, B., Yin, Q.-z., Moynier, F., et al. 2008, E&PSL, 272, 353

Jeans, J. H. 1919, Problems of Cosmogony and Stellar Dynamics (Cambridge: Cambridge Univ. Press)
Jenniskens, P., Baratta, G. A., Kouchi, A., et al. 1993, A&A, 273, 583
Johansen, A., Andersen, A. C., & Brandenburg, A. 2004, A&A, 417, 361
Johansen, A., & Klahr, H. 2005, ApJ, 634, 1353
Johansen, A., Mac Low, M.-M., Lacerda, P., & Bizzarro, M. 2015, SciA, 1, 1500109
Johansen, A., Oishi, J. S., Mac Low, M.-M., et al. 2007, Natur, 448, 1022
Johansen, A., Youdin, A., & Mac Low, M.-M. 2009, ApJL, 704, L75
Kama, M., Min, M., & Dominik, C. 2009, A&A, 506, 1199
Kamp, I., Thi, W.-F., Woitke, P., et al. 2017, A&A, 607, A41
Kataoka, A. 2017, Astrophysics and Space Science Library, 143
Kataoka, A., Tanaka, H., Okuzumi, S., et al. 2013, A&A, 557, L4
Kelling, T., Wurm, G., & Köster, M. 2014, ApJ, 783, 111
Keppler, M., Benisty, M., Müller, A., et al. 2018, A&A, 617, A44
Kessler-Silacci, J., Augereau, J.-C., Dullemond, C. P., et al. 2006, ApJ, 639, 275
Kida, S. 1981, JFM, 112, 397
Klaassen, P. D., Juhasz, A., Mathews, G. S., et al. 2013, A&A, 555, A73
Klahr, H., & Bodenheimer, P. 2006, ApJ, 639, 432
Klahr, H. H., & Henning, T. 1997, Icar, 128, 213
Kokubo, E., & Ida, S. 1996, Icar, 123, 180
Kokubo, E., & Ida, S. 1998, Icar, 131, 171
Kokubo, E., & Ida, S. 2000, Icar, 143, 15
Kolmogorov, A. 1941, DoSSR, 30, 301
Kornet, K., Stepinski, T. F., & Różyczka, M. 2001, A&A, 378, 180
Kothe, S. 2016, Mikrogravitationsexperimente zur Entwicklung eines empirischen Stoßmodells frprotoplanetare Staubagglomerate, PhD thesis, TU Braunschweig
Kretke, K. A., & Levison, H. F. 2014, AJ, 148, 109
Kretke, K. A., & Levison, H. F. 2015, Icar, 262, 9
Kretke, K. A., & Lin, D. N. C. 2007, ApJL, 664, L55
Krijt, S., Ormel, C. W., Dominik, C., et al. 2015, A&A, 574, A83
Krijt, S., Ormel, C. W., Dominik, C., et al. 2016, A&A, 586, A20
Kristensen, L. E., & Dunham, M. M. 2018, A&A, 618, A158
Krivov, A. V., Ide, A., Löhne, T., et al. 2018, MNRAS, 474, 2564
Kruijer, T. S., Kleine, T., & Borg, L. E. 2020, NatAs, 4, 32
Kruss, M., Demirci, T., Koester, M., et al. 2016, ApJ, 827, 110
Kruss, M., & Wurm, G. 2018, ApJ, 869, 45
Kuwahara, A., Kurokawa, H., & Ida, S. 2019, A&A, 623, A179
Lambrechts, M., & Johansen, A. 2012, A&A, 544, A32
Landry, R., Dodson-Robinson, S. E., Turner, N. J., et al. 2013, ApJ, 771, 80
Langkowski, D., Teiser, J., & Blum, J. 2008, ApJ, 675, 764
Lecar, M., Podolak, M., Sasselov, D., et al. 2006, ApJ, 640, 1115
Levison, H. F., Kretke, K. A., & Duncan, M. J. 2015, Natur, 524, 322
Li, X.-Y., Brandenburg, A., Haugen, N. E. L., & Svensson, G. 2017, JAMES, 9, 1116
Lin, M.-K. 2019, MNRAS, 485, 5221
Lin, M.-K., & Papaloizou, J. C. B. 2011, MNRAS, 415, 1426
Lorek, S., Lacerda, P., & Blum, J. 2018, A&A, 611, A18

Lovascio, F., & Paardekooper, S.-J. 2019, MNRAS, 488, 5290
Lyra, W., Johansen, A., Klahr, H., et al. 2009a, A&A, 493, 1125
Lyra, W., Johansen, A., Zsom, A., et al. 2009b, A&A, 497, 869
Lyra, W., & Klahr, H. 2011, A&A, 527, A138
Lyra, W., & Umurhan, O. M. 2019, PASP, 131, 072001
Lyttleton, R. A. 1961, MNRAS, 122, 399
MacGregor, M. A., Weinberger, A. J., Wilner, D. J., et al. 2018, ApJL, 855, L2
MacGregor, M. A., Wilner, D. J., Chandler, C., et al. 2016, ApJ, 823, 79
MacPherson, G. J., Bullock, E. S., Janney, P. E., et al. 2010, ApJL, 711, L117
Marcus, P. S., Pei, S., Jiang, C.-H., et al. 2015, ApJ, 808, 87
Marcy, G. W., & Butler, R. P. 1996, ApJL, 464, L147
Markiewicz, W. J., Mizuno, H., & Voelk, H. J. 1991, A&A, 242, 286
Martin, R. G., & Livio, M. 2012, MNRAS, 425, L6
Mathis, J. S., Mezger, P. G., & Panagia, N. 1983, A&A, 500, 259
Mathis, J. S., Rumpl, W., & Nordsieck, K. H. 1977, ApJ, 217, 425
Mayor, M., & Queloz, D. 1995, Natur, 378, 355
McElroy, D., Walsh, C., Markwick, A. J., et al. 2013, A&A, 550, A36
McKee, C. F., & Ostriker, J. P. 1977, ApJ, 218, 148
Meeus, G., Montesinos, B., Mendigutía, I., et al. 2012, A&A, 544, A78
Meijerink, R., Pontoppidan, K. M., Blake, G. A., et al. 2009, ApJ, 704, 1471
Meisner, T., Wurm, G., Teiser, J., et al. 2013, A&A, 559, A123
Min, M., Bouwman, J., Dominik, C., et al. 2016, A&A, 593, A11
Morbidelli, A., Bitsch, B., Crida, A., et al. 2016, Icar, 267, 368
Morbidelli, A., Bottke, W. F., Nesvorný, D., et al. 2009, Icar, 204, 558
Morbidelli, A., Chambers, J., Lunine, J. I., et al. 2000, M&PS, 35, 1309
Morbidelli, A., Lambrechts, M., Jacobson, S., et al. 2015, Icar, 258, 418
Morbidelli, A., Lunine, J. I., O'Brien, D. P., et al. 2012, AREPS, 40, 251
Mordasini, C., Alibert, Y., & Benz, W. 2009, A&A, 501, 1139
Morrison, D., & Simon, T. 1973, ApJ, 186, 193
Mulders, G. D., Ciesla, F. J., Min, M., et al. 2015, ApJ, 807, 9
Musiolik, G., Teiser, J., Jankowski, T., et al. 2016a, ApJ, 818, 16
Musiolik, G., Teiser, J., Jankowski, T., et al. 2016b, ApJ, 827, 63
Nauta, M. D. 2000, PhD thesis, Universiteit Utrecht
Neugebauer, G., Habing, H. J., van Duinen, R., et al. 1984, ApJL, 278, L1
Nixon, C. J., King, A. R., & Pringle, J. E. 2018, MNRAS, 477, 3273
Nozawa, T., Kozasa, T., Umeda, H., et al. 2003, ApJ, 598, 785
Oberc, P. 2007, Icar, 186, 303
Öberg, K. I., Boogert, A. C. A., Pontoppidan, K. M., et al. 2011, ApJ, 740, 109
Oka, A., Nakamoto, T., & Ida, S. 2011, ApJ, 738, 141
Okuzumi, S. 2009, ApJ, 698, 1122
Okuzumi, S., & Hirose, S. 2011, ApJ, 742, 65
Okuzumi, S., Momose, M., Sirono, S.-i., et al. 2016, ApJ, 821, 82
Okuzumi, S., Tanaka, H., Kobayashi, H., et al. 2012, ApJ, 752, 106
Ormel, C. W., & Cuzzi, J. N. 2007, A&A, 466, 413
Ormel, C. W., & Spaans, M. 2008, ApJ, 684, 1291
Paardekooper, S.-J. 2012, MNRAS, 421, 3286

Paardekooper, S.-J., & Mellema, G. 2004, A&A, 425, L9
Pan, L., & Padoan, P. 2013, ApJ, 776, 12
Paraskov, G. B., Wurm, G., & Krauss, O. 2007, Icar, 191, 779
Paruta, P., Hendrix, T., & Keppens, R. 2016, A&C, 16, 155
Pérez, L. M., Chandler, C. J., Isella, A., et al. 2015, ApJ, 813, 41
Petersen, M. R., Julien, K., & Stewart, G. R. 2007, ApJ, 658, 1236
Pinilla, P., Klarmann, L., Birnstiel, T., et al. 2016, A&A, 585, A35
Pinilla, P., Birnstiel, T., Ricci, L., et al. 2012, A&A, 538, A114
Podolak, M. 2003, Icar, 165, 428
Podolak, M., & Zucker, S. 2004, M&PS, 39, 1859
Pollack, J. B., Hollenbach, D., Beckwith, S., et al. 1994, ApJ, 421, 615
Pollack, J. B., Hubickyj, O., Bodenheimer, P., et al. 1996, Icar, 124, 62
Pontoppidan, K. M., Salyk, C., Blake, G. A., et al. 2010, ApJ, 720, 887
Popova, O., Borovička, J., Hartmann, W. K., et al. 2011, M&PS, 46, 1525
Poulet, F., Lucchetti, A., Bibring, J.-P., et al. 2016, MNRAS, 462, S23
Provenzale, A. 1999, AnRFM, 31, 55
Rafikov, R. R. 2004, AJ, 128, 1348
Raettig, N., Klahr, H., & Lyra, W. 2015, ApJ, 804, 35
Raymond, S. N., & Izidoro, A. 2017, Icar, 297, 134
Ricci, L., Rome, H., Pinilla, P., et al. 2017, ApJ, 846, 19
Riols, A., & Lesur, G. 2018, A&A, 617, A117
Rivkin, A. S., Howell, E. S., Vilas, F., et al. 2002, in Asteroids III, ed. W. F. Bottke, et al. (Tucson, AZ: Univ. Arizona Press), 235
Rogers, L. A. 2015, ApJ, 801, 41
Ros, K., & Johansen, A. 2013, A&A, 552, A137
Ros, K., Johansen, A., Riipinen, I., et al. 2019, A&A, 629, A65
Rózyczka, M., Kornet, K., Bodenheimer, P., et al. 2004, Revista Mexicana De Astronomia Y Astrofisica Conf. Series, 91
Ruge, J. P., Flock, M., Wolf, S., et al. 2016, A&A, 590, A17
Sadavoy, S. I., Myers, P. C., Stephens, I. W., et al. 2018, ApJ, 869, 115
Salyk, C., Lacy, J. H., Richter, M. J., et al. 2015, ApJL, 810, L24
Salyk, C., Lacy, J., Richter, M., et al. 2019, ApJ, 874, 24
Salyk, C., Pontoppidan, K. M., Blake, G. A., et al. 2008, ApJL, 676, L49
Schlichting, H. E., Fuentes, C. I., & Trilling, D. E. 2013, AJ, 146, 36
Schräpler, R., & Blum, J. 2011, ApJ, 734, 108
Schräpler, R., Blum, J., Krijt, S., et al. 2018, ApJ, 853, 74
Schräpler, R., Blum, J., Seizinger, A., et al. 2012, ApJ, 758, 35
Schmidt, O. 1959, A Theory of the Earth's Origin (London: Lawrence and Wishart)
Seizinger, A., & Kley, W. 2013, A&A, 551, A65
Semenov, D., Hersant, F., Wakelam, V., et al. 2010, A&A, 522, A42
Semenov, D., & Wiebe, D. 2011, ApJS, 196, 25
Sengupta, D., Dodson-Robinson, S. E., Hasegawa, Y., et al. 2019, ApJ, 874, 26
Shimaki, Y., & Arakawa, M. 2012, Icar, 221, 310
Shinbrot, T., Sabuwala, T., Siu, T., et al. 2017, PhRvL, 118, 111101
Sicilia-Aguilar, A., Hartmann, L., Calvet, N., et al. 2006, ApJ, 638, 897
Simon, J. B., Armitage, P. J., Li, R., et al. 2016, ApJ, 822, 55

Smith, B. A., & Terrile, R. J. 1984, Sci, 226, 1421
Smoluchowski, M. V. 1916, ZPhy, 17, 557
Stammler, S. M., Drążkowska, J., Birnstiel, T., et al. 2019, ApJL, 884, L5
Steinpilz, T., Joeris, K., Jungmann, F., et al. 2019, NatPh, 16, 225
Stern, S. A., Weaver, H. A., Spencer, J. R., et al. 2019, Sci, 364, aaw9771
Stevenson, D. J., & Lunine, J. I. 1988, Icar, 75, 146
Stone, J. M., & Gardiner, T. A. 2010, ApJS, 189, 142
Stoyanovskaya, O. P., Vorobyov, E. I., & Snytnikov, V. N. 2018, ARep, 62, 455
Strom, R. G., Malhotra, R., Ito, T., et al. 2005, Sci, 309, 1847
Surville, C., & Mayer, L. 2019, ApJ, 883, 176
Takeuchi, T., & Lin, D. N. C. 2002, ApJ, 581, 1344
Tanga, P., Babiano, A., Dubrulle, B., et al. 1996, Icar, 121, 158
Tanigawa, T., & Ohtsuki, K. 2010, Icar, 205, 658
Teiser, J., Engelhardt, I., & Wurm, G. 2011a, ApJ, 742, 5
Teiser, J., Küpper, M., & Wurm, G. 2011b, Icar, 215, 596
Teiser, J., & Wurm, G. 2009, MNRAS, 393, 1584
Testi, L., Birnstiel, T., Ricci, L., et al. 2014, in Protostars and Planets VI, ed. H. Beuther, et al. (Tucson, AZ: Arizona Univ. Press), 339
Thommes, E. W., Duncan, M. J., & Levison, H. F. 2003, Icar, 161, 431
Thommes, E. W., Matsumura, S., & Rasio, F. A. 2008, Sci, 321, 814
Todini, P., & Ferrara, A. 2001, MNRAS, 325, 726
Touboul, M., Kleine, T., Bourdon, B., et al. 2007, Natur, 450, 1206
Trapman, L., Facchini, S., Hogerheijde, M. R., et al. 2019, A&A, 629, A79
Tripathi, A., Andrews, S. M., Birnstiel, T., et al. 2018, ApJ, 861, 64
Uribe, A. L., Klahr, H., Flock, M., et al. 2011, ApJ, 736, 85
Turner, N. J., Willacy, K., Bryden, G., et al. 2006, ApJ, 639, 1218
Valiante, R., Schneider, R., Bianchi, S., et al. 2009, MNRAS, 397, 1661
van der Marel, N., van Dishoeck, E. F., Bruderer, S., et al. 2013, Sci, 340, 1199
van Dishoeck, E. F., Bergin, E. A., Lis, D. C., et al. 2014, in Protostars and Planets VI, ed. H. Beuther, et al. (Tucson, AZ: Arizona Univ. Press), 835
van Dishoeck, E. F., Herbst, E., & Neufeld, D. A. 2013, ChRv, 113, 9043
Viganò, D., Aguilera-Miret, R., & Palenzuela, C. 2019, PhFl, 31, 105102
Visser, R., van Dishoeck, E. F., Doty, S. D., et al. 2009, A&A, 495, 881
Vorobyov, E. I., Akimkin, V., Stoyanovskaya, O., et al. 2018, A&A, 614, A98
Wada, K., Tanaka, H., Suyama, T., et al. 2009, ApJ, 702, 1490
Wada, K., Tanaka, H., Suyama, T., et al. 2011, ApJ, 737, 36
Wakelam, V., Herbst, E., Loison, J.-C., et al. 2012, ApJS, 199, 21
Walsh, C., Millar, T. J., Nomura, H., et al. 2014, A&A, 563, A33
Ward-Thompson, D., Motte, F., & Andre, P. 1999, MNRAS, 305, 143
Weidenschilling, S. J. 1977, MNRAS, 180, 57
Weidling, R., Güttler, C., Blum, J., et al. 2009, ApJ, 696, 2036
Werner, M. W., Roellig, T. L., Low, F. J., et al. 2004, ApJS, 154, 1
Wetherill, G. W., & Stewart, G. R. 1989, Icar, 77, 330
Whipple, F. L. 1972, in From Plasma to Planet ed. A. Evlius (New York: Wiley Interscience Division), 211
Windmark, F., Birnstiel, T., Güttler, C., et al. 2012a, A&A, 540, A73

Windmark, F., Birnstiel, T., Ormel, C. W., et al. 2012b, A&A, 544, L16
Woitke, P., Kamp, I., & Thi, W.-F. 2009, A&A, 501, 383
Woodall, J., Agúndez, M., Markwick-Kemper, A. J., et al. 2007, A&A, 466, 1197
Woolfson, M. M. 1969, RPPh, 32, 135
Wurm, G., Paraskov, G., & Krauss, O. 2005, PhRvE, 71, 021304
Xiang, C., Matthews, L. S., Carballido, A., et al. 2020, ApJ, 897, 182
Youdin, A. N., & Goodman, J. 2005, ApJ, 620, 459
Youdin, A. N., & Lithwick, Y. 2007, Icar, 192, 588
Yu, M., Willacy, K., Dodson-Robinson, S. E., et al. 2016, ApJ, 822, 53
Zhang, K., Blake, G. A., & Bergin, E. A. 2015, ApJL, 806, L7
Zhang, K., Pontoppidan, K. M., Salyk, C., et al. 2013, ApJ, 766, 82
Zhdankin, V., Walker, J., Boldyrev, S., et al. 2017, MNRAS, 467, 3620
Zhu, Z., Stone, J. M., Rafikov, R. R., et al. 2014, ApJ, 785, 122
Zsom, A., Ormel, C. W., Güttler, C., et al. 2010, A&A, 513, A57

Chapter 4

From Pebbles to Planetesimals

4.1 Introduction

In Chapter 3, we mentioned that the meter-to-kilometer part of the solid size ladder might be backfilled from above by destructive collisions instead populated from below by inelastic collisions (see the inset box 3.1 entitled "Planet formation: The size scale"). With electrostatic sticking unlikely to produce meter-size rocks, we need a way to jump over the meter-size rung on the size ladder. Gravitationally collapsing pebble clouds can solve our problem: they become rubble-pile planetesimals with low material strength, similar to asteroid Itokawa and comet Churyumov-Gerasimenko (Fujiwara et al. 2006; Herique et al. 2019). Differentiated planetesimals such as Vesta come from rubble piles that form early and melt due to radiogenic heating (e.g., McSween et al. 2011). In this chapter, we learn how to concentrate pebbles into self-gravitating clouds and collapse the clouds into planetesimals. By moving from pebbles to planetesimals, we enter a realm in which the gravitational tugs from solid bodies compete with the gravitational pull of the star.

Before we dive in, we must emphasize the crucial role planetesimals play in forming planetary systems. Meteorites, asteroids, and comets provide a fossil record of planetesimals forming all across the solar system, and debris disks point to planetesimals orbiting stars with spectral types from B to M. Regardless of whether planet cores grow by collisions, pebble accretion, or a combination of both, planetesimals must be part of the process. While we can (and perhaps must) leapfrog the meter to decameter steps on the size ladder, there is no viable planet-formation theory that will let us bypass the 1–1000 km planetesimal size domain.

We begin in Section 4.2 by using the meteoritic record to assemble a history of planetesimal growth in the solar system. A successful planetesimal formation theory must tie in with the meteoritic clock. In Section 4.3 we present an analytical treatment of planetesimal-forming instabilities. There we will assess the gravitational stability of the midplane pebble layer, study the "reverse drag" solids exert on gas, calculate the conditions required for the reverse drag to concentrate pebbles via the

doi:10.1088/2514-3433/ac1db7ch4

streaming instability, and estimate the resulting planetesimal mass. We review the results from numerical simulations of planetesimal formation in Section 4.4, which also contains a discussion of alternatives to the streaming instability theory (Section 4.4.4). We present our conclusions in Section 4.5.

Insets that describe concepts that may be new to nonexpert readers include meteorite-finding techniques (inset box 4.1) and an introduction to the Kuiper Belt (inset box 4.2), which contains the most pristine collection of remnant planetesimals in the observable solar system. Table 4.1 defines new variables introduced in this chapter; previously introduced variables will retain their definitions from Chapters 2 and 3.

4.2 The Planetesimal Formation Timeline

A disk's likelihood of generating giant planets depends on the timing of planetesimal formation. If planetesimals form late in the disk's evolution, there is simply no time for solids to progress much further up the size ladder before the gas dissipates (Thommes et al. 2008). Though we lack unambiguous signatures of planetesimal formation in protoplanetary disks, meteorites give us a detailed chronology of planetesimal assembly in the inner solar nebula. One to a few supernovae, Wolf–Rayet stars, and/or asymptotic giant branch (AGB) stars seeded the solar nebula with short-lived radioisotopes (SLRIs) such as ^{26}Al, ^{182}Hf, and ^{60}Fe, which were incorporated into the growing solids (Lugaro et al. 2018). In the words of Lee et al. (1977), SLRIs are both fossil and fuel: daughter-product abundances and thermal evolution models can be combined to age-date the meteorite parent bodies, while SLRI decay generated enough heat to melt the earliest-forming planetesimals. In this section, we use the meteoritic record to reconstruct the planetesimal formation history of the inner solar nebula. Our discussion will follow the chronological order shown in Figure 4.1, an accretion timeline that includes information about Mars, the asteroids, Jupiter, and Saturn.

4.2.1 The Beginning

The chemical $t = 0$ is the formation of calcium-rich, aluminum-rich inclusions (CAIs; Figure 4.1, gray), the first solids that condensed out of hot gas near the inner edge of the solar nebula (Brearley et al. 1998; Amelin et al. 2010). Although meteorites, comets, and interplanetary dust particles contain presolar grains that are older than the solar nebula (Bradley 1994; Messenger et al. 2003; Heck et al. 2020), they do not include the same SLRI raw ingredients as the rest of the solids shown in Figure 4.1 and must be age-dated using different chronometers. The error bar on the CAI marker in Figure 4.1 shows that CAI condensation was not a single, global event; instead, ^{26}Al daughter-product abundances suggest a spread of at least 0.2 Myr in formation times, and maybe even 0.4 Myr (Kawasaki et al. 2019, 2020). Dauphas & Chaussidon (2011) place the CAI formation epoch near the boundary between the solar nebula's Stages I and II (Section 2.2), based on the presence of short-lived ^{10}Be implanted by the active young Sun. The Stardust mission recovered a CAI-like grain from Comet Wild 2 (McKeegan et al. 2006; Ciesla 2007), suggesting that winds were able to transport dust from the inner solar nebula to at least the asteroid belt, and maybe the Kuiper Belt (e.g., Ishii et al. 2008; Bai & Stone 2013; Section 2.4.6).

Table 4.1. Definitions of Variables

Variable	Units	Definition
Chemical and spectral properties of solids		
ϵ_W	parts per 10^4	Tungsten isotopic anomaly
$f^{Hf/W}$		Hafnium excess
λ	Myr^{-1}	Radioactive decay time constant
γ	% reflectance × μm^{-1}	Slope of reflectance spectra at wavelengths 0.4–1.0 μm
Disk properties		
Q_p		Gravitational stability criterion for pebble layer
S/G		Solid/gas mass ratio
$\tilde{G}$		Gravity parameter defined by Simon et al. (2016)
Properties of dust-gas two-fluid system		
$\vec{v}_p$, v_p, v_{pR}, $v_{p\phi}$	cm s^{-1}	Pebble-fluid velocity/speed; (R, ϕ) components
$\langle \vec{v}_p \rangle$	cm s^{-1}	Average pebble velocity
$\vec{v}_g$, v_g, v_{gR}, $v_{g\phi}$	cm s^{-1}	Gas velocity/speed; (R, ϕ) components
$\Delta\vec{v}_p$, Δv_p, $\Delta v_{p\phi}$	cm s^{-1}	Steady-state, drag-induced deviation from Keplerian velocity/speed; ϕ component
$\Delta\vec{v}_g$, Δv_g, $\Delta v_{g\phi}$	cm s^{-1}	Steady-state, drag-induced deviation from Keplerian velocity/speed; ϕ component
$\delta\rho_p$	g cm^{-3}	Pebble volume density perturbation
$\delta\vec{v}_p$, δv_p, δv_{px}, δv_{py}	cm s^{-1}	Pebble velocity/speed perturbation; shearing-box x and y components
$\delta\vec{v}_g$, δv_g, δv_{gx}, δv_{gy}	cm s^{-1}	Gas velocity/speed perturbation; shearing-box x and y components
k_y	cm^{-1} or au^{-1}	Radial wave number of density/velocity perturbations
k_z	cm^{-1} or au^{-1}	Vertical wave number of density/velocity perturbations
ω_r	s^{-1} or yr^{-1}	Frequency of Fourier perturbations
s	s^{-1} or yr^{-1}	Growth rate of Fourier perturbations
ℓ_c	au	Radius of the gravitationally bound pebble cloud
M_c	g or Ceres mass	Mass of gravitationally bound pebble cloud
$\mathcal{C}$		Pebble concentration factor

(*Continued*)

Table 4.1. (*Continued*)

Variable	Units	Definition
$\bar{St}$		Average Stokes number of particle size distribution
Planetesimal/embryo properties		
t_c	Myr	Core formation time
R_{pl}	km	Planetesimal radius
M_{pl}	g or Ceres mass	Planetesimal mass
a	au	Orbit semimajor axis
e		Orbit eccentricity
q	au	Perihelion distance

The dust velocities and the pebble concentration factor are functions of r_{eff}.

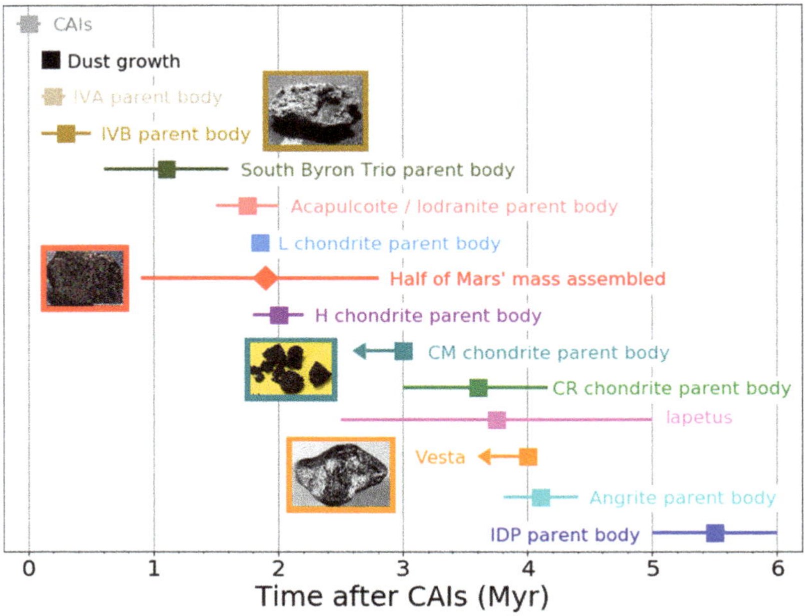

Figure 4.1. Timeline of planetesimal and planet formation in the inner 10 au of the solar nebula. Meteorites, asteroid surfaces, and moons record growth processes beginning with the condensation of the first microscopic solids (Section 4.2.1), progressing to the emergence of planetary embryos (Section 4.2.5), and culminating with the formation of giant planets (Section 4.2.7). This figure also includes multiple generations of planetesimals, from Ceres-size, early-forming collapsed pebble clouds (Section 4.2.2) to collisional fragments that reaccumulate after the solar nebula gas dissipates (Section 4.2.9). Outlined images show meteorites that come from the parent body plotted in the color that matches the outline.

We next move to electrostatic growth of dust grains (Figure 4.1, black, second from top)—the physics discussed in Chapter 3. Here we compare millimeter- to centimeter-size, coarse-grained, igneous CAIs—which melted after forming and so had their isotopic clocks "reset"—with their precursor particles: microscopic, fine-grained CAIs that resemble pristine nebular condensates. ^{26}Al daughter-product abundances tell us that it took between 20,000 and 50,000 years for fine-grained CAIs to grow to pebble size, melt, and recrystallize, forming coarse-grained CAIs (Jacobsen et al. 2008; MacPherson et al. 2010). Melting and sintering may have aided dust growth (Scott 2007). While there are many unanswered questions about how pebbles form, the chemical evidence shows that they can form quickly. The CAIs also show us that however advantageous H_2O ice may be for pebble formation (see Chapter 3), it's not a requirement. Pebble formation may have been even faster outside ~1 au, where lower temperatures would have improved the survival rate of macroscopic particles formed in the protosolar envelope (Stage 0/1).

4.2.2 The First Planetesimals

Magmatic iron meteorites, represented in Figure 4.1 by types IVA and IVB, present a conundrum. The isotopic clock records the time when atoms fractionate between different minerals, which happens during either a gas → solid or a liquid → solid phase change. Large planetesimals with molten interiors can retain their heat for millions of years, so there may be a considerable time lag between accretion and solidification-driven fractionation. For iron meteorites and stony achondrites—which come from the cores and mantles of differentiated planetesimals, respectively—one must combine SLRI-based crystallization ages with thermal models to find the accretion time. While irons and achondrites were originally thought to be younger than the "primitive" chondrites, which come from undifferentiated planetesimals, we now know that the magmatic irons are leftovers of some of the largest, earliest-forming planetesimals in the solar nebula (Kleine et al. 2005; Scherstén et al. 2006; Qin et al. 2008; Blichert-Toft et al. 2010). Forming early, the magmatic iron parent bodies contained large amounts of live ^{26}Al and other SLRIs; forming large, with radii >100 km, their low surface area/mass ratios allowed them to retain heat from SLRI decay and accretion, driving differentiation and magmatic activity for up to 10 Myr (Horan et al. 1998; Kleine et al. 2009; Neumann et al. 2012). Current models show that the IVA, IIAB, and IIIAB parent bodies formed between 0.1 and 0.3 Myr after CAI formation (Kruijer et al. 2014), while the IVB parent body accreted 0.1–0.5 Myr after CAI formation (Neumann et al. 2018b).

The IVA, IIAB, and IIIAB parent bodies formed in the inner solar system isotopic reservoir, which includes Earth, Mars, and Vesta (e.g., Poole et al. 2017; Section 4.2.8). However, the IVB parent body formed in a more distant reservoir (Trinquier et al. 2009)—possibly beyond the orbit of proto-Jupiter (Warren 2011)—that also spawned carbonaceous chondrites and the asteroid Ceres (Budde et al. 2016; Kruijer et al. 2017; McSween et al. 2018). (We will discuss the carbonaceous chondrites in greater detail in 4.2.6.) Like the IVB meteorite group, the South Byron Trio (Figure 4.1; olive green) has carbonaceous chondrite-like isotopic composition. Hilton et al. (2019) combined measurements of the $^{182}Hf \rightarrow ^{182}W$ system (half-life 8.9 Myr) with thermal modeling to show that the SBT parent body formed 1.1 ± 0.5 Myr after CAI formation.

Kruijer et al. (2017) suggest that on average, iron parent bodies in the Earth–Mars–Vesta region formed 0.5 Myr before similar parent bodies in the more distant isotopic reservoir. With a gas disk lifetime of 4 Myr (Wang et al. 2017; Alexander et al. 2018) and the pessimistic assumption that instead of the IVB meteorites, the later-forming SBT traces the emergence of Jupiter-building planetesimals, Jupiter would still have had at least 2.5 Myr to grow before the dissipation of the gas (see Section 2.2 for a discussion of the timescales of disk evolution). Our unfolding picture of the solar nebula features early planetesimal emergence in all regions sampled by meteorites (Libourel & Krot 2007).

4.2.3 Chondrules and Collisions

Here we will pause in our progress through Figure 4.1 to consider chondrules, the primary constituents of meteorites called chondrites. The left panel of Figure 4.2 shows the exterior structure of a chondrule from the type LL4 Bjurböle meteorite (Loesche et al. 2014), while the right panel shows a contrast-enhanced X-ray

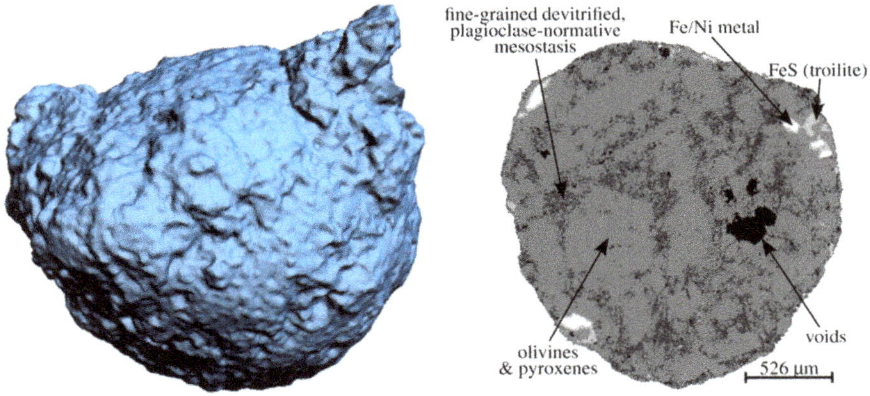

Figure 4.2. Left: The exterior structure of a chondrule from the Bjurböle meteorite as rendered by Loesche et al. (2014). Right: X-ray tomogram of a chondrule cross section from Loesche et al. (2013).

tomogram of a chondrule cross section (Loesche et al. 2013). Chondrules are roughly millimeter-size, solidified droplets of molten silicates that cooled on a timescale of minutes to hours (Hewins 1983; Scott 2007). Their formation mechanism has long been a puzzle: while chondrule textures suggest that they reached peak temperatures of 1400–1800 K when molten, their precursor aggregates grew in regions with ambient temperatures definitely below 1000 K (Russell et al. 2005) and probably below 650 K given the presence of troilite (FeS) in some chondrules (the Rubin et al. 1999 results were inconclusive for most chondrites in the sample). Furthermore, the Mg, O, Si, Fe, and K isotopes in chondrules are unfractionated, even though disk-model gas pressures are too low to prevent evaporation-driven fractionation (Clayton 1993; Kehm et al. 2003; Mullane et al. 2005). The flash-heated chondrules must have either cooled in localized high-pressure regions or clustered together in areas with high dust/gas ratios (Alexander et al. 2008). Chondrules began forming near $t = 0$: the oldest chondrules have age estimates that overlap with CAIs (Krot et al. 2004; Bizzarro et al. 2004; Connelly et al. 2012; Bollard et al. 2019). But unlike CAIs, chondrules have a wide age range; this can even be true for chondrules embedded in the same meteorite (e.g., Bizzarro et al. 2004; Wadhwa et al. 2007). Using uranium-corrected lead isotope dating, Connelly et al. (2012) found that the chondrule formation epoch lasted 3 Myr (later extended to 4 Myr by Connelly et al. 2017). While Bollard et al. (2017) revised the chondrule formation time interval downward to only 1 Myr based on new Pb–Pb age estimates from 22 chondrules, there's no getting around the fact that any localized chondrule-forming event had to be repeated over and over.

One theory is that chondrules formed in shocks, which could be powered by planetesimal collisions (Hood et al. 2009), large bodies on eccentric orbits (Ciesla et al. 2004; Hood et al. 2009; Morris et al. 2012; Boley et al. 2013), or spiral density waves (Desch & Connolly 2002; Boss & Durisen 2005; Boley & Durisen 2008). It is reasonable to theorize that the solar nebula had chondrule-forming spiral density waves, as such spirals commonly appear in high-resolution images of disks (e.g., Pérez et al., 2016; Dong et al., 2018; Lee et al. 2020; Johnston et al., 2020,

Section 2.4.7). However, spirals triggered by giant planets (e.g., Hall et al. 2018; Huang et al. 2018; Phuong et al. 2020; Uyama et al. 2020), rather than disk instability, cannot fully explain chondrule formation: the solar nebula started spawning chondrules before any planets appeared. How about planetesimal-driven shocks? From iron meteorites, we know that early planetesimal formation and late CAI formation may have overlapped (Amelin et al. 2010; Kawasaki et al. 2020), so a planetesimal origin for even the oldest chondrules is plausible. Unfortunately, models by Hood et al. (2009) suggest that impact-driven shocks can only produce a small fraction of the chondrule mass contained in the early asteroid belt. Bow shocks from planetesimals on eccentric orbits can produce the required chondrule quantity, but Hood et al. (2009) find that orbital eccentricities don't get high enough to drive shocks until Jupiter forms and its resonances sweep the inner solar system. Furthermore, Boley et al. (2013) calculate that chondrule-forming bow shocks can only be generated by an embryo with $r > 2000$ km—larger than the planetesimals produced by even the most optimistic streaming instability models (Section 4.4). While planetary embryos probably appeared at $t < 2$Myr (see the discussion of Mars in Section 4.2.5), there's no evidence that they predate the oldest chondrules.

Another planetesimal-related theory is that chondrules are "impact splash:" droplets splattered from colliding, molten planetesimals (Urey & Craig 1953; Asphaug et al. 2011; Sanders & Scott 2012; Dullemond et al. 2014; Lichtenberg et al. 2018). The impact splash model and a similar theory in which chondrule precursor droplets are launched from planetesimal surfaces by jets (Kieffer 1975; Johnson et al. 2015; Hasegawa et al. 2016; Wakita et al. 2017; Oshino et al. 2019) explain the presence of troilite in chondrules: the disk midplane at 1–4 au would have stayed below the troilite sublimation temperature (Section 2.4.5, Figure 2.19). Furthermore, planetesimals forming less than 2 Myr after CAIs contained enough live ^{26}Al to melt completely (Greenwood et al. 2005; Hevey & Sanders 2006), creating subsurface magma oceans that could have provided source material for impact splash (Elkins-Tanton 2012). (Note that Bollard et al. 2017 and Bollard et al. 2019 have revisited the solar nebula's initial ^{26}Al abundance and concluded that melted and differentiated planetesimals formed only within the first 1 Myr after CAIs, a time frame they also associate with chondrule formation.) In fact, Scott et al. (2015) argue that all parent bodies of igneous meteorites were collisionally disrupted while completely or partially molten. Finally, there is evidence for the early onset of planetesimal collisions energetic enough to launch jets. Greenwood et al. (2015) point out the shortage of Dunites—olivine-rich asteroids and meteorites—compared to metallic asteroids and iron meteorites. Dunites and irons should have come from the same parent bodies (mantles versus cores, respectively); the Dunite underrepresentation suggests that the mantles of differentiated planetesimals were obliterated early on by catastrophic collisions. Likewise, the cooling histories of group IVA iron meteorites indicate that their parent body was stripped of its mantle soon after accretion at $t \sim 0.1$Myr (Yang et al. 2007, 2008 Section 4.2.2). However, Budde et al. (2016) find that impact splash or jetting is unlikely to produce the complementary ^{183}W anomalies observed in chondrules and the fine-grained matrix dust that surrounds them, and Lichtenberg et al. (2018) argue that collisions between fully melted planetesimals would have produced basaltic spherules that are not observed in the meteoritic record. Jacquet (2014) suggests that impact-based models produce too much size diversity among chondrules, when in fact chondrules within a single meteorite typically cover an effective radius range of $\lesssim 6$.

The most appealing feature of the impact splash/jet theory of chondrule formation is its basis in events that definitely happened. Other flash-heating processes such as lightning (Whipple 1966; Pilipp et al. 1998; Desch & Cuzzi 2000), magnetic reconnection (Levy & Araki 1989; Eisenhour et al. 1994; McNally et al. 2013), and X-ray flares (Shu et al. 1997; Feigelson et al. 2002) are physically possible, but planetesimal collisions are inevitable. In fact, splash from a giant impact is the most widely accepted model for an isolated batch of late-forming chondrules (t = 5–6 Myr) incorporated into the CB chondrites (Krot et al. 2005, 2010; Bollard et al. 2015). Impact-based chondrule formation models may also explain the size-sorting phenomenon. While chondrules in meteorites sourced from the inner solar nebula—such as the acapulcoites/lodranites (<3 au; Neumann et al. 2018b) and H chondrites (possibly from asteroid Hebe, currently at 2.4 au; Gaffey & Gilbert 1998)—reach millimeter sizes and make up most of their host meteorites' mass (e.g., King & King 1978), chondrules from carbonaceous chondrites sourced from outside Jupiter's orbit (Warren 2011: Section 4.2.6) are smaller (~0.1 mm; King & King 1978) and less abundant. Because ice melting isn't an energy "sink" in collisions between rocky planetesimals in the inner disk (Sanders & Scott 2012), that's where chondrule production should be both frequent and efficient in terms of mass yield. If chondrules are indeed impact relics, their preservation in unmelted meteorites whose parent bodies formed at $t \gtrsim 2$Myr indicates that the solar nebula's solids were recycled through multiple generations of planetesimals. From studying the chondrule formation timeline, we build physical intuition about the fate of solids in the solar nebula. The concept that emerges is a continuous cycle of coalescing, colliding, and breaking.

4.1 Meteorite hunters

Despite continual improvements in age-dating techniques (e.g., Jackson et al. 2004; Gehrels et al. 2008; Paton et al. 2010), meteorites provide only anecdotal information about the primordial planetesimal population unless we know how likely they are to travel to Earth, survive the trip through the atmosphere, and be found on Earth's surface. While the rarity of large chunks of metal on the ground makes iron meteorites relatively conspicuous, stony meteorites can hide in plain sight among Earth rocks. Thus iron meteorites, though accounting for only 5.7% of observed meteor falls, used to make up 60% of meteorite finds (Seeds & Backman 2019). Today's diverse meteorite collections have their roots in a 1969–1970 geodetic survey by the Japanese Antarctic Research Expedition. On 21 December 1969, the scientists found three isolated, exposed rocks on the southeastern marginal ice field (later renamed Meteorite Ice Field) of the Yamato Mountains (Yoshida et al. 1971). Remembering a half-joking suggestion from Prof. Masao Gorai of the Tokyo University of Education that the survey team bring him a meteorite as a souvenir, team member Dr. Masaru Yoshida suggested that his colleagues collect all stones that appeared in front of their snow car.

Figure 4.3. Left: Alex Meshik and Morgan Nunn Martinez, of the 2013–2014 NASA Antarctic Search for Meteorites (ANSMET) field party, recover a meteorite from Antarctica's Miller Range. Image credit: NASA/JSC/ANSMET.[1] Right: radar signatures from meteors falling over Glendale, Arizona, USA on the evening of 2018 July 26. The blue/gray rectangles show detections from the 2.4° elevation sweep by the KIWA radar, part of the NEXRAD weather radar network operated by the National Oceanic and Atmospheric Administration. Green stripes indicate meteor signatures identified by the Phoenix Sky Harbor Airport radar. Image credit: NASA Astromaterials Research and Exploration Science.[2]

They returned to Japan with nine meteorites, all of different petrologic types. Prof. Gorai later wrote to Dr. Yoshida:

> This is indeed a surprise. According to what you told me, these meteorites were collected from a single small area; however, it is impossible to imagine that several meteorites with different lithologies would occur together within such a small area of ordinary land. Perhaps their occurrence is due to concentration by a glacial process that worked to bring the meteorites together from a much wider area. This is indeed a natural "Meteorite Museum" that made me extremely amazed and shocked.

(Yoshida 2010). Since the 1969 Yamato meteorite discovery, teams from all over the world have mined ablated Antarctic and Greenland glaciers for macrometeorites (Figure 4.3; left) and harvested micrometeorites from melted ultraclean ice. While macroscopic carbonaceous chondrites are rare, accounting for only 2% of meteorite falls, their petrologic relatives—the Antarctic carbonaceous micrometeorites—represent the bulk of extraterrestrial mass accreted by Earth today (e.g., Maurette et al. 2000). Further diversity in extraterrestrial samples comes from the interplanetary dust particles (IDPs) collected by high-altitude aircraft and the cosmic spherules picked from deep-sea sediments, some of which have no macrometeorite analog (Brownlee 1985; Plane 2012).

Estimates of meteorite survival rates began with radar tracking of falling meteors. Once Hey & Stewart (1947) conclusively established that transient ionospheric radio echoes were, in fact, reflections off meteor trails, atmospheric scientists began using the "meteor bounce effect" to monitor winds in the upper atmosphere (e.g., Robertson et al. 1953; Elford & Robertson 1953). By the mid-1950s, astronomers were setting up radar receiving stations exclusively dedicated to meteor monitoring and orbit mapping (Hawkins et al. 1956). Today, meteor tracking has moved into the public sphere: the

[1] https://solarsystem.nasa.gov/resources/2255/antarctic-meteorite/
[2] https://ares.jsc.nasa.gov/meteorite-falls/events/glendale-arizona

National Oceanic and Atmospheric Administration (USA) maintains a web-based repository of weather radar imagery, including access software and tutorials (Figure 4.3, right).[3] Similarly, NASA's All-sky Fireball Network publishes recent images of falling meteors brighter than the planet Venus and extrapolates their trajectories to predict the likely fragment landing place.[4] With the help of the American Meteoritical Society, which collates eyewitness accounts of meteor sightings,[5] citizen scientists can mount their own meteorite hunting expeditions.[6] Dynamical simulations of Earth/asteroid/dust interaction (e.g., Nesvorný et al. 2003; Borovička 2007; Grün et al. 2019) help us map meteorites' paths from interplanetary space into our labs, providing the final ingredient in meteorite population studies.

4.2.4 Approaching 2 Myr

Continuing our progress through Figure 4.1, we consider the acapulcoites and lodranites. The two meteorite groups, which come from the rocky upper mantle of the same parent body, show significant thermal alteration but were never fully molten. Touboul et al. (2009) find that the acapulcoite/lodranite parent body formed between 1.5 and 2 Myr after CAIs (see also Neumann et al. 2018b). Also forming during the same time period, about 1.9 Myr after CAIs, are the L-chondrite and H-chondrite parent bodies (Figure 4.1; sky blue and purple; Gail & Trieloff 2019; Henke et al. 2012; Monnereau et al. 2013). Like the acapulcoites/lodranites, the H- and L-chondrite meteorite groups include samples with high thermal metamorphism (petrologic type 6), which were heated to nearly 1200 K but did not quite reach the melting point of a eutectic iron–sulfur mixture (Slater-Reynolds & McSween 2005). Strikingly, the H-chondrite parent body may have had a radius exceeding 250 km (note that this study also predicts a slightly later formation time of 2.15 Myr after CAIs; Blackburn et al. 2017), which is much larger than the $r \gtrsim 100$ km found for the magmatic iron parent bodies (Neumann et al. 2012; Kruijer et al. 2014; Neumann et al. 2018a). The acapulcoites, lodranites, L chondrites, and H chondrites show that planetesimal thermal evolution is mainly controlled by the amount of live ^{26}Al (half-life 0.7 Myr) available during accretion; size is a secondary consideration (Bottke et al. 2006; Touboul et al. 2009). As formation times approach 2 Myr after CAIs (1 Myr according to Bollard et al. 2017), the live ^{26}Al abundance drops enough that newly formed planetesimals can no longer melt and differentiate (Baker et al. 2005). Instead, planetesimals forming at $t \gtrsim 2$Myr have unmelted, primitive

[3] https://www.ncdc.noaa.gov/wct/
[4] https://fireballs.ndc.nasa.gov/
[5] https://fireball.amsmeteors.org/members/imo/report_intro//
[6] See tutorial at https://ares.jsc.nasa.gov/meteorite-falls/how-to-find-meteorites/

structure with the constituent dust grains preserved (Hevey & Sanders 2006; Sugiura & Fujiya 2014; Scott et al. 2015).

As we discuss the extent to which ^{26}Al controls planetesimal interior structure, it is worth considering how solar system planetesimals may differ from planetesimals in nearby disks. At the Sun's $t = 0$, 7.5 billion years after the formation of the Milky Way, the solar nebula abundances of ^{60}Fe (half-life 1.5 Myr) and ^{26}Al (half-life 0.7 Myr) were more than 20 times higher than the corresponding Galactic steady-state abundances and orders of magnitude higher than the predicted abundances from three-phase interstellar medium mixing models (Huss et al. 2009). The only way to deposit abundant SLRIs in the solar nebula is to place the proto-Sun near at least one SLRI-synthesizing source, such as a supernova, Wolf–Rayet star, or AGB star (Lugaro et al. 2018). More likely, more than one source contributed to the solar nebula SLRI inventory; in that case, the Sun must have been born in a cluster with at least 1000 other young stars (Gounelle & Meynet 2012). But the best-studied disks are located in smaller clusters. The Taurus-Auriga star-forming region has a characteristic distance of only 140 pc, which is near enough for high angular-resolution disk observations, but a recent census that is complete for spectral types earlier than M6–M7 finds only 427 member stars (Luhman 2018). The TW Hydra association, which hosts the nearest protoplanetary disk (TW Hydrae, 47 pc), has 43 or fewer members (Gagné et al. 2017). Given that the mass of the most massive star in a cluster is proportional to the number of members (Bonnell et al. 2004), we don't see many progenitors of supernovae, Wolf–Rayet stars, or AGB stars in the nearest star-forming regions. The Orion nebula, with about 3500 members (Megeath et al. 2012), may be similar to the Sun's natal cluster, but at a distance 414 pc (Menten et al. 2007), it's hard to resolve the giant planet-forming regions of the cluster's disks. In fact, Gounelle (2015) predicts that only 1% of the Galaxy's planetary systems formed ^{26}Al-rich environments.

Planetesimal melting and differentiation (or lack thereof) have consequences for planet formation. During giant impacts, planetesimal mantles and cores are vaporized at different rates, changing the core/mantle ratio and the long-lived radioisotope abundance in growing protoplanets (Bonsor et al. 2015; Carlson et al. 2018). Radioactive uranium, thorium, and potassium constitute a major internal heat source driving Earth's plate tectonics (Pollack et al. 1993), which can cool the climate by drawing down CO_2 from the atmosphere and oceans (Raymo et al. 1988). While early results from Bonsor et al. (2020) indicate that about 66% of exoplanetesimals swallowed by white dwarfs are differentiated, a robust examination of the planetesimal makeup in other star systems awaits the discovery of more polluted white dwarfs with high-precision Ca and Fe abundances.

4.2.5 Martian Meteorites and Embryo Formation

One major milestone on the road to giant-planet formation is the growth of embryos in the Pluto–Mars mass range ($1.3 \times 10^{25} - 6.4 \times 10^{26}$ g;

Wetherill & Stewart 1993).[7] Roughly speaking, planetesimals in the asteroid belt that were "born big" had radii and masses similar to Ceres (r = 476 km, $M = 9.5 \times 10^{23}$ g; Morbidelli et al. 2009; Johansen et al. 2007). It's not clear whether the first Kuiper Belt planetesimals had a similar size spectrum to the first asteroids (Simon et al. 2016, 2017) or started much smaller, with $r \lesssim 4$ km (Schlichting et al. 2013; see the inset box 4.2 on the Kuiper Belt). In this book, the word "embryo" will refer to an object with $M \gtrsim 10^{25}$ g—roughly Pluto's mass, and 10 times Ceres' mass—no matter the location in the disk. A planetesimal grows into an embryo by either colliding with other planetesimals or capturing drifting pebbles.

Mars is labeled a leftover embryo because simulations of terrestrial planet growth usually result in an Earth or super-Earth occupying Mars' position at 1.5 au from the Sun (Chambers 2001; Raymond et al. 2004; Kokubo et al. 2006; Clement et al. 2020), even when resonances from Jupiter and Saturn are taken into account (O'Brien et al. 2006). Theorists have developed many models to explain the "small Mars" problem. Hansen (2009), who points out that Mercury is also puzzlingly small given the number of close-in, massive exoplanets detected, suggests that terrestrial planet formation might have taken place in a narrow annulus of planetesimals covering R = 0.7–1 au. Mercury and Mars would then have faced arrested development after getting scattered out of the annulus by Venus and Earth (see also Walsh & Levison 2016). Other possibilities are viscously stirred pebble accretion, which produces a small Mars because pebble stopping time increases with distance from the star (Levison et al. 2015; though the authors did not model Mercury's formation), and incomplete condensation of solids in the inner solar nebula (Morgan & Anders 1980; Ebel & Alexander 2011; though this theory does not solve the small Mars problem). But even though we still don't know why Mars got stuck as an embryo (Lykawka & Ito 2019), we can use Martian meteorites—pieces of Mars' mantle that were launched by giant impacts and eventually fell to Earth (Figure 4.1; red box)—to estimate its accretion time.

To construct a growth chart for Mars, we analyze the behavior of tungsten, a siderophile ("metal lover") that sinks to a planet's iron–nickel core during differentiation, and hafnium, a lithophile ("rock lover") that remains in the mantle. (Unlike planetesimals, embryos are massive enough to melt and differentiate from accretion heating alone, though Mars also had some help from ^{26}Al.) Because core formation occurs when a protoplanet is nearing its final mass (Nimmo & Kleine 2007), we can roughly say that Hf–W fractionation traces Mars' emergence as a planetary embryo. We define the tungsten isotopic anomaly ϵ_W with respect to chondrites, which retain a record of the isotopic evolution of undifferentiated bodies:

$$\epsilon_{\rm W} = 10^4\left[\frac{(^{182}{\rm W}/^{183}{\rm W})}{(^{182}{\rm W}/^{183}{\rm W})_{\rm CHUR}} - 1\right], \tag{4.1}$$

[7] Descriptions of embryos usually quote a Moon–Mars mass range, but because the Moon is the result of a giant impact onto proto-Earth (Canup & Asphaug 2001; Cuk & Stewart 2012), Pluto is more representative of the types of objects that grow from planetesimals.

where CHUR stands for chondritic uniform reservoir. Assuming that core formation happens much faster than planet growth, the core formation time t_c is given by

$$\epsilon_{\mathrm{W}} = 10^4 \left(\frac{^{180}\mathrm{Hf}}{^{182}\mathrm{W}}\right)^{\mathrm{now}}_{\mathrm{CHUR}} \left(\frac{^{182}\mathrm{Hf}}{^{180}\mathrm{Hf}}\right)^{t_0} f^{\mathrm{Hf/W}} \exp[-\lambda t_c] \tag{4.2}$$

(Jacobsen 2005), where $\left(\frac{^{180}\mathrm{Hf}}{^{182}\mathrm{W}}\right)^{\mathrm{now}}_{\mathrm{CHUR}}$ is measured in today's chondrites, $\left(\frac{^{182}\mathrm{Hf}}{^{180}\mathrm{Hf}}\right)^{t_0}$ is the inner solar nebula's hafnium isotopic ratio at $t = 0$, and $\lambda = 0.078$ Myr^{-1} is the time constant for $^{182}\mathrm{Hf} \rightarrow {}^{182}\mathrm{W}$ decay (Vockenhuber et al. 2004). The most elusive quantity in Equation (4.2) is $f^{\mathrm{Hf/W}}$, a measure of hafnium excess in the Martian mantle:

$$f^{\mathrm{Hf/W}} = \frac{(\mathrm{H/W})}{(\mathrm{H/W})_{\mathrm{CHUR}}} - 1. \tag{4.3}$$

Luckily, because thorium and hafnium behave similarly during igneous processing (Nimmo & Kleine 2007), $(\mathrm{Th/Hf})_{\mathrm{CHUR}} = (\mathrm{Th/Hf})_{\mathrm{Marsmantle}}$. By combining new measurements of $(\mathrm{Th/Hf})_{\mathrm{CHUR}}$ with the known value of Th/W in Martian meteorites, Dauphas & Pourmand (2011) were able to calculate $f^{\mathrm{Hf/W}}$ and t_c. They found that half of Mars' mass was already assembled at $t = 1.9^{+0.9}_{-1.0}$ Myr after CAI formation.

Mars' early emergence tells us (1) that progress up the size ladder is rapid—we have moved from $r = 1$ mm to $r = 3400$ km, a factor of 3.4×10^9, in only 2 Myr (see the inset on the size scale in Section 3.1)—and (2) that progress up the size ladder is messy. The canonical picture of inelastic planetesimal collisions leading to embryo growth (e.g., Kokubo & Ida 1998; Ida & Lin 2004; Raymond et al. 2005) is too simplistic: planetesimals and embryos were forming in the same region of the solar nebula at the same time, and planetesimal collisions were causing at least as much destruction as growth (Bottke et al. 2005; Leinhardt & Stewart 2012). Embryo collisions can also be destructive (e.g., Agnor & Asphaug 2004); some dynamical models suggest that Mercury is a remnant embryo that had its mantle collisionally stripped (Benz et al. 1988). We encourage any reader who is looking for a challenging project to build a model of embryo growth that includes ongoing planetesimal building, collisional fragmentation, and dust production so that solids can move all the way up and down the size ladder. N-body codes that contain two of the three required ingredients—collisional fragmentation and dust production—already exist and have been used to model the observational signatures of embryo formation and stirring by unseen giant planets (Leinhardt et al. 2015; Dobinson et al. 2016).

4.2.6 Meteorites from Afar

Our discussion of the IVB and South Byron Trio parent bodies (Section 4.2.2) has already brought us into contact with carbonaceous chondrites (CCs) and, by extension, C-type asteroids. Broadly speaking, today's C-type asteroids occupy the outer asteroid belt: most have semimajor axes between 2.8 and 3.3 au (Gradie & Tedesco 1982), though dynamical mixing has left some interlopers in the inner asteroid belt (DeMeo & Carry 2014). Carbonaceous chondrites—victims of unfortunate taxonomy in that they are not necessarily carbon-rich (Jarosewich 1990; Pearson et al. 2006)—have similar ultraviolet, visible, and near-infrared spectra to Ceres and other C-type asteroids (Figure 4.4; Bus & Binzel 2002; DeMeo et al. 2009; Schafer 2018), which was the first clue that they might have C-type parent bodies (e.g., Larson et al. 1979; Feierberg et al. 1985; Wetherill & Chapman 1988). The discovery that most C-type asteroids and CCs show signs of aqueous alteration

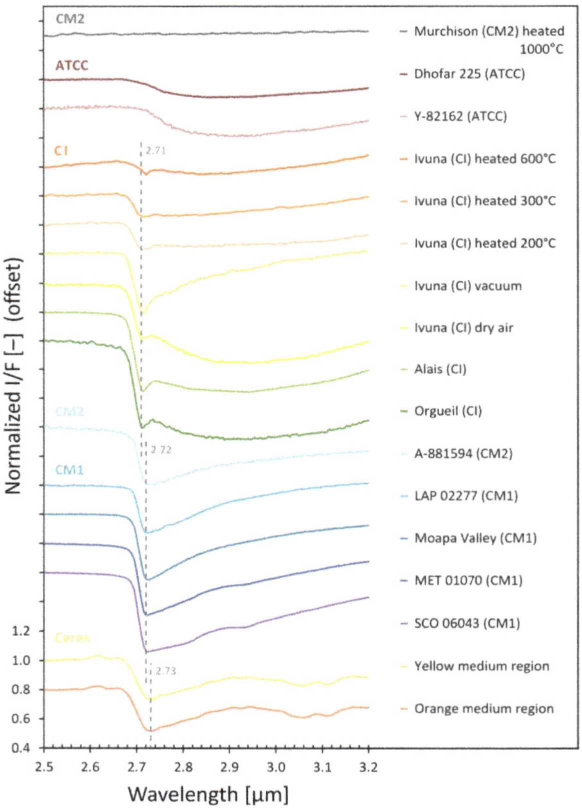

Figure 4.4. Spectral comparison between Ceres and carbonaceous chondrites from Schäfer et al. (2018), copyright John Wiley & Sons. The bottom two lines show composite infrared reflectance spectra from the VIR hyperspectral imager aboard the Dawn spacecraft. The spectrum labeled "Yellow medium region" (where "yellow" denotes a spectral slope of $-10 < \gamma < -3\%$ reflectance per μm at wavelengths 0.4–1.0 μm) was extracted from a surface region 20 × 85 km in size, located near the eastern rim of Ikapati crater. The "Orange medium region" spectrum ("orange" meaning 0.4–1.0 μm spectral slope of $\gamma \geqslant -3.0$ % reflectance per μm) comes from a 90 × 40 km region located 20 km west of Tafakula crater. All other entries in the figure are laboratory spectra of meteorites. The Ceres surface spectra are most similar to the four laboratory spectra of CM1 chondrites (LAP 02277, Moapa Valley, MET 01070, and SCO 06043), though none of the chondrites show Ceres' 3.0–3.1 μm ammoniated phyllosilicate absorption feature.

(Tomeoka & Buseck 1985, Vilas & Gaffey, 1989; Clayton & Mayeda, 1999, e.g.) led to the idea that C-type asteroids formed in a cold region of the solar nebula where they were able to accrete water ice (Dufresne & Anders 1962; Grimm & McSween 1989). Indeed, topographical models by Ruesch et al. (2016) and Fu et al. (2017) suggest that Ceres has subsurface liquid, and the best-fit differentiation and interior structure model by Neumann et al. (2020) has Ceres forming in the Kuiper Belt (though this is not yet a consensus scenario). Furthermore, Ceres has water vapor plumes similar to those of active comets from the Kuiper Belt and the main asteroid belt (Hsieh & Jewitt 2006; Küppers et al. 2014).

Further evidence that CCs grew beyond the terrestrial planets comes from titanium, chromium, and oxygen isotopic compositions (Trinquier et al. 2009; Qin et al. 2010; Warren 2011; Desch & Kalyaan 2018). Figure 4.5, taken from Burbine & Greenwood (2020), places Earth, Mars, the Moon, and meteorites on a plot of Δ^{17} O, a measure of departure from the terrestrial oxygen fractionation pattern, as a function of ϵ^{54} Cr, the parts-per-10,000 $^{54}Cr/^{52}Cr$ ratio referenced to a standard sample (replace $^{182}W/^{183}W$ with $^{54}Cr/^{52}Cr$ in Equation (4.1) to define ϵ^{54} Cr). The CCs and NCs (all non-CC solids, including Martian meteorites and Earth/Moon samples) land in two distinct parts of the plot. While it's easy to understand the connection between asteroid orbits and aqueous alteration (see the discussion of the water ice line in box 3.3), it's much less clear why the chromium or titanium isotopic

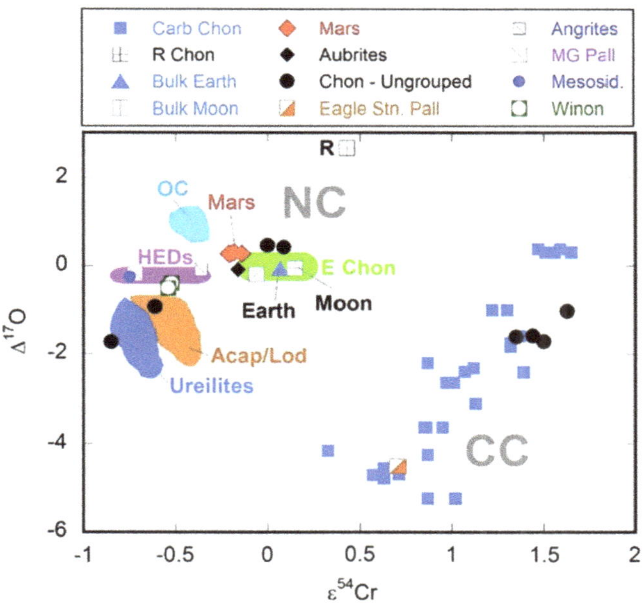

Figure 4.5. Δ^{17} O, which measures departure from the terrestrial oxygen fractionation pattern, as a function of ϵ^{54} Cr, which is defined as in Equation (4.1) but with ^{54}Cr and ^{52}Cr standing in for ^{182}W and ^{183}W, respectively. Noncarbonaceous samples—those from Earth, Moon, Mars, ordinary chondrites (OC), enstatite chondrites (E Chon), acapulcoites/lodranites, and ureilites—are clearly separated from samples of the carbonaceous chondrite isotopic reservoir. Image from Burbine & Greenwood (2020), with permission of Springer.

ratio should vary with distance from the young Sun. Revisiting Section 3.2.2 and Figure 3.6 provides some insight. Dauphas et al. (2010) showed that ^{54}Cr, a neutron-rich chromium isotope that was probably synthesized in a nearby supernova, found its way into meteorites through fine-grained particles with $r_{\rm eff} < 100$nm. Such particles are entrained in the gas (Equations (3.7) and (3.8)), while coarse-grained, isotopically normal particles settle to the midplane, grow, and start drifting inward. Even if supernova ejecta is initially distributed homogeneously throughout the solar nebula, either aerodynamic drag or the fine-grained dust's vulnerability to heat will fractionate the isotopes (Dauphas et al. 2010; Poole et al. 2017). Another possibility is that there was a compositional gradient in the solar system's parent molecular cloud that imprinted on the solar nebula (Nanne et al. 2019).

From the abundances of deuterium, ^{54}Cr, and ^{26}Mg (the ^{26}Al decay product), Alexander et al. (2012) and van Kooten et al. (2011) suggest that the CM parent body (Figure 4.1, teal; see the inset image of CM fragments) formed at the boundary between the NC and CC isotopic reservoirs. Unlike the other planetesimals whose provenance we have discussed, the CM parent body seems to have grown gradually, with accretion taking up to 3 Myr. In terms of formation time, the CM chondrite parent body is representative of many of the CCs: while planetesimal formation began early in all isotopic reservoirs (Section 4.2.2), it became more gradual with increasing distance from the star, with most CC mass assembling at t = 3–4 Myr (Neveu & Vernazza 2019). van Kooten et al. (2020) point out that the Bells meteorite, an anomalous CM chondrite that has some isotopic similarities with solids sourced from beyond Saturn, may have come from a surface layer on the CM parent body that was composed of captured pebbles (see Chapter 5 in Volume 2 for more on pebble accretion). With fewer, smaller chondrules and more gradual accretion than ordinary chondrites, the CM chondrites may be hinting that embryo growth becomes less violent as we move farther from the Sun and orbital speeds decrease. Of course, collisions are omnipresent throughout any planetesimal disk—witness the collisionally produced dust in the 96 resolved debris disks cataloged at https://circumstellardisks.org—but the Kuiper Belt appears less "broken down" than the asteroid belt (e.g., A'Hearn 2011; Movshovitz et al. 2016), and the asteroid belt's destructive hit-and-run collisions can turn into planet-building grazing and merging in the Kuiper Belt (Leinhardt et al. 2010; McKinnon et al. 2020; see box 4.2 for more on the Kuiper Belt).

Multiple lines of evidence suggest that the CR chondrites (Figure 4.1, green) formed even farther from the proto-Sun than the CM chondrites. Based on their ^{54}Cr and ^{26}Mg abundances, the CR chondrites must be made of 25%–50% primordial molecular cloud material, which cannot survive thermal processing in the warm inner solar nebula (van Kooten et al. 2011). CR chondrites are enriched in ^{15}N (Prombo & Clayton 1993; Budde et al. 2018), another hallmark of formation in the outer solar system (Füri & Marty 2015). Examining the formation times of meteorites from afar helps us piece together a timeline of Jupiter's evolution. If Jupiter scattered the CR chondrite parent body into the inner solar system, the scattering must have happened after CR parent formation at $t = 3.6 \pm 0.6$Myr

(Schrader et al. 2017, but note that Budde et al. 2018 estimate $t > 4.0^{+0.5}_{-0.3}$). The prevailing theory explaining the distinct isotopic compositions shown in Figure 4.5 posits that (1) proto-Jupiter opened a gap in the solar nebula that frustrated mixing of dust from the different isotopic reservoirs (see Sections 2.4.2 and 3.2.3), and (2) after planetesimal formation was well underway, Jupiter's growth and/or migration scattered fully formed planetesimals so that compositional mixing occurred on a macroscopic level (Warren 2011; Budde et al. 2016; van Kooten et al. 2011; Kruijer et al. 2017; Budde et al. 2018).

4.2.7 Iapetus and Saturn

Our progress on age-dating the outer solar system continues with Saturn's moon Iapetus (Figure 4.6). While we don't have physical samples of Iapetus in hand, we do have ground-breaking images from the *Cassini* spacecraft.[8] From a close flyby on 2005 December 31, the Cassini team measured equatorial × polar radii of 747.4 × 712.4 km (Thomas et al. 2007). The 35 km difference between the two radii indicates that Iapetus is rotationally flattened. What's odd, though, is that Iapetus' present orbit and rotation period should yield no more than a 10 m deviation from a spherical shape. According to Castillo-Rogez et al. (2007), the icy lithosphere solidified and strengthened while Iapetus was in the process of spinning down, preserving the extreme flattening and leaving the low-viscosity liquid interior to dissipate angular momentum. Castillo-Rogez et al. (2007) were able to back out the SLRI abundance required to keep the interior liquid long enough for Iapetus to spin down from a 16 hour rotation period—which is consistent with its topography—to the present value of 79.3 days (Yoder 1995). To match the ^{26}Al and ^{60}Fe abundances in the best-fit despinning models, Iapetus must have formed between 2.5 and 5 Myr after CAIs.

What does Iapetus tell us about Saturn's formation? Nothing, if Iapetus is a captured satellite like its sister moon Phoebe (Johnson & Lunine 2005), but Ward (1981) and Nesvorný et al. (2014) show that despite its high-inclination orbit, Iapetus could have formed in the Kronian subnebula (Saturn's circumplanetary disk; Mosqueira & Estrada 2003). Furthermore, compositional data link Iapetus' bright side with Saturn's regular satellites Dione and Rhea (Buratti et al. 2002), suggesting a common origin for all three moons. Finally, the despinning modeled by Castillo-Rogez et al. (2007) had to be induced by tidal torques from Saturn—it couldn't have happened to an isolated dwarf planet. If Iapetus formed in the Kronian subnebula, most of Saturn's mass had to have assembled before Iapetus started to grow. That means we can restrict estimates of Saturn's formation time to $t = 2.5$–5 Myr.

Because we already know the solar nebula's gas lasted about 4 Myr (Alexander et al. 2018, but note that Wang et al. 2017 quote a more precise gas lifetime of 3.8 Myr based on paleomagnetic studies of meteorites), the 5 Myr upper limit on Saturn's formation time isn't very useful. However, the possibility that Saturn

[8] https://solarsystem.nasa.gov/missions/cassini/overview/

Figure 4.6. Cassini images of Iapetus reveal a geoid shaped by a 16 hour rotation period, but the actual rotation period is 79.3 days. The icy trailing hemisphere has a similar composition to sister moons Dione and Rhea, while the dark leading hemisphere sweeps up debris from comets impacting the captured satellite Phoebe (Verbiscer et al. 2009; Denk et al. 2010). The equatorial ridge outlined on the right-hand side of the image is a product of spindown (Castillo-Rogez et al. 2007). Image credit: NASA/JPL/Space Science Institute.

accreted most of its mass within 2.5 Myr is intriguing.[9] As we saw in Section 4.2.6; planetesimals were still forming beyond the H_2O-ice line at $t > 3$ Myr. Progress up the size ladder (box 3.1) was far from orderly, with planetesimals, embryos (Section 4.2.5), and giant planets all forming at the same time.

[9] A similar time frame might apply to extrasolar giant planets as well: Najita & Kenyon (2014) conclude that in the Taurus-Auriga star-forming region, a large fraction of solids have already been converted into planets or planetesimals by 3 Myr.

4.2.8 Vesta and the Angrite Parent Body

Vesta is a prolific source of both meteorites (McCord et al. 1970; Larson & Fink 1975; Consolmagno & Drake 1977; Figure 4.1, orange box) and still-orbiting collisional fragments (Williams 1979; Binzel & Xu 1993; Zappalà et al. 1995). As one would expect from an aspherical object (see the image in Figure 3.1) that has showered the inner solar system with debris, Vesta is not fully intact (Consolmagno et al. 2015). To explain the varying crustal thickness—20 km in the northern hemisphere and 60–100 km in the southern hemisphere (Clenet et al. 2014)—Haba et al. (2019) suggest that a hit-and-run impact stripped Vesta's northern crust, after which the southern hemisphere reaccreted some of the crust plus other collisional debris. With only 27% of Ceres' mass (Konopliv et al. 2014), Vesta does not come close to meeting our definition of planetary embryo; it is solidly in the planetesimal category. Still, young Vesta experienced many of the same geophysical processes as the terrestrial planets. All samples sourced from Vesta contain igneous rock (Drake 2001; McSween et al. 2011), indicating that Vesta melted, differentiated, and went through a period of volcanic activity. Like other airless bodies such as Mercury and Earth's Moon, Vesta's crust has been significantly altered by "impact gardening" (Denevi et al. 2016). Vesta, Earth, and Mars are linked by more than just geology—they are all part of the same isotopic reservoir (Magna et al. 2006; Wiechert & Halliday 2007; Sarafian et al. 2014), which means many of the Earth-forming planetesimals were Vesta analogs (Kleine et al. 2004; Bottke et al. 2006).

The angrite parent body (Figure 4.1, cyan), another planetesimal that melted, differentiated, and played host to volcanoes, was larger than Vesta is today (Bryson et al. 2019; Zhu et al. 2019). Weiss et al. (2008) showed through paleomagnetic analysis that the angrite parent body's molten metallic core generated a dynamo, providing further evidence that parent bodies of basaltic meteorites had terrestrial planet-like interior structures. Interestingly, the angrites may contain remnants of radially drifting pebbles: while angrites overall have a dry, volatile-poor composition consistent with growth in the warm inner solar nebula (Collinet & Grove 2020), they are enriched in a few selected highly volatile elements such as hydrogen, carbon, and fluorine. Sarafian et al. (2017) suggest that the angrite parent body accreted pebbles that drifted inward from beyond the H_2O-ice line within the first 2 Myr after CAI formation. Like the iron meteorites (Section 4.2.2), the angrites cannot be traced to any known asteroid—their parent body was disrupted, indicating that for planetesimals, large size is no guarantee of survival.

Given their lengthy volcanic/magmatic epochs, the formation time estimates for Vesta and the angrite parent body are somewhat model dependent. The angrite parent body had an extended cooling history of 8 Myr, as demonstrated by the age range of various angrite meteorites (Zhu et al. 2019). The planetesimal must have assembled almost all of its mass by the crust formation age of $t = 4.1 \pm 0.3$Myr, as dated from chromium and magnesium isotopes (Zhu et al. 2019). Vesta, which retained enough internal heat to sustain magmatism for at least 30 Myr (Roszjar et al. 2016; Jourdan et al. 2020), had an even longer cooling history. Because meteorites from Vesta sample multiple layers of the crust and Vesta had multiple epochs of magmatism (Hublet et al. 2017), it is harder to give an absolute age. From the ^{26}Al–^{26}Mg chronometer, Hublet et al. (2017) find $t = 2.66$ Myr for the formation of Vesta's upper crust and $t = 2.88$ Myr for the differentiation of two subtypes of Vestan meteorites from each other. Tang & Dauphas (2012) quote a core formation time of $t = 3.5^{+2.5}_{-1.7}$ Myr from the ^{60}Fe–^{60}Ni isotopic system. Roszjar et al. (2016) simply quote a formation time upper limit of $t = 4$ Myr based on the amount of live ^{26}Al necessary for maintaining magmatism for tens of millions of years. In Figure 4.1, we plot the upper limit from Roszjar et al. (2016) to emphasize the uncertainty in thermal evolution models for large bodies that remain mostly intact, not because Vesta formed late. Part of the reason that the iron meteorites (Section 4.2.2) have more precise formation time estimates is because their parent bodies were disrupted shortly after formation, so the fragments cooled quickly (e.g., Yang & Johansen 2014).

4.2.9 Interplanetary Dust Particles

Our last entry in Figure 4.1 concerns the interplanetary dust particles (IDPs) collected in the stratosphere, one of which is pictured in Figure 3.1. The IDPs are tiny meteors with a size distribution peaking at 200 μm (Love & Brownlee 1993). Along with micrometeorites in the 0.5–2 mm size range, they contribute most of the extraterrestrial mass that hits Earth (Hughes 1978; Engrand & Maurette 1998). Based on spectroscopic similarities, Vernazza et al. (2015) come to the conclusion that P- and D-type asteroids found in the main belt and among the Jupiter Trojans are the primary IDP parent bodies, consistent with earlier suggestions by Bradley et al. (1996) and Dermott et al. (2002). There are also IDPs from short-period comets and Kuiper Belt objects (KBOs; Brownlee et al. 1995; Liou et al. 1996; Bradley 2003; Busemann et al. 2009). Indeed, the P- and D-type asteroids may themselves be interlopers from the outer solar system (Hartmann et al. 1987; Morbidelli et al. 2005; Fraser et al. 2014).

P/D asteroids are fragile. Material strength matters, even in planetesimals primarily held together by gravity, and the IDP parent bodies are ice/rock rubble piles with $\rho_m \approx 1$ g cm^{-3} (Castillo-Rogez et al. 2019) that can be catastrophically disrupted at much lower speeds than rocky asteroids like Vesta (Benz & Asphaug 1999; Stewart & Leinhardt 2009; Movshovitz et al. 2016). The friable[10] fragments in the 10–100 m size range that find their way to Earth do not survive their trip through the atmosphere (Ceplecha et al. 1998), and only remnant dust grains that enter the atmosphere at highly oblique angles make it into astrophysicists' collections. With such tiny specimens, it's difficult to do the same type of chemical analysis that underlies the age-dating of macroscopic meteorites. Often the best age constraints come from observations of the parent-body surfaces.

In contrast to C-type asteroids, most P and D types do not have aqueously altered surfaces, despite having abundant ice (Bell et al. 1989; Campins et al. 2010; Usui et al. 2019). The prevailing theory is that P- and D-type asteroids form late, stay

[10] Friable (adjective): capable of being easily crumbled or reduced to powder. From the Oxford English Dictionary.

cold, and don't melt (e.g., Bland & Travis 2017). (Not that C-type asteroids ever get hot: in the thermal evolution models of Travis & Schubert 2005, the C-types never reach interior temperatures above 450 K, even when the ^{26}Al abundance is artificially increased by 50%.) According to the thermal evolution simulations of Neveu & Vernazza (2019), P- and D-type parent bodies with $r \approx 100$ km must have finished accreting at least 5–6 Myr after CAI formation in order to have low enough SLRI abundance to keep their anhydrous surfaces. The same simulation set shows that the outer layers of C-type asteroids were assembled 3–4 Myr after CAI formation, in agreement with the timeline discussed in Section 4.2.6. The C/P/D asteroids may have begun accreting much earlier, but if they did their growth was gradual. Again, we may be seeing a gentler growth process as we move farther from the Sun. Neveu & Vernazza (2019) suggest that the P- and D-type asteroids formed at $R > 10$au.

One important point is that the final assembly of P- and D-type asteroids cannot have been assisted by gas, because the solar nebula gas was gone by $t = 5$ Myr (Wang et al. 2017; Alexander et al. 2018). The asteroids' outer layers must have been acquired by Bondi–Hoyle accretion or gravitational focusing, not by gas-driven pebble drift. Neveu & Vernazza (2019) hypothesize today's P- and D-type asteroids were preceded by at least one generation of planetesimals that formed at $R > 10$au; if so, today's surviving asteroids may be products of pairwise collisions between the previous generation's fragments.

With our timeline of planetesimal building in hand, we next move to mathematical models of planetesimal-forming instabilities. These models describe the gas-assisted clumping of pebble clouds that collapse to form large, first-generation planetesimals.

4.3 Mathematical Fundamentals of Planetesimal-forming Instabilities

In Chapter 3, we learned how difficult it is for collisional growth to produce particles beyond centimeter sizes. Given the abundant meteoritic evidence showing that the solar nebula spawned planetesimals with $r > 100$ km within $t \sim 10^5$ years, we require a theory in which "the fate of planetary accretion no longer...hinge[s] on the stickiness of the surfaces of dust particles" (Goldreich & Ward 1973). Mutual gravity is certainly insufficient to hold two colliding pebbles together, but what if we had a self-gravitating pebble cloud? One could imagine such a cloud forming from an overdense perturbation in a layer of pebbles concentrated in the midplane (e.g., Weidenschilling 1980; Youdin & Shu 2002). We will first calculate the surface density of solids required to produce such a self-gravitating cloud, then describe the "reverse drag" exerted by solids on gas, which enables pebble concentration by the streaming instability. Next we will examine the conditions required to trigger the streaming instability in a laminar disk. Finally, we will calculate a mass scale for the planetesimals formed by pebble-cloud collapse.

4.3.1 The Self-gravitating Pebble Layer

In Section 2.4.7, we calculated the Toomre Q gravitational stability criterion for a gaseous disk. Here we adapt Equation (2.85) to find Q_p for a thin disk of pebbles concentrated near the midplane:

$$Q_p = \frac{v_{pt}\Omega_K}{\pi G \Sigma_p}, \tag{4.4}$$

where v_{pt} is the expected speed of pebbles stirred by turbulence and Σ_p is the pebble surface density (see Chapter 3). At $R \gtrsim 1$ au, the largest pebbles that can grow by direct sticking have $St < 1$, so $v_{pt} = \sqrt{\alpha}\ c_s$ (Section 3.2.4). Because far more dust mass resides in pebbles than in microscopic grains after $t \sim 50{,}000$ years (Sections 3.3.3 and 4.2; Birnstiel et al. 2012), we can safely say that $\Sigma_p/\Sigma = S/G$, where S/G is the solid/gas mass ratio. We can then rewrite Equation (4.4) as

$$Q_p = \frac{G}{S}\ \sqrt{\alpha}\ Q, \tag{4.5}$$

where Q is the Toomre Q criterion for the gas disk. Figure 4.7 shows Q_p for our model disk with $G/S = 100$, the canonical value assuming radial pebble drift and photoevaporation have not substantially affected the mass inventory. Even with weak turbulence ($\alpha = 10^{-4}$, red dotted–dashed line), our model disk has a gravitationally stable pebble layer out to 100 au. Figure 4.7 also compares v_{pt} with v_{crit}, the critical rms pebble speed required to push down to $Q_p = 1$ and trigger instability in the pebble layer. Because $v_{\text{crit}} < v_{pt}$ throughout the planet-forming region, pebble-cloud fragmentation without some local particle-concentrating mechanism would require extreme damping of pebble random velocities. While Goldreich & Ward (1973) suggest damping by inelastic collisions, experiments show that collisions of both icy and rocky pebbles produce small fragments (Blum & Wurm 2008) that can be kicked high above the thin pebble layer by weak turbulence (Equation (3.51), Section 3.2.4). Furthermore, recall that our model disk resembles the maximum-mass solar nebula (MAXSN; Nixon et al. 2018; box 2.2): if its pebble layer is stable even with minimal turbulence (Nelson et al. 2013), the prospects for

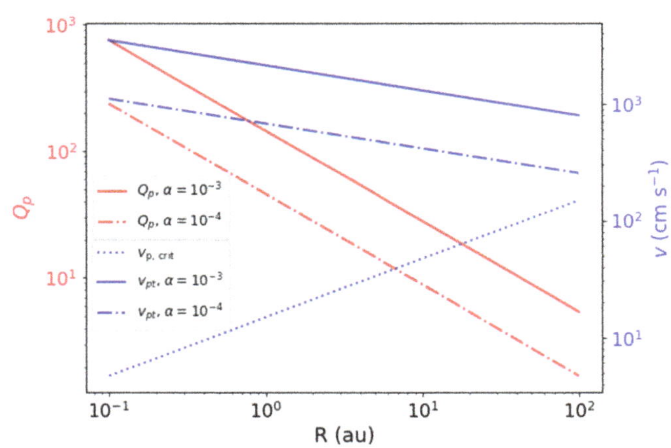

Figure 4.7. Q_p as a function of distance from the star (red, left-hand axis); rms particle velocities and critical rms velocity required for gravitational collapse of pebble layer (blue, right-hand axis). Even for weak turbulence with $\alpha = 10^{-4}$ (dashed–dotted lines), $Q_p > 1$ and $v_{pt} > v_{\text{crit}}$ (dotted blue line) throughout the disk.

planetesimal formation by direct gravitational collapse seem dismal in most astrophysical disks. Finally, a highly concentrated pebble layer can be self-destructive. By pulling its co-orbiting gas up to Keplerian speed (for more on the "reverse drag" exerted by solids on gas, see Section 4.3.2), the pebble layer sets up a strong vertical shear that can trigger the Kelvin–Helmholtz instability and stir up the solids (e.g., Chiang 2008; Barranco 2009; Lee et al. 2010).

To guarantee that gravitationally bound pebble clouds can form, we need (ironically) a particle concentration mechanism that doesn't rely on the self-gravity of the midplane pebble layer. Although we have already explored particle concentration by pressure bumps and vortices (Section 3.2.3), we have so far treated solids as tracer particles that don't affect the gas. Now we will expand our conceptual framework by allowing the particles to play an active role in the disk dynamics. We will begin by analyzing the reverse drag that highly concentrated solids exert on the gas. This reverse drag, often called a "backreaction" in the literature, is the key ingredient that enables particle concentration by the streaming instability.

4.3.2 Reverse Drag and Steady-state Flow

Imagine a swarm of pebbles clustered tightly enough for the solid density to approach the gas density. All pebbles have effective radius $r_{\rm eff} = 1$ cm and material density $\rho_m = 2$ g cm^{-3}. At 5 au in the midplane of our model disk, where $\rho = 4.1 \times 10^{-11}$ g cm^{-3}, the pebble number density required to reach $\rho_p/\rho = 1$ is 4.9×10^{-12} cm^{-3}. If we cut out a cube of disk material that contains, on average, just one pebble, that cube is 59 m on a side. It may seem counterintuitive that pebbles could affect the gas flow when they are far enough apart to each inhabit their own building on a university campus, but consider the gas–solid collision timescale. For $\alpha = 10^{-4}$–10^{-3}, $v_{pt} \ll \bar{v}$, the pebbles are essentially stationary targets for the fast-moving gas molecules. Each gas molecule encounters a pebble every $1/(n_p \sigma_p \bar{v}) \approx 0.027$ years, yielding almost 4400 gas–pebble collisions per molecule per orbit. Substantial gas–particle momentum exchange looks promising.

To quantify the reverse drag in a laminar disk, we assume the pebbles make up a continuous fluid with negligible pressure.[11] Pebbles and incompressible gas form an interpenetrating, two-fluid system.[12] We write coupled Equations of motion that express conservation of linear momentum. For pebble-fluid velocity $\vec{v}_p$ and gas velocity $\vec{v}_g$ in the (R, ϕ) plane,

$$\frac{\partial \vec{v}_p}{\partial t} + \left(\vec{v}_p \cdot \vec{\nabla}\right)\vec{v}_p = -\Omega_K^2 \vec{R} - \frac{\vec{v}_p - \vec{v}_g}{\tau_f}, \tag{4.6}$$

[11] If the disk midplane is turbulent, the pebble velocity dispersion provides a pressure analog, as in Equation (4.4).

[12] To circumvent the need for ever-higher spatial and temporal resolution in numerical simulations as the particle size decreases, Laibe & Price (2014) derived a one-fluid treatment in which a dust/gas mixture has a single center-of-mass velocity $\vec{v}_c$ with a cooling function.

$$\frac{\partial \vec{v}_g}{\partial t} + \left(\vec{v}_g \cdot \vec{\nabla}\right)\vec{v}_g = -\Omega_K^2 \vec{R} + \frac{\rho_p}{\rho}\left(\frac{\vec{v}_p - \vec{v}_g}{\tau_f}\right) - \frac{1}{\rho}\vec{\nabla}P. \tag{4.7}$$

The first term on the right-hand side of Equation (4.6) encodes stellar gravity, while the second represents gas drag in the form of Equation (3.1). On the right-hand side of Equation (4.7), we reverse the sign of the drag term to account for the fact that encounters with particles speed up the gas, rather than slowing it down; multiplying the drag term by ρ_p/ρ boosts the reverse drag according to the pebble concentration factor. All of Chapter 3 is formulated for the test-particle limit, $\rho_p/\rho \to 0$. Gas pressure support is included in the final term on the right-hand side of Equation (4.7).

Because both the gas and dust fluids have nearly Keplerian azimuthal velocity, we will rewrite $v_{p\phi}$ and $v_{g\phi}$ as a Keplerian component minus a small deviation:

$$v_{p\phi} = v_K - \Delta v_{p\phi} \tag{4.8}$$

$$v_{g\phi} = v_K - \Delta v_{g\phi}, \tag{4.9}$$

where $\Delta v_{p\phi}$, $\Delta v_{g\phi} \ll v_K$. The gas and dust radial speeds are also highly sub-Keplerian: v_{pR}, $v_{gR} \ll v_K$. Note that $\Delta v_{p\phi}$ and $\Delta v_{g\phi}$ are different than Δv from Chapter 3, which is defined as $v_p - v_g$. We will assume the dust and gas density distributions are axisymmetric ($\partial/\partial\phi = 0$) and the velocity gradients are smooth ($\partial \Delta v/\partial R \sim \Delta v/R$, $\partial v_R/\partial R \sim v/R$). We will also ignore accretion velocity and assume the only gas radial motion is driven by reverse drag. When we linearize Equations (4.6) and (4.7) by ignoring second-order terms (i.e., terms containing products of sub-Keplerian speeds), then write a separate scalar Equation for each (R, ϕ) vector component, we find

$$\frac{\partial v_{pR}}{\partial t} = -2\Omega_K \Delta v_{p\phi} - \left(\frac{v_{pR} - v_{gR}}{\tau_f}\right) \tag{4.10}$$

$$\frac{\partial \Delta v_{p\phi}}{\partial t} = \frac{1}{2}\Omega_K v_{pR} - \left(\frac{\Delta v_{p\phi} - \Delta v_{g\phi}}{\tau_f}\right) \tag{4.11}$$

$$\frac{\partial v_{gR}}{\partial t} = -2\Omega_K \Delta v_{g\phi} + \frac{\rho_p}{\rho}\left(\frac{v_{pR} - v_{gR}}{\tau_f}\right) - \Delta g \tag{4.12}$$

$$\frac{\partial \Delta v_{g\phi}}{\partial t} = \frac{1}{2}\Omega_K v_{gR} + \frac{\rho_p}{\rho}\left(\frac{\Delta v_{p\phi} - \Delta v_{g\phi}}{\tau_f}\right). \tag{4.13}$$

By setting all $\partial/\partial t = 0$, we can find the steady-state solution to our system of linear equations,[13] originally derived by Nakagawa et al. (1986):

[13] This author solved for the steady-state solutions using the `linsolve` function in `Python`'s `sympy` package.

$$v_{pR} = \frac{\Delta g \ \tau_f}{\left(1 + \frac{\rho_p}{\rho}\right)^2 + St^2} \tag{4.14}$$

$$\Delta v_{p\phi} = -\frac{\Delta g \left(1 + \frac{\rho_p}{\rho}\right)}{2\Omega_K \left[\left(1 + \frac{\rho_p}{\rho}\right)^2 + St^2\right]} \tag{4.15}$$

$$v_{gR} = -\frac{\Delta g \ \tau_f \ \rho_p/\rho}{\left(1 + \frac{\rho_p}{\rho}\right)^2 + St^2} \tag{4.16}$$

$$\Delta v_{g\phi} = -\frac{\Delta g \left(1 + \frac{\rho_p}{\rho} + St^2\right)}{2\Omega_K \left[\left(1 + \frac{\rho_p}{\rho}\right)^2 + St^2\right]}. \tag{4.17}$$

Figure 4.8 shows the steady-state flow at 5 au for $0.01 \leqslant \rho_p/\rho \leqslant 100$, assuming all particles have $r_{\rm eff} = 20$ cm ($St = 0.22$). In the test-particle limit, there is not enough solid mass to trigger strong reverse drag and the disk behaves as in Chapter 3, with

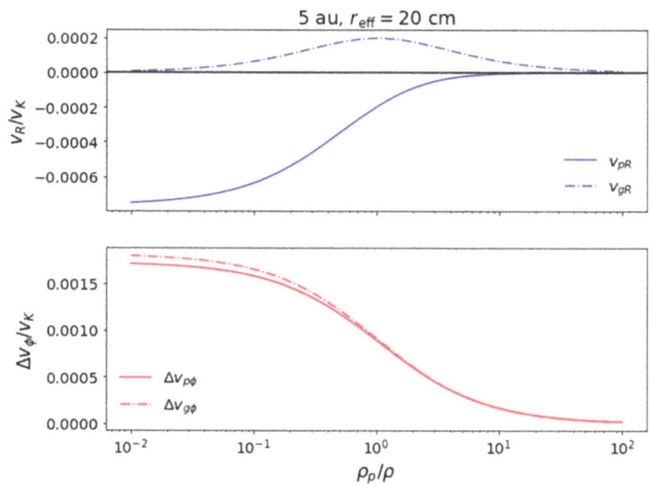

Figure 4.8. Steady-state flow at 5 au for a disk with all solids incorporated into rocks with $r_{\rm eff} = 20$ cm ($St = 0.22$). Top: radial motion. Bottom: azimuthal motion. Solid lines show pebble-fluid speeds and dashed–dotted lines show gas speeds. Both dust and gas deviate from Keplerian motion under the influence of drag and reverse drag.

inspiraling solids ($v_{pR} < 0$) and pressure-supported, sub-Keplerian gas ($\Delta v_{gR} \approx 0$, $\Delta v_{p\phi} < \Delta v_{g\phi}$). As $\rho_p/\rho \to 1$, the drag force on any individual particle decreases as reverse drag speeds up the gas ($|\Delta v_{p\phi} - \Delta v_{g\phi}| \to 0$). However, the strong solid mass-loading means the total solid-to-gas angular momentum transfer rate is high, which forces the gas to conserve angular momentum by spiraling outward ($v_{gR} > 0$). When $\rho_p/\rho \gg 1$, solids dominate the dynamics and the entire system is in Keplerian motion. Figure 4.8 is a compelling illustration of how reverse drag can alter the gas dynamics. Bai & Stone (2010) derive generalized steady-state flow solutions for a particle size distribution, not just a single size (note that their numerical simulations also predict reduced radial speeds for high-metallicity disks).

When we examine the Nakagawa et al. (1986) steady-state radial motion, we can see the potential for a two-fluid instability. In Chapter 2, we learned how relative motion between two fluid layers with similar densities—a.k.a. shear—can drive turbulence (Section 2.3.3). The top panel of Figure 4.8 shows a comparable situation: we have two substances with similar densities and strong relative motion.[14] Furthermore, Youdin & Goodman (2005) spotted the parallels between dust-rich protoplanetary disk midplanes and interpenetrating plasmas (Spitzer 1965). When two plasmas with relative motion are coupled by electric fields, the flow is unstable, meaning small perturbations will grow exponentially. Figure 4.8 shows exactly the same situation that leads to two-fluid plasma instability—coupled, interpenetrating fluids with relative motion. There is even an energy source available to drive an instability, which is the pressure work done on the outward-flowing gas (Youdin & Johansen 2007; Pinilla & Youdin 2017). In the next section, we will show that the Nakagawa et al. (1986) steady-state laminar flow is unstable, particularly when the gas and dust mass densities are similar.

4.3.3 The Streaming Instability

Suppose we have a small perturbation to the steady-state flow, so that $\vec{v}_g \to \vec{v}_g + \delta\vec{v}_g$, $\vec{v}_p \to \vec{v}_p + \delta\vec{v}_p$, and $\rho_p \to \rho_p + \delta\rho_p$. To find out if the perturbation grows, we will perform a linear stability analysis. Our disk is governed by Equations (4.6) and (4.7), plus the continuity equation for pebbles:

$$\frac{\partial \rho_p}{\partial t} + \vec{\nabla} \cdot (\rho_p \vec{v}_p) = 0. \tag{4.18}$$

While we ignored vertical motion in Section 4.3.2, we need to consider it here. However, we can simplify our treatment of the z dimension by assuming that $g_z \approx 0$ and $\partial P/\partial z \approx 0$—reasonable assumptions given that it's only possible to reach $\rho_p/\rho \sim 1$ near the midplane (e.g., Youdin & Chiang 2004). With $g_z \approx 0$, there is no steady-state vertical motion of either pebbles or gas ($v_{pz} = v_{gz} = 0$), but the

[14] The analogy between gas–dust relative motion and shear breaks down when we compare the resulting instabilities. The streaming instability segregates gas from solids, while the shear-driven instabilities mix adjacent fluid layers.

perturbed velocities $\delta\vec{v}_p$ and $\delta\vec{v}_g$ may have nonzero z components. The combination of Equations (4.6), (4.7), and (4.18) gives us seven scalar equations that describe seven properties of the perturbation: δv_{pR}, $\delta v_{p\phi}$, δv_{pz}, δv_{gR}, $\delta v_{g\phi}$, δv_{gz}, $\delta\rho_p$. We will assume the perturbation does not compress the gas; instead, the clumping pebbles merely push the gas out of their way. As in Youdin & Goodman (2005), we consider only axisymmetric perturbations.

Because planetesimal formation is a local phenomenon that operates on small scales, we can trade cylindrical coordinates for the shearing-box reference frame ($y = R_0 - R$, $x = R(\phi - \phi_0)$, where ϕ_0 is the azimuthal coordinate at the box center; see the inset box 3.2 on the shearing box). Moving into the shearing box simplifies our mathematics: we can use sinusoids instead of Bessel functions as the radial basis for our wavelike perturbations. We recast our governing equations for the shearing-box frame by including Coriolis deflection and advection by the background shear flow:[15]

$$\frac{\partial\rho_p}{\partial t} + \vec{\nabla}\cdot\left(\rho_p\vec{v}_p\right) + \frac{3}{2}\Omega_K\frac{\partial\rho_p}{\partial y}y = 0 \tag{4.19}$$

$$\frac{\partial\vec{v}_p}{\partial t} + \left(\vec{v}_p\cdot\vec{\nabla}\right)\vec{v}_p + \frac{3}{2}\Omega_K\frac{\partial\vec{v}_p}{\partial y}y = -2\Omega_K v_{px}\hat{y} - \frac{1}{2}\Omega_K v_{py}\hat{x} - \frac{\vec{v}_p - \vec{v}_g}{\tau_f} \tag{4.20}$$

$$\begin{aligned}\frac{\partial\vec{v}_g}{\partial t} + \left(\vec{v}_g\cdot\vec{\nabla}\right)\vec{v}_g + \frac{3}{2}\Omega_K\frac{\partial\vec{v}_g}{\partial y}y = &- 2\Omega_K v_{gx}\hat{y} - \frac{1}{2}\Omega_K v_{gy}\hat{x}\\ &+ \frac{\rho_p}{\rho}\left(\frac{\vec{v}_p - \vec{v}_g}{\tau_f}\right) - \frac{1}{\rho}\frac{\partial P}{\partial y}\hat{y}.\end{aligned} \tag{4.21}$$

In Equations (4.19)–(4.21), all velocities are measured relative to the Keplerian flow: $v_{px} = -\Delta v_{p\phi}$, $v_{gx} = -\Delta v_{g\phi}$, $v_{py} = -v_{pR}$, and $v_{gy} = -v_{gR}$, where $\Delta v_{p\phi}$, $\Delta v_{g\phi}$, v_{pR}, and v_{gR} are given by Equations (4.14)–(4.17). Three constraints help us simplify the equations of motion: (1) axisymmetry in both the steady-state flow and perturbations (all $\partial/\partial x = 0$), (2) no steady-state vertical motion ($\partial v/\partial z = 0$), and (3) the fact that for a given particle size, τ_f changes slowly with R (Figure 3.3), so $\tau_f(r_{\rm eff})$ is approximately constant throughout the shearing box. The background value of ρ_p/ρ, also assumed constant throughout the box, is a free parameter. As in Section 4.3.2, we linearize the governing equations by neglecting terms of order $(\delta v)^2$ and $\delta v\ \delta\rho$. As an example of how the linearization works, we present the perturbed equation governing gas radial speed $v_{gy} + \delta v_{gy}$, derived from

[15] Youdin & Johansen (2007) include local pressure gradients from gas density fluctuations in Equation (4.21), which we ignore here.

Equation (4.21), in raw form. Figuring out why each term marked $\nearrow^{0}$ can be ignored is a good exercise for the reader:

$$\frac{\partial(v_{gy}+\delta v_{gy})}{\partial t}+\cancelto{0}{(v_{gx}+\delta v_{gx})\frac{\partial(v_{gy}+\delta v_{gy})}{\partial x}}+(v_{gy}+\delta v_{gy})\frac{\partial(v_{gy}+\delta v_{gy})}{\partial y}$$
$$+\cancelto{0}{(v_{gz}+\delta v_{gz})\frac{\partial(v_{gy}+\delta v_{gy})}{\partial z}}+\frac{3}{2}\Omega_K y\frac{\partial(v_{gy}+\delta v_{gy})}{\partial y}$$
$$=-2\Omega_K(v_{gx}+\delta v_{gx})+\frac{\rho_p+\delta\rho_p}{\rho\tau_f}(v_{py}+\delta v_{py}-v_{gy}-\delta v_{gy})-\frac{1}{\rho}\frac{\partial P}{\partial y}.$$

We use the fact that the background flow is in steady state to subtract off the unperturbed terms. The linearized equation for gas momentum conservation in the radial direction is then

$$\begin{aligned}&\frac{\partial\delta v_{gy}}{\partial t}+\frac{\partial v_{gy}}{\partial y}\delta v_{gy}+v_{gy}\frac{\partial\delta v_{gy}}{\partial y}+\frac{3}{2}\Omega_K\,y\frac{\partial\delta v_{gy}}{\partial y}+2\Omega_K\delta v_{gx}\\&-\frac{1}{\rho\tau_f}\Big[(v_{py}-v_{gy})\delta\rho_p+\rho_p\delta v_{py}-\rho_p\delta v_{gy}\Big]=0.\end{aligned}\tag{4.22}$$

To get the other six governing equations for the perturbed two-fluid disk midplane, we linearize each scalar equation contained in the vector Equations (4.19)–(4.21).

Now we are ready to specify the mathematical form of our perturbations. As in Youdin & Goodman (2005), each perturbed quantity is described by a wave function:

$$\delta f=\tilde{\delta f}\exp[i(k_y y+k_z z-[\omega_r+is]t)],\tag{4.23}$$

where $\tilde{\delta f}$ is the amplitude of the perturbation, k_y is the radial wave number, k_z is the vertical wave number, ω_r is the wave frequency, and s is the growth/damping rate. The space and time derivatives of the Fourier perturbations are simple: $\partial/\partial y=ik_y$, $\partial/\partial z=ik_z$, and $\partial/\partial t=-i[\omega_r+is]$. Continuing with Equation (4.22), we find

$$\begin{aligned}&\left[ik_y\left(v_{gy}+\frac{3}{2}\Omega_K y\right)+\frac{\partial v_{gy}}{\partial y}+\frac{\rho_p}{\rho\tau_f}\right]\tilde{\delta v}_{gy}\\&-\frac{\rho_p}{\rho\tau_f}\tilde{\delta v}_{py}+2\Omega_K\tilde{\delta v}_{gx}-\left(\frac{v_{py}-v_{gy}}{\rho\tau_f}\right)\tilde{\delta\rho}_p\\&=i(\omega_r+is)\tilde{\delta v}_{gy}.\end{aligned}\tag{4.24}$$

Because our goal is to find combinations of parameters ρ_p/ρ, τ_f and wavenumbers k_y, k_z that will lead to exponentially growing perturbations, we need to collect our governing equations into an eigenvalue problem of the form

$$\delta\mathbf{f} = \begin{bmatrix} \delta\tilde{v}_{px} \\ \delta\tilde{v}_{py} \\ \delta\tilde{v}_{pz} \\ \delta\tilde{v}_{gx} \\ \delta\tilde{v}_{gy} \\ \delta\tilde{v}_{gz} \\ \delta\tilde{\rho}_p \end{bmatrix} \tag{4.25}$$

$$\mathbf{A}\ \delta\mathbf{f} = i[\omega_r + is]\ \delta\mathbf{f}.$$

If s is positive, $\partial\delta\mathbf{f}/\partial t$ is also positive and the perturbations grow. The entries in matrix **A** include contributions from the wavenumbers k_y, k_z, the free parameters ρ_p/ρ, τ_f, the steady-state gas/dust velocities from Equations (4.14)–(4.17), and their derivatives $\partial/\partial y$.

Figure 4.9 shows the streaming instability growth rates in our model disk at 5 au, given wavenumbers set by the pressure gradient: k_y, $k_z = g/R|\Delta g|$ (similar to K_y, $K_z = 1$ as defined by Youdin & Johansen 2007). We do not consider rocks with $r_{\rm eff} > 30$cm because Equations (4.19)–(4.21) are not valid in the Stokes drag regime, where τ_f depends on $\vec{v}_p$ (see Section 3.2.1). Growing modes exist for all combinations of ρ_p/ρ and $r_{\rm eff}$, but the instability grows fastest at $\rho_p/\rho \sim 1$ and at $r_{\rm eff} > 10$ cm. The transition between the Epstein and Stokes drag laws occurs at $r_{\rm eff} = 20$ cm at 5 au in our model disk, so the results for the largest particles in

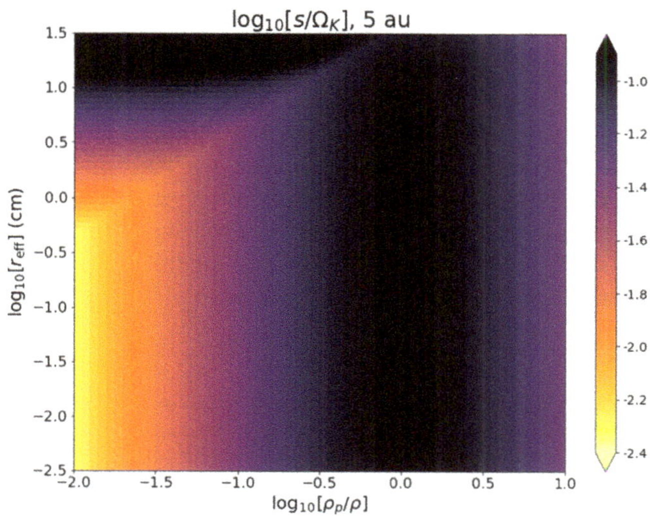

Figure 4.9. Streaming instability growth rates at $R = 5$ au for perturbation wavenumbers k_y, $k_z = g/R|\Delta g|$. The instability growth timescale is <3 orbits for $r_{\rm eff} \gtrsim 10$ cm and $\rho_p/\rho \sim 1$. The results for the largest particles in Figure 4.9 are approximate, as particles with $r_{\rm eff} > 20$ cm are in the Stokes drag regime.

Figure 4.9 are approximate; see Stoyanovskaya et al. (2020) for a generalized treatment that includes particles in the Stokes regime. If $S/G \sim 0.01$ (which can be globally true in a disk young enough not to have lost much mass to radial drift, or locally true in a more evolved disk with a dust trap) and all solid mass is incorporated into rocks with $r_{\rm eff} = 10$ cm, then settling–stirring equilibrium[16] gives $\rho_p/\rho = 0.33$ for $\alpha = 10^{-4}$, $(R, z)=$ (5 au, 0) and near-maximum growth rate is achieved. However, we know from Chapter 3 that it's difficult to grow solids beyond ~1 cm via direct sticking. For $r_{\rm eff} = 1$cm and $\alpha = 10^{-4}$, $\rho_p/\rho = 0.11$ at $(R, z) = (5$ au, $0)$. The instability can still grow, but the growth timescale is a factor of three longer than one would get if the solid mass were concentrated in decimeter-size rocks (see also Gole et al. 2020; Umurhan et al. 2020). For more information on the streaming instability, such as examinations of the eigenvectors from Equation (4.25) and the gas density perturbations, which are not included in this analysis, see Youdin & Goodman (2005) and Youdin & Johansen (2007).

Squire & Hopkins (2018) break the parameter space illustrated in Figure 4.9 into "low-metallicity" and "high-metallicity" parts, with the boundary at $\rho_p/\rho = 1$. They argue that at $\rho_p/\rho < 1$, the streaming instability is one of a family of related resonant drag instabilities (RDIs). In an RDI, a propagating pressure wave has a phase velocity equal to the dust drift velocity, and the resulting resonance couples the dust to the pressure wave. The dust drifts into the gas pressure maxima, the reverse drag on the gas increases the gas pressure, and a positive feedback loop forms. RDIs may be triggered by any type of pressure wave, such as an acoustic wave or magnetosonic wave (Squire & Hopkins 2018; Hopkins & Squire 2018). However, the fastest-growing RDI is the settling instability, in which epicyclic gas motion resonates with the particles' vertical motion.[17] The low-metallicity streaming instability, which has slower growth rates, is triggered when epicyclic gas motion resonates with $-u$ and/or w, the radial and azimuthal dust drift speeds in the midplane (Section 3.2.2). Squire & Hopkins (2018) argue that the high-metallicity streaming instability ($\rho_p/\rho \gtrsim 1$) is not an RDI; instead, it is more akin to a destabilized harmonic oscillator. They also posit that dust ring formation in the disk midplane, as seen in the DSHARP survey (Andrews et al. 2018), is an inevitable result of the settling instability and may provide the requisite dust density enhancement to trigger the Youdin & Goodman (2005)-type streaming instability we explore here. However, numerical simulations by Krapp et al. (2020) show that the settling instability is viable only when turbulence is unrealistically weak ($\alpha \lesssim 10^{-6}$). Krapp et al. also found that settling

[16] Note that there is some physical inconsistency between assuming settling–stirring equilibrium and computing streaming instability growth rates using Equation (4.25), as the Nakagawa et al. (1986) steady-state drift equations that describe the background flow are derived for laminar disks. Furthermore, a settled dust layer may self-destruct by triggering the Kelvin–Helmholtz instability (Barranco 2009), so the analytical expressions that predict the scale height of the dust layer (Equation (3.51)) and the midplane number density (Equation (2.14), with $\rho_0 \rightarrow \rho_{p,0}$ and $\Sigma \rightarrow \Sigma_p$) might not apply.

[17] A similar phenomenon, in which particles would spontaneously "unmix" from a homogeneous gas–dust blend and form swarms with 10 times the background solid density, was found in numerical simulations by Lambrechts et al. (2016).

proceeds faster than instability-driven particle clumping, in which case the settling instability is unlikely to drive planetesimal formation on its own.

A non-RDI alternative to the classical streaming instability described in this section comes from Lin & Youdin (2015) and Lin (2021), who document a two-fluid instability driven by vertical shear $R\partial\Omega_K/\partial z$ instead of radial shear $R\partial\Omega_K/\partial R$. The mathematical setup is similar to Equations (4.19)–(4.21) except that the disk is stratified, so that $\partial P/\partial z$ and g_z are nonzero. The vertical shear streaming instability growth rates are faster than those of the classical streaming instability, though the growing modes are confined to shorter radial length scales. Vertical buoyancy and turbulent viscosity are efficient stabilizers, and previous simulations by Ishitsu et al. (2009) suggest that the growing modes may lead to turbulence instead of clumping if the particles are pebble size ($St \ll 1$). Further numerical work is needed to determine the role of vertical streaming instabilities in planetesimal formation.

4.3.4 Mass Scale

For now, let us assume that the streaming instability grows rapidly enough to produce gravitationally bound pebble clouds that collapse into planetesimals.[18] (We will review the numerical simulations that test this assumption in Section 4.4.) How big will the planetesimals be? To find out, we can use similar mathematics to Section 2.4.7 to find a planetesimal Jeans mass. Here we will examine a spherical perturbation embedded in a three-dimensional disk instead of a circular perturbation in an infinitely thin disk because bound pebble clouds are much smaller than the scale height of the midplane pebble layer. Our reasoning loosely follows that of Klahr & Schreiber (2020), who derived density- and diffusion-based planetesimal collapse criteria.

Consider a pebble cloud with radius ℓ_c and mass $M_c = (4/3)\pi\ell_c^3\rho_p$. The two processes that could stabilize the cloud are Keplerian shear and diffusion. The cloud will collapse if its freefall time $\tau_{\rm ff}$ is smaller than both the shear time τ_s and the time it takes a pebble to random-walk its way out of the cloud τ_d. The freefall time is the time it takes a particle at the outer edge of the cloud to fall a distance ℓ_c:

$$\ell_c = \frac{1}{2}\frac{GM_c}{\ell_c^2}\tau_{\rm ff}^2 \tag{4.26}$$

$$\tau_{\rm ff} = \sqrt{\frac{3}{2\pi G\rho_p}}. \tag{4.27}$$

Diffusion, the particle analog of gas pressure, is best at stabilizing small-scale perturbations. To find the minimum size of a pebble cloud in which self-gravity defeats diffusion—the pebble-cloud Jeans length—we set $\tau_{\rm ff} = \tau_d = \ell^2/D_p$ (see the

[18] Kuiper (1952, p. 362) anticipated the possibility of planet(esimal) formation through gravitational collapse of critical density material, writing "When the contraction had proceeded to the point where the density locally exceeded the critical Roche density, gravitationally stable clouds could form which enlarged into the protoplanets." Kuiper was envisioning gas clouds, but his ideas apply equally well to pebble clouds.

discussion of particle diffusion coefficients in Section 3.2.4). Recall that our mathematical description of the streaming instability only holds for particles in the Epstein drag regime, for which $D_p \approx \nu$. Then

$$\ell_c \geqslant \left(\frac{3\nu^2}{2\pi G \rho_p} \right)^{1/4}. \tag{4.28}$$

Not all pebble overdensities that meet the Jeans length criterion in Equation (4.28) are self-gravitating, however. To collapse, a cloud must exceed the critical pebble mass density $\rho_{p,c}$, which we find by incorporating the shear criterion. For a spherical pebble cloud, the critical density is

$$\rho_{p,c} = \frac{9\Omega_K^2}{4\pi G} \tag{4.29}$$

(Roche 1847; Goldreich & Ward 1973). Substituting Equation (4.29) into Equation (4.28), we find that for pebble concentrations exactly at the critical density, the Jeans length criterion simplifies to

$$\ell_c \geqslant \left(\frac{2}{3} \right)^{1/4} \sqrt{\alpha}\, H. \tag{4.30}$$

If we assume each gravitationally bound pebble cloud collapses into a single planetesimal without fragmenting, a critical-density cloud yields a planetesimal with mass $M_{pl} = (4/3)\pi \ell_c^3 \rho_{p,c}$, or

$$M_{pl} \simeq 2.2\, \alpha^{3/2} \frac{c_s^3}{G\Omega_K}. \tag{4.31}$$

The quantity c_s^3/G is the characteristic accretion rate of any object forming via the collapse of an isothermal cloud, derived in a star formation context by Shu (1977). Equation (4.31) states that the planetesimal mass is proportional to both the cloud-collapse accretion rate and the orbital timescale, and that the proportionality constant is determined by the turbulent efficiency α (Gole et al. 2020). The top panel of Figure 4.10 shows the expected planetesimal mass from Equation (4.31) for our model disk with $\alpha = 10^{-4}$ (blue solid line) and $\alpha = 10^{-3}$ (red dashed line), while the bottom panel of Figure 4.10 shows the expected planetesimal radius R_{pl} assuming the pebble material density is the same as the bulk density of Ceres, $\rho_m = 2$ g cm^{-3}.[19] If turbulence is weak, as in a magnetically dead zone (see Section 2.4.6), the gravitational collapse of planetesimal clouds produces planetesimal sizes consistent with both the meteoritic record (Section 4.2) and the asteroid belt size distribution (Morbidelli et al. 2009; Delbo' et al. 2017). Again, we emphasize that

[19] While Earth is massive enough to be gravitationally compressed, Ceres' mass is low enough for compression to be negligible. We expect planetesimals like Ceres to have bulk density similar to the material density of the pebbles from which they formed.

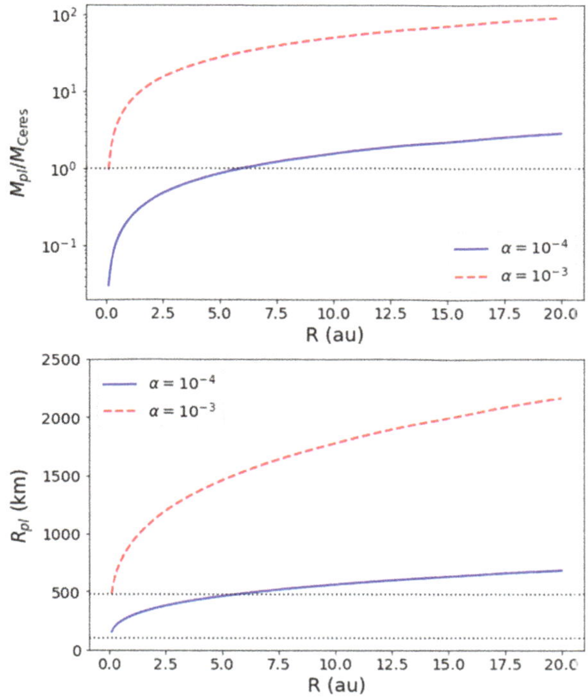

Figure 4.10. Top: planetesimal mass predicted by Equation (4.31), in Ceres-mass units, for $\alpha = 10^{-3}$ (red dashed line) and $\alpha = 10^{-4}$ (blue solid line). Bottom: planetesimal radius given $\rho_m = 2$ g cm^{-3}. Black dotted lines mark Ceres' radius (476 km) and $R_{pl} = 100$ km, the oft-quoted typical size of the first planetesimals in the inner solar system (Morbidelli et al. 2009; Delbo' et al. 2017; e.g.,).

Equation (4.31) was derived under the assumptions that the collapsing pebble cloud neither fragments nor accretes additional solids as it collapses. Section 4.4 discusses numerical simulations that test both of those assumptions.

How efficiently does the streaming instability have to concentrate solids in order to form bound pebble clouds at the midplane? If we assume all particles have the same size r_{eff}, we can combine Equation (2.14) with Equation (3.51) to find $\rho_p(r_{\text{eff}}, z = 0)$ given by settling–stirring equilibrium. Self-gravitating pebble clouds can only form if we boost the equilibrium pebble mass density by a concentration factor of

$$\mathcal{C}(R, r_{\text{eff}}) = \rho_{p,c}(R)/\rho_{p,0}(R, r_{\text{eff}}). \tag{4.32}$$

Figure 4.11 shows $\mathcal{C}(R, r_{\text{eff}})$ for our model disk with $\alpha = 10^{-4}$. Focusing on cm-size pebbles, which are the constituents of rubble-pile asteroids and comets (Section 3.4), we need concentration factors of $\mathcal{C} = 200$–900 between 5 and 20 au. For decimeter-size rocks, higher equilibrium values of $\rho_p(z = 0)$ lead to lower required concentration factors. We will examine the pebble concentration factors predicted by numerical simulations in Section 4.4.2.

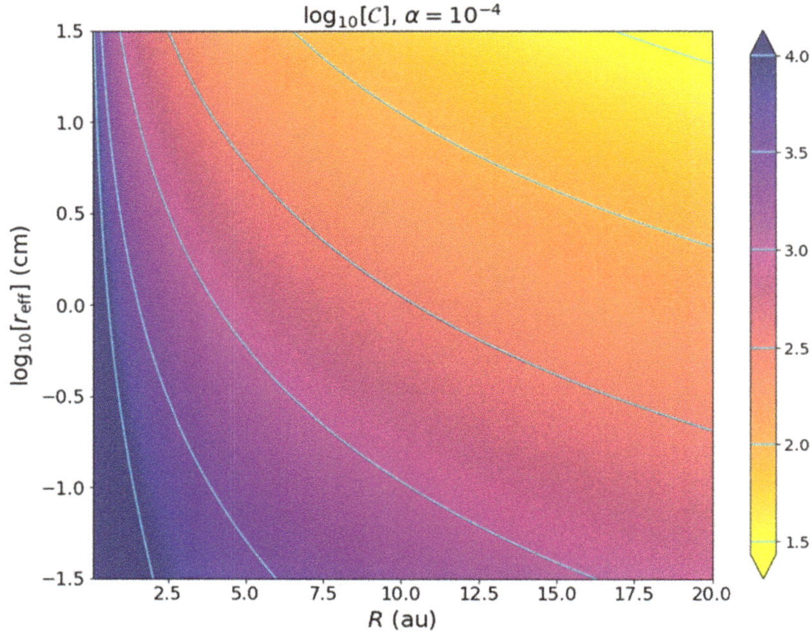

Figure 4.11. Midplane particle concentration factor $\mathcal{C}$ required for pebble-cloud collapse in our model disk with $\alpha = 10^{-4}$. Throughout the giant-planet-forming region, centimeter-size pebbles must be concentrated over and above the density of the settled midplane layer by factors of 200–900 in order to collapse into planetesimals.

Pebble-cloud collapse is a good candidate for delivering the "born early, born big" planetesimal formation scenario indicated by the meteoritic timeline (Section 4.2) as long as some particle concentration mechanism can deliver $\mathcal{C} \sim 500$. In the next section, we will review numerical models of the streaming instability.

4.4 Numerical Models of Planetesimal-forming Instabilities

Though we have identified a possible pathway from pebbles to Ceres, we haven't solved the planetesimal formation problem yet. Equation (4.25) tells us the conditions that trigger the streaming instability, but it doesn't tell us the fate of the exponentially growing perturbations. Does the instability saturate, or does it concentrate solids efficiently enough to produce self-gravitating pebble clouds? Do the pebble clouds fragment, or does each one collapse into a single planetesimal, as presumed in Section 4.3.4? To answer these questions, we review the literature describing numerical simulations of planetesimal formation. Sections 4.4.1–4.4.3 focus on the streaming instability and related RDIs, while Section 4.4.4 presents numerical results exploring other types of instabilities.

4.4.1 Foundational Model: Boulders, Pressure Maxima, and the Streaming Instability

Based on laboratory experiments and observational studies of meteorites and comets, we have repeatedly emphasized that pebbles are the building blocks of planetesimals (Sections 3.3.3, 3.4, & 4.2.1). But in the minimum-mass solar nebula (mmsn; box 2.2), instead of the MAXSN (Nixon et al. 2018) used in this work, the gas mean free path is large enough for meter-size boulders to be in the Epstein drag regime at $R \gtrsim 1$au. Given that the streaming instability growth rate increases with particle size (Figure 4.9), we expect planetesimal formation models to favor the largest particles to which the mathematics in Section 4.3.3 apply. The foundational numerical analysis by Johansen et al. (2007; hereafter J07), which uses a disk with $2\Sigma_{\rm mmsn}$ at 5 au, therefore focuses on boulders. Another crucial difference between the J07 models and the analytical framework presented in Section 4.3 is that the latter is formulated for laminar disks. J07 present a more complete model by specifying turbulence-induced local pressure maxima as the "prestreaming" particle concentration mechanism that drives the streaming instability (Johansen et al. 2006) while noting that large vortices and spiral arms could also work. In fact, all of the radial drift-halting structures mentioned in Section 3.2.3 are good candidates for triggering the streaming instability, as is vertical settling in massive disks (Section 4.3.3).

J07 generate MRI turbulence with $\alpha \approx 10^{-3}$ using the ideal-MHD module of the `PENCIL` code (Brandenburg & Dobler 2002; Section 2.4.2). With $S/G = 0.01$ and all solids incorporated into boulders with $St = 1$ ($r_{\rm eff} = 1$ m at 5 au), vertical gravity g_z creates a midplane sedimentary layer with time-averaged $\rho_p/\rho = 0.5$. MRI turbulence then seeds the streaming instability by forming transient pressure maxima that trap boulders, yielding $\rho_p/\rho > 10$ (see Figure 1(d) of J07). Finally, the streaming instability concentrates the boulders enough to form gravitationally bound clusters with $\rho_p/\rho > 300$. J07 find that even when the boulders have a Stokes number distribution $0.25 < St < 1$ ($15 < r_{\rm eff} < 60$ cm at 5 au), all boulder sizes can still find their way into the same bound cluster—an important result, given that particles with different sizes reach different drift speeds (Figure 3.6) and therefore have a nonnegligible relative motion that could handicap cluster formation.

Planetesimal formation is not done yet, however—the velocity dispersion of rocks within the bound clusters is too high for gravitational collapse without some kind of damping. As suggested by Goldreich & Ward (1973), inelastic collisions are effective at damping out the solids' relative velocities: J07 show that overdensities created by the streaming instability in the 2× mmsn disk with $S/G = 0.01$ evolve into Ceres-mass collapsing clusters in less than 10 orbits after the self-gravity and collision modules are turned on (Figure 4.12, Figure 3 of J07; see also Wahlberg Jansson & Johansen 2014). However, collisional cooling is not required for planetesimal formation. When the boulders' drag-induced loss of kinetic energy is taken into account, gravitational collapse proceeds even without collisions as long as the total (gas+solid) surface density is increased to $3\Sigma_{\rm mmsn}$. The resulting planetesimal may be more massive than the initial overdensity formed by the streaming instability, as the boulder clump grows by gravitationally accreting nearby solids. In fact, Johansen

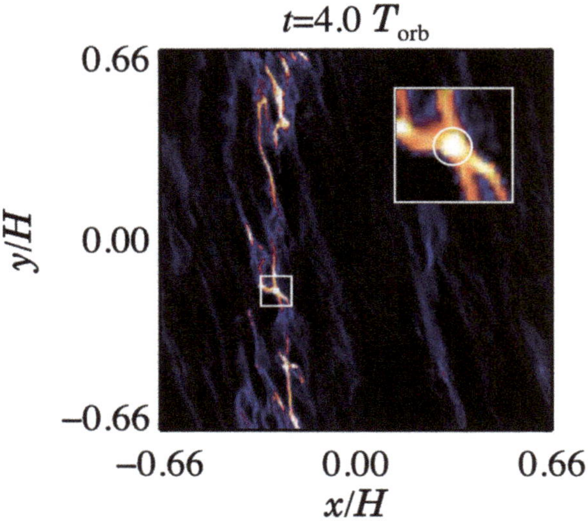

Figure 4.12. A subpanel of Figure 2 from J07 showing a gravitationally bound boulder cluster embedded in unstable, interpenetrating solid-gas flows. Drag-coupled gas and solids evolve in a shearing box for 20 orbits before self-gravity and collisional cooling are turned on. Four orbits after the onset of self-gravity, bound clusters of boulders have formed in overdense perturbations generated by the streaming instability. The inset shows the most massive boulder clump, with the gravitationally bound region circled in white. At the end of the simulation, the clump has reached 3.5 Ceres masses. Image: Johansen et al. (2007), with permission of Springer.

et al. (2015) posit that only a fraction of the solids in the early solar nebula were boulder-size. Their model suggests that after the streaming instability concentrated the boulders into "seed" planetesimals, the planetesimals went on to accrete surface layers of chondrules (Section 4.2.3), forming the parent bodies of the ordinary and carbonaceous chondrites.

There are two ways in which the original J07 model may not be consistent with observational or experimental results. First, while successful J07-style planetesimal formation requires some turbulence to be present *after* boulders have formed, it's not clear that pebbles could settle to the midplane and grow into boulders in a disk with $\alpha \approx 10^{-3}$ (Birnstiel et al. 2012; Drażkowska & Dullemond 2014; Sengupta et al. 2019). Doing so would require an efficient target-to-projectile mass transfer in collisions with $m_1 \gg m_2$ (Teiser & Wurm 2009; Deckers & Teiser 2014; Section 3.3.2). To complicate the picture, the streaming instability itself can drive turbulence that increases collision speeds and stirs particles out of the solid-rich midplane (Johansen & Youdin 2007; Youdin & Johansen 2007; Bai & Stone 2010). Second, while giant planets have a strong tendency to orbit metal-rich host stars (Gonzalez 1997; Fischer & Valenti 2005), the planet–metallicity correlation for Neptune-size and smaller planets orbiting sunlike stars is weak to nonexistent (Buchhave et al. 2012; Winn & Fabrycky, 2015; Lu et al. 2020, though low-mass hosts of super-Earths are generally metal rich). Debris disks—circumstellar dust rings produced by colliding planetesimals—are similarly indiscriminate about host-star metallicity, indicating that planetesimal formation is a robust process for star systems with [Fe/H] >-0.5 (Beichman et al. 2006; Maldonado et al. 2012, 2015). But when the solids are bound up in decimeter-size "cobbles" rather than meter-size boulders, triggering the streaming instability in mmsn-type disks requires supersolar metallicity, $S/G \gtrsim 0.02$ (Johansen et al. 2009; Bai & Stone 2010; Johansen et al. 2014; Drażkowska & Dullemond 2014). While Carrera et al. (2017) suggest gas photoevaporation as a possible way to increase S/G, it takes millions of years for photoevaporation to begin to substantially modify the gas surface density (Alexander et al. 2006; Wise & Dodson-Robinson 2018), though a magnetically driven disk wind (e.g., Bai & Stone 2013) might be able to deplete the gas more quickly. The photoevaporation-triggered planetesimal formation scenario is therefore inconsistent with the meteoritic evidence that planetesimals formed within the first 0.3 Myr of solar nebula evolution (Section 4.2.2).

A simple solution to the metallicity problem is to model a more massive disk, as indicated by several lines of observational evidence (Section 2.3.1, inset box 2.2). While the higher gas density in our model disk leads to a lower St and therefore a slower instability growth rate at any particular (R, $r_{\rm eff}$) combination than in the mmsn, many numerical and analytical models have shown that the streaming instability nevertheless occurs even when particles have $St \ll 1$ (Section 4.4.2; below). Slower growth rates in massive disks are compensated by higher background solid densities ρ_p, which reduce the value of $\mathcal{C}$ needed for self-gravitating pebble clouds to form. Indeed, Gerbig et al. (2020) show that planetesimals can form in the weakly turbulent dead zones of massive disks with solar metallicity. We conclude that the solar-metallicity (or even mildly subsolar) MAXSN is a viable environment for planetesimal formation by the streaming instability.

4.4.2 Chondrules, Pebbles, and Mixed-size Particles

Shi & Chiang (2013) emphasized that a universally applicable planetesimal formation model cannot rely on the availability of boulders, given the difficulty of forming meter-size particles in pairwise collisions (Section 3.3.3). Here we describe streaming instability simulations that treat realistic particle sizes and/or size distributions ($r_{\rm eff,max} \sim 1$ mm–1 cm). Drażkowska & Dullemond (2014) linked the collision and streaming regimes by adding a semianalytical streaming instability model based on the numerical results of Bai & Stone (2010) to their Monte Carlo dust coagulation code. Their mmsn-based disk was capable of producing planetesimals beyond the H_2O-ice line at 3 au, but not in the ice-poor inner disk (recall from Section 3.3.3 that icy grains are more resistant to fragmentation than silicates and can reach larger sizes; Okuzumi et al. 2012; Kataoka et al. 2013; Aumatell & Wurm 2014). Their conclusions were echoed by Krijt et al. (2016), Drażkowska & Alibert (2017), and Schoonenberg & Ormel (2017). Yet the asteroid belt and the meteoritic record are replete with planetesimal relics composed primarily of silicates. Ida & Guillot (2016) proposed a solution: ice sublimates from pebbles that drift inside the H_2O-ice line, leaving behind the remnant silicate grains on which the ice mantles originally deposited.[20] Being much smaller with their ice mantles removed, the remnant silicates slow down their radial drift, forming a dust-rich "traffic jam" just inside the H_2O-ice line in which the streaming instability can create rocky planetesimals. The Ida & Guillot (2016) model could explain why Earth, Mars, and Vesta appear to come from the same isotopic reservoir (Section 4.2.8 Sarafian et al. 2014)—their constituent planetesimals would all have formed in the same dust annulus (which, according to Hansen 2009 and Walsh et al. 2011, would subsequently have to be truncated at 1 au by Jupiter's migration in order to explain Mars' small mass).

The preponderance of chondrules in meteorites sourced from the inner solar nebula (Section 4.2.3) prompted Carrera et al. (2015) to investigate whether the streaming instability could turn a settled layer of millimeter-size particles into planetesimals. The top panel of Figure 4.13 summarizes their results for the mmsn at 2.5 au: if $S/G \gtrsim 0.065$, chondrule-size particles ($St \sim 0.003$) can concentrate themselves into gravitationally bound clumps. Of course, $S/G = 0.065$ is quite high given that the most metal-rich stars in the Galaxy only have about three times the heavy-element abundance as the Sun (e.g., Brewer et al. 2016). Assuming protostars and their accretion disks have the same atomic inventory, even a disk with [Fe/H] = 0.5 would need some kind of dust trap (Section 3.2.3) in order to locally enhance S/G and create the conditions for a chondrule-driven streaming instability. However, Yang et al. (2017) updated the Carrera et al. (2015) study with a new algorithm that facilitated longer simulation times and higher resolution (Yang & Johansen 2016). The bottom panel of Figure 4.13 shows their findings. Millimeter-size particles can trigger the streaming instability at 2.5 au in the mmsn given $S/G \gtrsim 0.04$, and more importantly from a giant-planet perspective, low-porosity centimeter-size particles with $St \sim 0.01$ can form collapsing clumps even at $S/G \lesssim 0.02$. Such a solid/gas ratio is easily reached even in solar-composition disks given that ice freezeout (including water, ammonia, CO, and CH_4) can double the canonical value of $S/G = 0.01$ (e.g., Dodson-Robinson et al. 2009) without particle trapping. The Yang et al. (2017) results are promising; we eagerly await simulations of pebble streaming in disks more massive than the mmsn, such as the Nixon et al. (2018) MAXSN adopted in this work and the Most Appealing Solar Nebula model (MAXN) of Lenz et al. (2020).

Nesvorný et al. (2019) used the inclination distribution of binary KBOs to test whether the streaming instability was responsible for seeding dwarf-planet formation in the outer solar nebula. (For more on the Kuiper Belt, box 4.2.) Their simulation parameter space covered $0.3 < St < 2$—equivalent to subcentimeter particles at 45 au in the mmsn if gas density has been reduced by photoevaporation—and $0.02 < S/G < 0.1$. (In our model disk, a 1 cm particle in the midplane has $St = 0.1$ at 45 au.) Nesvorný et al. (2019) identified the emerging gravitationally bound pebble clumps using a tree code, computed the total angular momentum J of each clump, and calculated the ratio J/J_c, where J_c is the angular momentum of a

[20] Water ice does not nucleate its own solid grains in the low vapor-pressure conditions found in molecular clouds or protoplanetary disks; rather, monolayers of ice build on silicate and metallic seeds. See Fraser et al. (2001) for more information on the conditions required for the desorption of icy grain mantles.

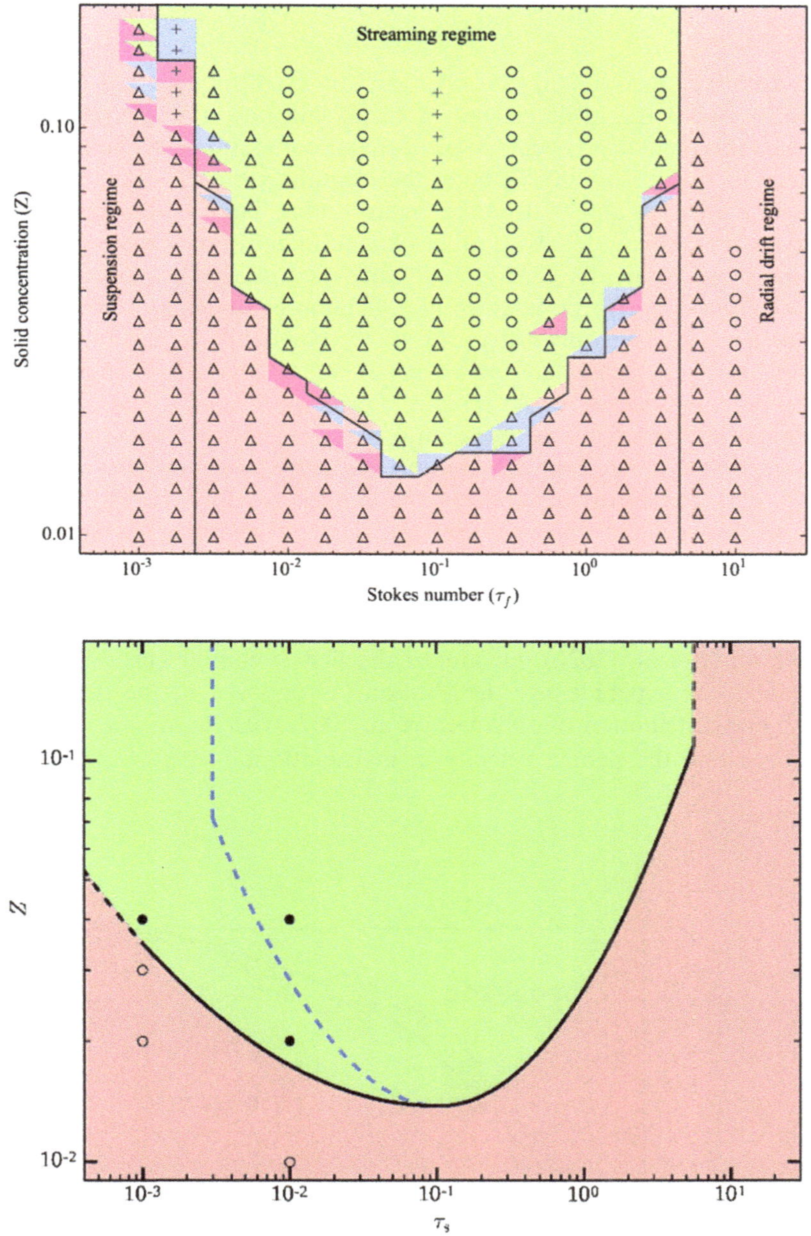

Figure 4.13. Top: summary of the simulations by Carrera et al. (2015, their Figure 8; reproduced with permission © ESO) showing the parts of the *S*/*G* (*y*-axis) and *St* (*x*-axis) parameter space at which streaming instability is possible in the minimum-mass solar nebula at 2.5 au. Particles in the suspension regime are tightly coupled to the gas ($St \ll 1$), while particles in the radial drift regime are weakly coupled ($St \gtrsim 1$). Particle concentration by the streaming instability is very likely in the green shaded region, but unlikely in the red shaded region. Blue and magenta areas represent combinations of *S*/*G* and *St* that led to mixed simulation outcomes, depending on the randomly chosen initial particle positions. Bottom: results from updated simulations by Yang et al. (2017; reproduced with permission © ESO) push the critical solid-to-gas ratio required to trigger the streaming instability down to $S/G \sim 0.04$ for millimeter-size particles in the mmsn. The solid black line shows the revised separatrix between unstable (green) and stable (red) regimes, while the blue dashed line shows the original separatrix from Carrera et al. (2015).

critically rotating Jacobi ellipsoid. Almost all clumps were supercritical, with $J/J_c \gg 1$, so would fragment instead of collapsing into a single, isolated planetesimal. Such fragmentation occurs in protostellar cores and yields binary or multiple stars (e.g., Goodwin et al. 2007); here, the fragmentation produces binary planetesimals (Robinson et al. 2020). Like binary stars, binary planetesimals can be disrupted by close encounters with other planetesimals, and widely separated "soft" binaries are more vulnerable than close "hard" binaries (Parker & Kavelaars 2010). The recent discovery of a population of soft binaries that may have originated at 38 au and been pushed outward to 45 au by Neptune's migration suggests that nearly 100% of the planetesimals grown at ~38 au must have formed in binary pairs; otherwise, no soft binaries would remain (Fraser et al. 2017). The Nesvorný et al. (2019) simulations predict a pebble-clump obliquity distribution that is consistent with the observed inclination distribution of Kuiper Belt binaries (Figure 4.14), indicating that streaming instability models can produce realistic planetesimal populations. The authors' choice to use a gas-depleted mmsn as the initial conditions makes sense if the Kuiper Belt planetesimals formed well after the gas giants, when photoevaporation and/or MHD winds had already carried away much of the gas. For more on the Kuiper Belt, see box 4.2.

Paardekooper et al. (2020) introduced the first streaming instability models to take into account a pebble size distribution (as opposed to the decimeter–meter "boulder" size distribution of Johansen et al. 2007) with $dn_p/dr_{\rm eff} \propto r_{\rm eff}^{-3.5}$ (Equation (3.68)). Replacing the pebble velocity $\vec{v}_p$ in Equations (4.19)–(4.21) with average velocity

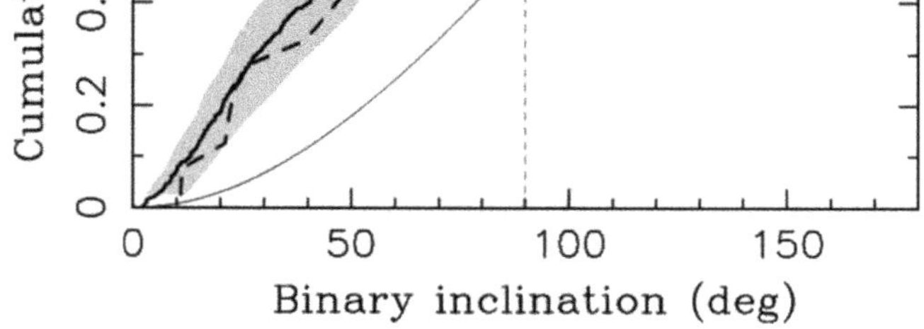

Figure 4.14. Cumulative obliquity distribution of gravitationally bound pebble clumps (solid line) and 68% confidence interval (gray shading) from the simulations of Nesvorný et al. (2019), with permission of Springer. Their clump obliquity distribution is a good match for the observed Kuiper Belt binary inclination distribution (dashed line). The thin solid line shows a cumulative inclination distribution of randomly oriented orbits.

$$\langle \vec{v}_p \rangle = \frac{\int n_p(r_{\rm eff})\vec{v}_p(r_{\rm eff})dr_{\rm eff}}{\int n_p(r_{\rm eff})dr_{\rm eff}}, \tag{4.33}$$

and assuming that all pebbles (1) move at terminal speed (Garaud & Lin 2004; Jacquet et al. 2011) and (2) are tightly coupled to the gas (i.e., the largest particle in the size distribution has $St \ll 1$; our model disk meets this criterion at $R < 20$au as long as $r_{\rm eff,max} \lesssim 1$ cm), they used linear stability analysis to find the wavenumbers and growth rates of exponentially growing modes. They found that for $\rho_p/\rho \ll 1$, the polydisperse streaming instability operates at wavelengths longer by a factor of $1/\bar{St}$ (where $\bar{St}$ is an average Stokes number) than the Youdin & Goodman (2005) predictions. When $\rho_p/\rho \gtrsim 1$, the polydisperse streaming instability behaves much like the classical model presented in Section 4.3.3, producing modes that grow on an orbital timescale. The Paardekooper et al. (2020) results show that the pebble size need not be uniform for the streaming instability to produce pebble-pile planetesimals.

Recall that our model disk requires pebble concentration factors $\mathcal{C} = 200$–900 in order to generate self-gravitating pebble clumps (Section 4.3.4). Can pebble streaming models deliver such efficient concentration? The answer depends on both S/G and $v_K - v_\phi$, which is the maximum possible pebble drift speed (see Section 3.2.2); our model disk has $S/G = 0.01$ and $v_K - v_\phi = 0.05c_s$ at 5 au. Though some relative radial motion between pebbles and gas is required in order to trigger the streaming instability, high values of $v_K - v_\phi$ inhibit particle concentration (see Figure 5 of Drażkowska & Dullemond 2014).The mmsn-based models of Bai & Stone (2010), which include a particle radius distribution (albeit one that tops out at boulder sizes), find a maximum $\mathcal{C} \sim 1200$ with $S/G = 0.03$, $St_{\rm min} = 10^{-3}$, and $v_K - v_\phi = 0.05c_s$,[21] but inefficient clumping with $\mathcal{C} \sim 10$ for canonical $S/G = 0.01$. Their results can be viewed with cautious optimism given that (1) ice freezeout in a solar-composition disk naturally yields $S/G \approx 0.02$ (e.g., Dodson-Robinson et al. 2009), and (2) the high vapor pressure just outside the H_2O-ice line leads to efficient ice deposition, possibly forming decimeter-size snowballs (Ros & Johansen 2013). Johansen et al. (2012), Yang & Johansen (2014), and Johansen et al. (2015) also present simulations in which $\mathcal{C} > 1000$ in some overdense filaments. Carrera et al. (2015) and Yang et al. (2017) do not report maximum values of $\mathcal{C}$ in their simulations.

4.4.3 Size and Mass Distribution

So far, numerical simulations have shown us that the streaming instability can produce planetesimals from a range of raw materials—boulders, pebbles, chondrules, or a mixture of all three. Now we want to examine the characteristics of the resulting planetesimals. Our order-of-magnitude mass estimate from Section 4.3.4 is built on the assumptions that (1) all pebble clouds are exactly at critical density $\rho_{p,c}$,

[21] Our estimate of $\mathcal{C}$ is based on their figures 4 and 5.

(2) no pebble cloud accretes extra solids during collapse, and (3) none of the collapsing clouds fragment into multiple planetesimals. Reality is undoubtedly messier—for example, we already know from Section 4.4.2 that fragmentation is likely. What do numerical simulations tell us about the planetesimal size and mass distribution?

Simon et al. (2016) ran a suite of two-fluid simulations using the `Athena` code (Gardiner & Stone 2005, 2008; Stone et al. 2008), with a fast-Fourier-transform-based Poisson solver added to treat pebble-cloud self-gravity. They computed planetesimal size distributions in shearing boxes of size L_x, L_y, $L_z = 0.2H$, with varying grid resolution and gravity parameter $\tilde{G}$:

$$\tilde{G} = \frac{4\pi G\rho_0}{\Omega_K^2}. \tag{4.34}$$

All simulations had $St = 0.3$, $v_K - v_\phi = 0.05c_s$, and $h_p/H = 0.02$. When reporting results from their fiducial simulation set, they translated the dimensionless code units into physical units by specifying an mmsn disk model with $R = 3$ au and $\tilde{G} = 0.05$. Different values of $\tilde{G}$ correspond to different locations in the disk; their parameter study covered $0.02 \leqslant \tilde{G} \leqslant 0.1$. While planetesimals with $R_{pl} \lesssim 80$ km ($M_{pl} \lesssim 0.004$ Ceres masses) only emerged in the highest-resolution simulation (512^3 grid zones), the maximum planetesimal radius of ~300 km was consistent even among simulations with varying resolution. (For comparison, Vesta and Ceres have effective radii of 250 km and 476 km, respectively.) The smallest planetesimals formed had $R_{pl} \simeq 50$ km. In a similar set of simulations using the `Pencil` code, identical shearing-box dimensions, and a 512^3 grid, Johansen et al. (2015) produced a pebble-pile planetesimal population with radii $30 \lesssim R_{pl} \lesssim 120$ km. By allowing the newly formed planetesimals to accrete drifting chondrules, they were able to produce the large sizes seen in the asteroid belt today.

Johansen et al. (2015) fit the planetesimal initial mass function with an exponentially tapered power law, written in cumulative form as

$$N_{>M_{pl}} = N_0 \left(\frac{M_{pl}}{M_0} \right)^{1-\zeta} \exp\left(-M_{pl}/M_1\right). \tag{4.35}$$

The power-law portion of Equation (4.35) yields differential mass distributions with $dN/dM_{pl} \propto M_{pl}^{-\zeta}$ below some characteristic mass M_1, similar to the grain size distributions in Chapter 3. Johansen et al. (2015) found a best-fit value of $\zeta = 1.6$ for a variety of grid resolutions. Simon et al. (2016), who did not include the exponential taper at large sizes in their fit, also found $\zeta \simeq 1.6$, with only a minor dependence on $\tilde{G}$ and resolution. Abod et al. (2019) showed that ζ only depends weakly on $v_K - v_\phi$. For $0.037\,5c_s < v_K - v_\phi < 0.1c_s$, they found $\zeta = 1.5$–1.6 for a simple power-law mass distribution without the exponential taper. However, adding the exponential taper yielded $\zeta = 1.3$. Simon et al. (2017) argued that a universal value of $\zeta \simeq 1.6$ emerges from the small-scale structure of streaming-induced turbulence and applies to almost all disk conditions and particle sizes, but

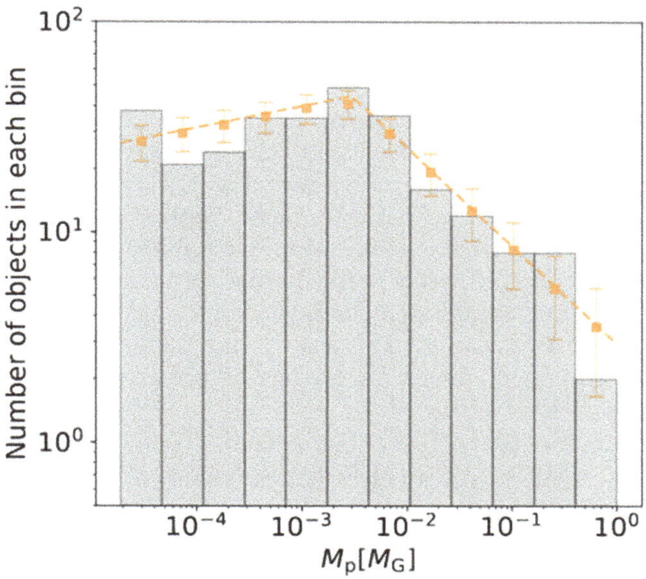

Figure 4.15. The planetesimal mass distribution reprinted from Li et al. (2019), Run I, which has L_x, L_y, $L_z = 0.1H, 0.1H, 0.2H$, $512 \times 512 \times 1024$ grid cells, $St = 2$, and $S/G = 0.1$. While the high-mass end of the mass distribution is consistent with $dN/dM_{pl} \propto M_{pl}^{-1.3}$, as in Abod et al. (2019), there is a turnover at the low-mass end.

Li et al. (2019) disagreed, finding that both the functional form of the planetesimal initial mass function (simple power law, broken power law, exponentially tapered power law, etc.) and the best-fit parameters depended on the simulation setup. Their best-fit exponentially tapered power-law model had $\zeta = 1.3$, in agreement with Abod et al. (2019). Li et al. (2019) also showed evidence of a turnover at low masses in which ζ becomes negative (Figure 4.15), which is necessary if the total planetesimal mass is to remain finite.

Let us assume that some part of the initial mass function can be described by $dN/dM_{pl} \propto M_{pl}^{-1.6}$. The corresponding power-law component of the planetesimal radius distribution, $dN/dR_{pl} \propto R_{pl}^{-q}$, has $q = 2.8$, which is similar to the observational findings of the Outer Solar Systems Origins Survey for KBOs with diameters under 100 km ($q \simeq 3$; Lawler et al. 2018). The observed size distributions for objects with $R \gtrsim 100$ km are steeper in both the Kuiper Belt and the asteroid belt (Jedicke et al. 2002; Bottke et al. 2005; Fraser et al. 2014), possibly indicating an exponential cutoff in the initial mass function but also perhaps a consequence of postcollapse pebble accretion (Johansen et al. 2015) or collisions (e.g., Schlichting et al. 2013). Benchmarking size distributions predicted by streaming instability models against observations is nontrivial as the solar system's remnant planetesimals have been collisionally altered since their formation (e.g., Schlichting et al. 2013).

While Johansen et al. (2015) and Simon et al. (2016) found that the planetesimal initial mass function was relatively independent of grid resolution, they both used shearing boxes with azimuthal and radial dimensions L_x, $L_y = 0.2H$, which were too small to allow more than one overdense axisymmetric filament to form. Schäfer et al. (2017) tested larger simulation boxes—L_x, $L_y = 0.4H$ and L_x, $L_y = 0.8H$—to make room for multiple filaments. Because of their large box sizes, their highest resolution of 256 × 256 × 128 yielded grid cells with side length $H/640$, which is actually the largest grid cell used by Johansen et al. (2015) and Simon et al. (2016). Since the minimum planetesimal radius decreases with grid cell size, Schäfer et al. (2017) did not produce the smaller planetesimals that make up the power-law part of the initial mass function. However, their largest simulation box (L_x, $L_y = 0.8H$) yielded a 620 km planetesimal, and the box with L_x, $L_y = 0.4H$ produced seven planetesimals larger than Ceres. The size of the simulation box is clearly an important consideration in numerical models of planetesimal formation, a conclusion echoed by Li et al. (2019).

Figure 4.16 helps us extrapolate the Simon et al. (2016) results to our model disk, which has $v_K - v_\phi = 0.05c_s$ at 5 au. The corresponding value of $\tilde{G}$ is 0.11, which is slightly higher than the maximum value of 0.1 explored in their numerical simulations. For $\alpha = 10^{-4}$, particles with $h_p/H = 0.02$ at 5 au in our model disk have $St = 0.25$, compared to $St = 0.3$ in Simon et al. (2016). Such particles are snowballs with $r_{\rm eff} = 22.5$ cm, which are larger than either the constituent pebbles of comet 67P (e.g., Poulet et al. 2016) or the size that can readily form via two-body collisions of icy particles (Gundlach & Blum 2015; Lorek et al. 2018), but perhaps within reach of the Ros & Johansen (2013) vapor deposition model. $S/G = 0.02$ is consistent with our model disk if abundant water ice is part of the solid inventory. From Figure 12 of Simon et al. (2016), the largest planetesimals in the $\tilde{G} = 0.1$ simulations have $M_{pl} \simeq 0.4$ Ceres masses, which is only a factor of 2 less massive than the characteristic planetesimal predicted by the order-of-magnitude estimate in Section 4.3.4, Figure 4.10. Though Equation (4.31) does not capture the full complexity of pebble-cloud formation and collapse, it nevertheless provides useful

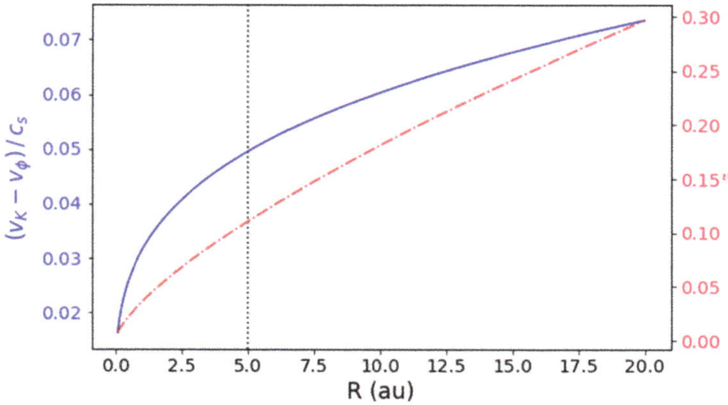

Figure 4.16. $(v_K - v_\phi)/c_s$ (blue solid line, left-hand axis) and $\tilde{G}$ (red dashed–dotted line, right-hand axis) as a function of R in our model disk. The dotted black line marks $R = 5$ au, which gives a good match to the parameters in simulation set SI128-G0.1 of Simon et al. (2016; see their table 1).

physical intuition about the largest planetesimals produced by the streaming instability.

4.4.4 Numerical Simulations Based on Alternative Theories

Though most theorists broadly agree that self-gravitating pebble clouds form via streaming instability and/or related RDIs, there are some caveats worth mentioning. In some parts of the disk, turbulence can increase collision speeds to the point where the particle size distribution becomes fragmentation limited with $r_{\text{eff,max}}$ below the threshold needed to trigger the streaming instability (e.g., Drążkowska & Dullemond 2014; see Figure 3.17). Even when turbulence-triggered collisional fragmentation is not considered, a background turbulent velocity field (i.e., turbulence not caused by the streaming instability itself) strongly suppresses the instability growth rate (Chen & Lin 2020; Umurhan et al. 2020). One possibility is that the high concentration factors needed to form gravitationally bound pebble clouds come from spiral arms instead of the streaming instability (Rice et al. 2004; Dipierro et al. 2015; Vorobyov et al. 2018). In simulations by Rice et al. (2004), the overdense arms of a marginally stable disk acted as traps for drifting particles, in some cases increasing the local value of S/G by a factor of 50. Vertical settling can then provide the high midplane values of ρ_p/ρ necessary to trigger pebble-cloud collapse. However, it's unclear whether a disk can retain enough mass to keep its spiral arms for up to 3 Myr—the shortest possible time frame over which pebble-pile planetesimals formed in the solar nebula, according to the meteoritic record (Figure 4.1; Section 4.2.6). Gibbons et al. (2015) argue that vortices are at least as efficient as spiral arms at locally increasing S/G, if not more so. In either the spiral-arm or the vortex scenario, streaming instability need not play a direct role (though there is a model of vortex particle concentration followed by streaming instability by Raettig et al. 2015), but the resulting planetesimals are still described by the pebble-cloud collapse model in Section 4.3.4.

Another mechanism that could lead to planetesimal formation via pebble-cloud collapse is turbulent concentration. Cuzzi et al. (2008) suggest that a weakly turbulent disk produces clouds of size-sorted particles similar to those observed in chondrites (Section 4.2.3) and that cloud densities can reach $\rho_p/\rho = 100$. The smaller particle clouds are vulnerable to disruption by ram pressure, but the largest ones will be gravitationally bound and will collapse into "sandpile" planetesimals. Cuzzi et al. (2008) predict initial planetesimal sizes ranging from 10 to 100 km. Highlighting the fact that pebble clouds need to reside in regions of low vorticity to avoid rotational breakup, Chambers (2010) and Hopkins (2016) present turbulent concentration models that are benchmarked against the vorticity distribution of different-size eddies in the disk (see Section 3.2.4 for more on turbulent eddies). The Chambers model produces large planetesimals with initial sizes of ~100 km in the asteroid belt and ~400 km in the Kuiper Belt. The asteroid belt predictions are a good match for the meteoritic record (Section 4.2.2), but robust observational constraints on the initial Kuiper Belt size distribution are still being derived (see the inset on the Kuiper Belt below). Hartlep & Cuzzi (2020) correct the turbulent cascade model used in

Cuzzi et al. (2008) and update the turbulent concentration model, finding that instead of a power law (as usually seen in streaming instability simulations), the initial planetesimal size distribution has a well-defined peak. The exact peak size depends on the location in the disk and the turbulent efficiency but is generally between 10 and 100 km. Hartlep & Cuzzi emphasize that the new turbulent concentration model requires centimeter-size seed particles (a.k.a. pebbles) instead of millimeter-size chondrules.

4.2 The Kuiper Belt

In *The origin and evolution of the solar system*, Edgeworth (1949, p. 609) speculated that:

> It would be unreasonable to suppose that the original rotating disk … came to an abrupt end outside the orbit of Neptune. There must have been a gradual thinning out of the material at the outer boundary. …It is not unreasonable to suppose that this outer region …is in fact a vast reservoir of potential comets.

Kuiper (1951, p. 13) agreed, adding that

> "The planet Pluto, which sweeps through the whole zone from 30 to 50 astronomical units, is held responsible for having started the scattering of the comets throughout the solar system."[22]

Kuiper's suggestion that Pluto (discovered by Clyde Tombaugh in 1930) was responsible for scattering comets throughout the solar system made sense given that Pluto's original mass estimate was $M > 7M_\oplus$, based on erroneous measurements of anomalies in Neptune's orbit. Neither Edgeworth nor Kuiper considered that Pluto itself could be the vanguard of their elusive comet reservoir. But there were hints that the reservoir existed. For example, after McKinnon (1989) hypothesized that Charon (discovered by Christy & Harrington 1978) was a reaccreted byproduct of a catastrophic collision between two minor planets, Stern (1991) calculated that such a collision would be likely only if the early solar system contained hundreds of Plutos.[23] Exploring another line of reasoning, Fernandez (1980) and Duncan et al. (1988) pointed out that a trans-Neptunian planetesimal belt would be a more dynamically efficient source of short-period comets than the Oort cloud.

[22] The scenario in which dwarf planets dynamically excite smaller neighboring planetesimals is called self-stirring. The planetesimal belts in the solar system are planet-stirred by Jupiter (asteroid belt) and Neptune (Kuiper Belt; e.g., Wyatt 2008).

[23] Shannon & Wu (2011) came to a similar conclusion about extrasolar debris disks, pointing out that their high dust luminosities could only be sustained if each disk contained ~1000 Plutos apiece that dynamically excite collisions between smaller neighbors.

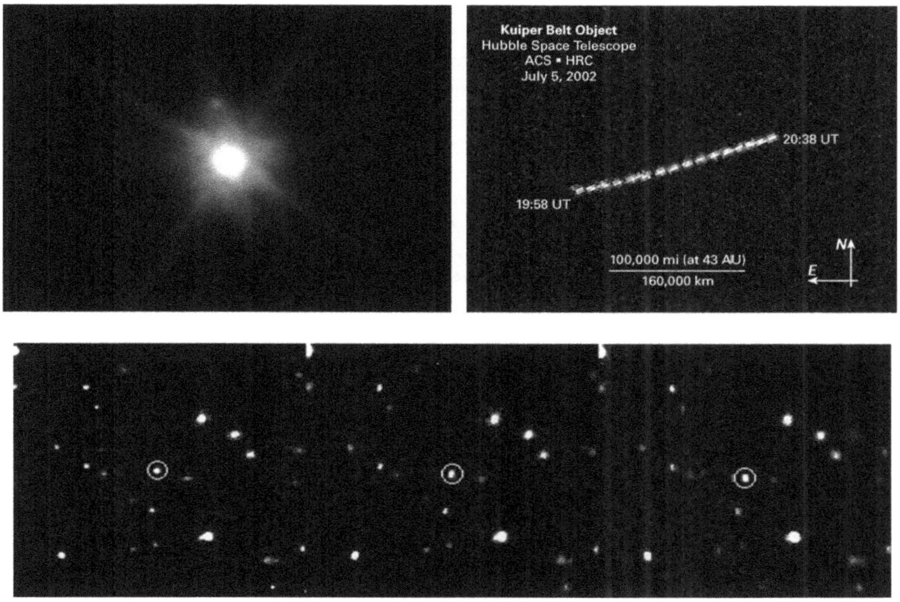

Figure 4.17. Top left: Makemake, the third-largest known Kuiper Belt object at $R_{pl} \simeq 715\ km$, and its moon MK2 observed by Hubble in 2015. Image credit: NASA, ESA, A. Parker, and M. Buie (Southwest Research Institute). Top right: a composite of 16 Hubble Advanced Camera for Surveys images reveals the track of Quaoar across the sky. Most Kuiper Belt objects are identified based on their motion with respect to background stars and galaxies. Image credit: NASA and M. Brown (Caltech). Bottom: discovery images of Eris, the most massive known Kuiper Belt object, from the Samuel Oschin Telescope at Palomar Observatory. Eris, circled in white, moves southward against the fixed background objects. Images were taken 90 minutes apart on 2003 October 21. Image credit: NASA/JPL/Caltech.

Then, five years after launching their search for distant "SMOs" (slow-moving objects) in the ecliptic plane, Jewitt & Luu (1993) discovered 1992 QB_1/Albion. Two more SMOs—1993 FW (Luu et al. 1993) and 1993 RO (Jewitt et al. 1993)—followed in quick succession. By 1995, there were 24 such candidate KBOs, enough for astronomers to consider their ensemble properties (Duncan et al. 1995; Irwin et al. 1995; Morbidelli et al. 1995). KBOs are identified in surveys based on their motion against fixed background stars and galaxies (Figure 4.17); as of this writing, there are 2679 known trans-Neptunian objects with semimajor axes that place them, on average, beyond Neptune.[24] If a KBO is bright enough, it may be imaged at high resolution in order to search for moons or binary companions, which enable accurate mass measurements.

We now know that Neptune, not Pluto, is primarily responsible for shaping the dynamics of the small bodies in the outer solar system. Figure 4.18 shows the eccentricity–semimajor axis distribution of all objects with 25 au $<a<$ 80 au. Only

[24] Source: IAU Minor Planet Center, https://minorplanetcenter.net/.

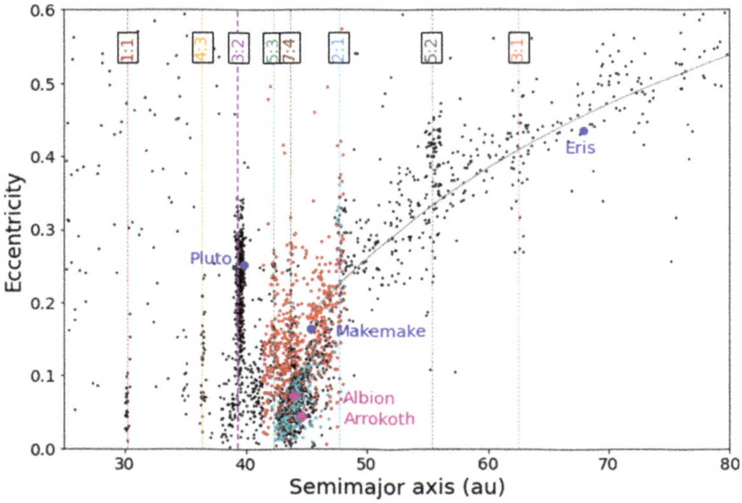

Figure 4.18. Eccentricity as a function of semimajor axis for objects with $25 < a < 80$ au. Members of the cold classical Kuiper Belt are shown in green, while objects in the dynamically hot classical Kuiper Belt are shown in red. Black data points show centaurs with $a < 30$ au, resonant KBOs, members of the scattering disk, detached KBOs, and classical KBOs with $2° < i < 6°$ whose cold/hot dynamical status is ambiguous. Well-populated resonances with Neptune are marked with vertical dashed lines. The gray curve traces out the high-a portion of the scattering disk.

the cold classical Kuiper Belt, a low-inclination, low-eccentricity population centered at $a \simeq 44$ au (Petit et al. 2011), whose members are plotted in green on Figure 4.18, remains relatively unaffected by giant-planet migration, i.e., "dynamically cold" (Batygin et al. 2011; Dawson & Murray-Clay 2012; Wolff et al. 2012). The cold classical objects have redder colors (Tegler & Romanishin 1998; Trujillo & Brown 2002; Marsset et al. 2019) and a different size distribution (Bernstein et al. 2004; Peixinho et al. 2012; Fraser et al. 2014) than the rest of the KBOs, suggesting a different formation and collisional evolution history. The leading theory is that the cold classical KBOs formed in situ and experienced minimal orbit alteration, while the other KBOS either got trapped sweeping resonances (shown in Figure 4.18 by dashed vertical lines) or were scattered onto eccentric, highly inclined orbits by a migrating Neptune (e.g., Malhotra 1995; Hahn & Malhotra 2005; Parker & Kavelaars 2012). Inhabitants of the hot classical Kuiper Belt (Figure 4.18, red), the resonances, and the scattering disk (see below) formed interior to the cold classical Kuiper Belt (Levison et al. 2008; Nesvorný & Vokrouhlický 2016).

Even today, Neptune is still scattering leftover planetesimals that venture near its orbit during perihelion passage. From a model of the orbital evolution of known KBOs, Gladman et al. (2008) picked out the objects with rapid semimajor axis jumps of $\Delta a > 1.5$ au within a 10 Myr time frame; these objects form the "scattering disk." In Figure 4.18, the scattering disk roughly follows a gray curve that shows eccentricity as a function of semimajor axis for objects with perihelion distance $q = a(1 - e) = 37$ au. While Gladman et al. (2008) emphasize that no single value of q defines the scattering disk, the gray curve encodes the concept that planetesimals get "kicked" onto high-eccentricity orbits during close encounters with Neptune. Thus, the scattering disk can be understood as a class of objects with aphelia beyond the classical Kuiper Belt and perihelia near Neptune. The Gladman et al. models highlight the fact that KBO orbits are not necessarily stable, even if some KBOs have remained in their current orbits for approximately the age of the solar system. On billion-year timescales, some planetesimals drop out of resonance with Neptune (Morbidelli 1997; Tiscareno & Malhotra 2009; Lawler et al. 2019), others undergo slow Kozai oscillations between high-eccentricity and high-inclination states (Gomes 2003), and high-perihelion objects experience successive weak interactions with Neptune that drive random walks in the semimajor axis (Bannister et al. 2017). On the flip side, Neptune's migration scattered some planetesimals into stable, nonresonant orbits in the hot classical belt (Petit et al. 2011).

Despite its intricate and highly evolved dynamical structure, the Kuiper Belt holds a relatively pristine planetesimal population. As stated in Section 4.2.6, the protracted parent-body formation and scarcity of chondrules in carbonaceous chondrites hint at gradual, gentle accretion in the outer solar nebula. Based on orbital timescales and velocities (4.8 km s^{-1} for a circular orbit at $a = 40$ au versus 20 km s^{-1} at $a = 3$ au), we expect collisions to be less frequent and destructive in the Kuiper Belt than in the asteroid belt (Abedin et al. 2021). Indeed, while the asteroid belt boasts at least 122 collisional families (Nesvorný 2015), there is only one collisional family in the Kuiper Belt (Brown et al. 2007). Furthermore, the collisional families in the asteroid belt have a high ratio of velocity dispersion to parent-asteroid mass, indicating that the family-forming collisions tended to be catastrophic (e.g., Michel et al. 2004; Nesvorný et al. 2006)—as opposed to the Kuiper Belt's Haumea event, which was a gentle graze and merge that probably also gave Haumea its elongated shape (Leinhardt et al. 2010). Certainly, the planetesimals in the Kuiper Belt are collisionally altered: Pluto and Charon host impact craters from KBOs with 0.3 km $< R_{pl} <$ 40 km (Singer et al. 2019), and Pluto/Charon, Eris, Makemake, Haumea, Orcus, and Quaoar all have moons that are probably collision fragments (though see Brown & Butler 2018 for caveats about Eris' moon Dysnomia; Barranco 2016). Successful models of the current KBO size distribution rely on the theory of collisional cascades (e.g., Pan & Schlichting 2012; Kenyon & Bromley 2020). But compared to the asteroid belt, the Kuiper Belt is a much better representation of the solar system's primordial planetesimal population.

It is not yet clear whether the Kuiper Belt size distribution primarily encodes building up, breaking down, or both. Schlichting et al. (2013) presented a combined fragmentation–coagulation model based on an initial population of 1 km planetesimals that, when evolved forward in time, produced a good match for the observed size distribution. Weidenschilling (1997) and Kenyon & Luu (1999) also argued in favor of bottom-up growth, and Krivov & Wyatt (2021) present evidence that extrasolar planetesimals should be "born small" if high-luminosity debris disks are to maintain their prodigious dust production. (The minimum planetesimal size that the streaming instability/gravitational collapse model can produce is still unknown, thanks to the fact that small planetesimals only emerge in ultra-high-resolution simulations; Li et al. 2019). If these models are correct, then instability-based planetesimal formation models that predict $R_{pl} \sim 100$ km need to be reexamined (Section 4.4.3).

On the other hand, Morbidelli & Rickman (2015) suggest that the kilometer-size Jupiter family comet nuclei are fragments from collisions between larger KBOs of the type that may have formed via the streaming instability (though they caution that their results depend on the timing of both gas disk dispersal and dynamical instability among the giant planets). Furthermore, the binary fraction in the Kuiper Belt—~30%

among the cold classical population, a high number given that binaries can become unbound in gravitational interactions with third bodies—is much more consistent with a streaming instability origin (which naturally forms binaries; Section 4.4.2) than with collisional coagulation (Robinson et al. 2020). The bilobed structure of Arrokoth (Figure 4.18, magenta; Figure 3.1), a contact binary in the cold classical Kuiper Belt that was visited by the New Horizons spacecraft in 2019, suggests low-speed accumulation within a collapsing particle cloud (McKinnon et al. 2020). Finally, the KBO size distribution cannot be modeled by a single power law; it is instead a "broken" power law with a flatter slope (lower ζ) at small sizes and a steeper slope (higher ζ) at large sizes.[25] (In fact, Schlichting et al. 2013 show that there may be as many as four breaks in the observed size distribution.) According to collisional cascade theory, the "knee" of the power law—i.e., the break radius in a two-component model—is approximately the characteristic initial planetesimal size (Hartlep & Cuzzi 2020). There is some uncertainty in the KBO break radius because the conversion from observed magnitude to R_{pl} depends on the assumed albedo, but current estimates place the break between 20 km and 100 km (H-band absolute magnitude of 7.7–8.5; Fraser et al. 2014; Shankman et al. 2016; Lawler et al. 2018), in rough agreement with the streaming instability models of Johansen et al. (2015) and Simon et al. (2016).

Statisticians trying to infer the size, semimajor axis, eccentricity, and argument of perihelion distributions of observed KBOs have a difficult job: the detection biases against faint, distant objects are formidable, and the selection biases that can lead to spurious orbital clusterings among the discoveries are nonintuitive (Shankman et al. 2017; Lawler et al. 2018; Napier et al. 2021). But with the completion of the Outer Solar System Origins Survey (OSSOS) (Bannister et al. 2016), the publication of the survey catalog featuring more than 800 new detections (Bannister et al. 2018), and the OSSOS team's progress toward deriving a debiased KBO size distribution (e.g., Shankman et al. 2016; Lawler et al. 2018), the Kuiper Belt is gaining importance as a testbed for planetesimal formation models.

4.5 Conclusions

Observational studies of asteroids and comets are converging on the idea that the solar system's first planetesimals were collapsed pebble clouds. The Hayabusa and Rosetta missions both revealed the rubble-pile structure of their targets, asteroid Itokawa and comet 67P/Churyumov-Gerasimenko, and the high binary fraction in the Kuiper Belt is inconsistent with "bottom-up" growth from small seed particles. One feature of pebble-cloud collapse is that it tends to produce large planetesimals, tens to hundreds of meters (Sections 4.3.4 and 4.4.3). The extended magmatic histories of Vesta, the angrite parent body, and the iron meteorites can only be explained if the original planetesimals had $R_{pl} \gtrsim 100$ km (Vesta has $R = 250$ km). The fact that the knee or divot of the KBO scattering-disk size distribution is at 20–100 km is also consistent with some numerical simulations of the streaming instability, though simulated size distributions depend strongly on shearing-box size and resolution. Though the streaming instability is the leading candidate for concentrating pebbles into gravitationally bound clouds (e.g., Youdin & Goodman

[25] There may even be a "divot" in the differential size distribution, instead of a knee; see Shankman et al. (2016).

2005; Johansen et al. 2007; Carrera et al. 2017; Nesvorný et al. 2019; Lin 2021), there are other instability types, such as RDIs (Squire & Hopkins 2018). It is possible that vortices, spiral arms, and/or turbulence could concentrate particles enough to form bound pebble clouds directly, without any need for a two-fluid instability (Section 4.4.4).

The meteoritic record tells us that planetesimal formation was both faster and more violent in the inner solar nebula than in the giant-planet-forming region and the Kuiper Belt (Section 4.2.6). Some meteorites sourced from beyond $R \sim 3$ au suggest that their parent bodies accreted slowly, possibly taking up to 3 Myr (e.g., van Kooten et al. 2011), though there were rapidly accreting planetesimals beyond 3 au as well (e.g., the IVB parent body; Neumann et al. 2018b). (Why some pebble-based planetesimals accrete slowly while others collapse rapidly is an interesting research question for a motivated reader.) In contrast, the inner solar nebula was still forming planetesimals at $t \sim 2$Myr, such as the acapulcoite/lodranite, H-chondrite, and L-chondrite parent bodies, but their growth was presumably rapid (e.g., Neumann et al. 2018b; Gail & Trieloff 2019). The low densities of P- and D-type asteroids—parent bodies of the interplanetary dust particles that are assumed to have formed beyond 10 au—indicate formation in a "gentle" environment, as repeated high-speed collisions would have compacted the growing planetesimals. Likewise, the paucity of collisional families in the Kuiper Belt points toward pebble-cloud collapse followed by infrequent, low-speed collisions, as expected based on the long dynamical timescale in the outer solar system. It is likely that solids from the inner solar were "recycled" through more generations of planetesimals than solids from the giant planet-forming region and beyond.

It is essential that streaming instability and RDI simulations are applied to massive disks like the MAXSN or MASN (Nixon et al. 2018; Lenz et al. 2020). With the exception of Gerbig et al. (2020), all of the numerical work reviewed in Section 4.4 has been applied to disks with at most two to three times the mass of the mmsn (though it is sometimes possible to reinterpret simulations conducted in dimensionless units using a different disk model, as we have done with the results from Simon et al. 2016 in Section 4.4.3). Furthermore, there is now enough statistically robust information about the size distribution, binary fraction, and orbits of KBOs to make the Kuiper Belt a vital testbed of planetesimal formation theory. Given that (1) planetesimals from the outer solar nebula have been emplaced in the asteroid belt (e.g., Hsieh & Jewitt 2006; Budde et al. 2018) and (2) pebble drift from beyond the ice line may have provided surface layers of planetesimals from the inner solar nebula (Sarafian et al. 2017), it's important to work toward a global model of planetesimal formation. In particular, the timing of planetesimal accretion beyond the ice line affects the abundance of drifting pebbles and the timescale over which they could have been accreted by asteroid parent bodies. Given the computational expense even of local shearing-box simulations, a solar-nebula-wide planetesimal formation model will be difficult to construct; as a first step, we suggest constructing a global semianalytical model that includes radial drift and a prescription for planetesimal formation based on local simulations.

This concludes Volume 1 of "Origins of Giant Planets." In Volume 2, we will transition from gas-dominated dynamics to solid-dominated dynamics. We will begin with the growth of solid planet cores (Chapter 5), then explore the accretion of gaseous envelopes (Chapter 6), and finally finish with migration and dynamical instabilities among growing planets (Chapter 7).

References

Abedin, A. Y., Kavelaars, J. J., Greenstreet, S., et al. 2021, AJ, 161, 195
Abod, C. P., Simon, J. B., Li, R., et al. 2019, ApJ, 883, 192
Agnor, C., & Asphaug, E. 2004, ApJL, 613, L157
A'Hearn, M. F. 2011, ARA&A, 49, 281
Alexander, C. M. O., Bowden, R., Fogel, M. L., et al. 2012, Sci, 337, 721
Alexander, C. M. O., Grossman, D., Ebel, J. N., & Ciesla, F. J. 2008, Sci, 320, 1617
Alexander, C. M. O., McKeegan, K. D., & Altwegg, K. 2018, SSRv, 214, 36
Alexander, R. D., Clarke, C. J., & Pringle, J. E. 2006, MNRAS, 369, 229
Amelin, Y., Kaltenbach, A., Iizuka, T., et al. 2010, E&PSL, 300, 343
Andrews, S. M., Huang, J., Pérez, L. M., et al. 2018, ApJL, 869, L41
Asphaug, E., Jutzi, M., & Movshovitz, N. 2011, E&PSL, 308, 369
Asphaug, E., & Reufer, A. 2014, NatGe, 7, 564
Aumatell, G., & Wurm, G. 2014, MNRAS, 437, 690
Bai, X.-N., & Stone, J. M. 2010, ApJ, 722, 1437
Bai, X.-N., & Stone, J. M. 2013, ApJ, 769, 76
Baker, J., Bizzarro, M., Wittig, N., et al. 2005, Natur, 436, 1127
Bannister, M. T., Gladman, B. J., Kavelaars, J. J., et al. 2018, ApJS, 236, 18
Bannister, M. T., Kavelaars, J. J., Petit, J.-M., et al. 2016, AJ, 152, 70
Bannister, M. T., Shankman, C., Volk, K., et al. 2017, AJ, 153, 262
Barr, A. C., & Schwamb, M. E. 2016, MNRAS, 460, 1542
Barranco, J. A. 2009, ApJ, 691, 907
Batygin, K., Brown, M. E., & Fraser, W. C. 2011, ApJ, 738, 13
Beichman, C. A., Bryden, G., Stapelfeldt, K. R., et al. 2006, ApJ, 652, 1674
Bell, J. F., Davis, D. R., Hartmann, W. K., et al. 1989, in Asteroids II, ed. R. P. Binzel, et al. (Tucson, AZ: Univ. Arizona Press), 921
Bernstein, G. M., Trilling, D. E., Allen, R. L., et al. 2004, AJ, 128, 1364
Benz, W., & Asphaug, E. 1999, Icar, 142, 5
Benz, W., Slattery, W. L., & Cameron, A. G. W. 1988, Icar, 74, 516
Binzel, R. P., & Xu, S. 1993, Sci, 260, 186
Birnstiel, T., Klahr, H., & Ercolano, B. 2012, A&A, 539, A148
Bizzarro, M., Baker, J. A., & Haack, H. 2004, Natur, 431, 275
Blackburn, T., Alexander, C. M. O., Carlson, R., & Elkins-Tanton, L. T. 2017, GeoCoA, 200, 201
Bland, P. A., & Travis, B. J. 2017, SciA, 3, e1602514
Blichert-Toft, J., Moynier, F., Lee, C.-T. A., Telouk, P., & Albarede, F. 2010, E&PSL, 296, 469
Blum, J., & Wurm, G. 2008, ARA&A, 46, 21
Boley, A. C., & Durisen, R. H. 2008, ApJ, 685, 1193
Boley, A. C., Morris, M. A., & Desch, S. J. 2013, ApJ, 776, 101
Bollard, J., Connelly, J. N., & Bizzarro, M. 2015, M&PS, 50, 1197

Bollard, J., Connelly, J. N., Whitehouse, M. J., et al. 2017, SciA, 3, e1700407
Bollard, J., Kawasaki, N., Sakamoto, N., et al. 2019, GeoCoA, 260, 62
Bonnell, I. A., Vine, S. G., & Bate, M. R. 2004, MNRAS, 349, 735
Bonsor, A., Carter, P. J., Hollands, M., et al. 2020, MNRAS, 492, 2683
Bonsor, A., Leinhardt, Z. M., Carter, P. J., et al. 2015, Icar, 247, 291
Borovička, J. 2007, in Proc. IAU 236, Near Earth Objects, Our Celestial Neighbors: Opportunity and Risk (Cambridge: Cambridge Univ. Press), 107
Boss, A. P., & Durisen, R. H. 2005, ApJL, 621, L137
Bottke, W. F., Durda, D. D., Nesvorný, D., et al. 2005, Icar, 175, 111
Bottke, W. F., Nesvorný, D., Grimm, R. E., Morbidelli, A., & O'Brien, D. P. 2006, Natur, 439, 821
Bradley, J. P. 1994, Sci, 265, 925
Bradley, J. P. 2003, TrGeo, 1, 711
Bradley, J. P., Keller, L. P., Brownlee, D. E., & Thomas, K. L. 1996, M&PS, 31, 394
Brandenburg, A., & Dobler, W. 2002, CoPhC, 147, 471
Brearley, A. J., & Jones, R. H. 1998, in Reviews in Mineralogy 36, Planetary Materials, ed. J.J. Papike (Washington, DC: Mineralogical Society of America), 3-1
Brewer, J. M., Fischer, D. A., Valenti, J. A., & Piskunov, N. 2016, ApJS, 225, 32
Brown, M. E., Barkume, K. M., Ragozzine, D., & Schaller, E. L. 2007, Natur, 446, 294
Brown, M. E., & Butler, B. J. 2018, AJ, 156, 164
Brownlee, D. E. 1985, AREPS, 13, 147
Brownlee, D. E., Joswiak, D. J., Schlutter, D. J., et al. 1995, LPSC, 26, 183
Bryson, J. F. J., Neufeld, J. A., & Nimmo, F. 2019, E&PSL, 521, 68
Buchhave, L. A., Latham, D. W., Johansen, A., et al. 2012, Natur, 486, 375
Budde, G., Burkhardt, C., Brennecka, G. A., et al. 2016, E&PSL, 454, 293
Budde, G., Kleine, T., Kruijer, T. S., Burkhardt, C., & Metzler, K. 2016, PNAS, 113, 2886
Budde, G., Kruijer, T. S., & Kleine, T. 2018, GeoCoA, 222, 284
Buratti, B. J., Hicks, M. D., Tryka, K. A., Sittig, M. S., & Newburn, R. L. 2002, Icar, 155, 375
Burbine, T. H., & Greenwood, R. C. 2020, SSRv, 216, 59
Burke, C. J., Christiansen, J. L., Mullally, F., et al. 2015, ApJ, 809, 8
Bus, S. J., & Binzel, R. P. 2002, Icar, 158, 146
Busemann, H., Nguyen, A. N., Cody, G. D., et al. 2009, E&PSL, 288, 44
Campins, H., Hargrove, K., Pinilla-Alonso, N., et al. 2010, Natur, 464, 1320
Canup, R. M., & Asphaug, E. 2001, Natur, 412, 708
Carlson, R. W., Brasser, R., Yin, Q.-Z., Fischer-Godde, M., & Qin, L. 2018, SSRv, 214, 121
Carrera, D., Gorti, U., Johansen, A., & Davies, M. B. 2017, ApJ, 839, 16
Carrera, D., Johansen, A., & Davies, M. B. 2015, A&A, 579, A43
Cassan, A., Kubas, D., Beaulieu, J.-P., et al. 2012, Natur, 481, 167
Castillo-Rogez, J. C., Matson, D. L., Sotin, C., et al. 2007, Icar, 190, 179
Castillo-Rogez, J., Vernazza, P., & Walsh, K. 2019, MNRAS, 486, 538
Ceplecha, Z., Borovička, J., Elford, W. G., et al. 1998, SSRv, 84, 327
Chambers, J. E. 2001, Icar, 152, 205
Chambers, J. 2006, Icar, 180, 496
Chambers, J. 2008, Icar, 198, 256
Chambers, J. E. 2010, Icar, 208, 505
Chawner, H., Gomez, H. L., Matsuura, M., et al. 2020, MNRAS, 493, 2706

Chen, K., & Lin, M.-K. 2020, ApJ, 891, 132
Chiang, E. 2008, ApJ, 675, 1549
Christy, J. W., & Harrington, R. S. 1978, AJ, 83, 1005
Ciesla, F. J. 2007, Sci, 318, 613
Ciesla, F. J., Hood, L. L., & Weidenschilling, S. J. 2004, M&PS, 39, 1809
Clayton, R. N. 1993, AREPS, 21, 115
Clayton, R. N., & Mayeda, T. K. 1999, GeoCoA, 63, 2089
Clement, M. S., Kaib, N. A., & Chambers, J. E. 2020, PSJ, 1, 18
Clenet, H., Jutzi, M., Barrat, J.-A., et al. 2014, Natur, 511, 303
Collinet, M., & Grove, T. L. 2020, GeoCoA, 277, 358
Connelly, J. N., Bizzarro, M., Krot, A. N., et al. 2012, Sci, 338, 651
Connelly, J. N., Bollard, J., & Bizzarro, M. 2017, GeoCoA, 201, 345
Consolmagno, G. J., & Drake, M. J. 1977, GeoCoA, 41, 1271
Consolmagno, G. J., Golabek, G. J., Turrini, D., et al. 2015, Icar, 254, 190
Cuk, M., & Stewart, S. T. 2012, Sci, 338, 1047
Cuzzi, J. N., Hogan, R. C., & Shariff, K. 2008, ApJ, 687, 1432
Dauphas, N., & Chaussidon, M. 2011, AREPS, 39, 351
Dauphas, N., & Pourmand, A. 2011, Natur, 473, 489
Dauphas, N., Remusat, L., Chen, J. H., et al. 2010, ApJ, 720, 1577
Davis, A. M., & McKeegan, K. D. 2014, in Treatise on Geochemistry, Vol. 1, Meteorites and Cosmochemical Processes (Amsterdam: Elsevier), 361
Dawson, R. I., & Murray-Clay, R. 2012, ApJ, 750, 43
Deckers, J., & Teiser, J. 2014, ApJ, 796, 99
Delbo', M., Walsh, K., Bolin, B., et al. 2014, Sci, 357, 1026
DeMeo, F. E., Binzel, R. P., Slivan, S. M., et al. 2009, Icar, 202, 160
DeMeo, F. E., & Carry, B. 2014, Natur, 505, 629
Denevi, B. W., Beck, A. W., Coman, E. I., et al. 2016, M&PS, 51, 2366
Denk, T., Neukum, G., Roatsch, T., et al. 2010, Sci, 327, 435
Dermott, S. F., Kehoe, T. J. J., Durda, D. D., et al. 2002, in Proc. Asteroids, Comets, and Meteors: ACM 2002 (Noordwijk: ESA Publications), 319
Desch, S. J., & Connolly, H. C. 2002, M&PS, 37, 183
Desch, S. J., & Cuzzi, J. N. 2000, Icar, 143, 87
Desch, S. J., & Kalyaan, A. 2018, ApJS, 238, 11
Dipierro, G., Pinilla, P., Lodato, G., et al. 2015, MNRAS, 451, 974
Dobinson, J., Leinhardt, Z. M., Lines, S., et al. 2016, ApJ, 820, 29
Dodson-Robinson, S. E., Willacy, K., Bodenheimer, P., et al. 2009, Icar, 200, 672
Dong, R., Liu, S.-y., Eisner, J., et al. 2018, ApJ, 860, 124
Drake, M. J. 2001, M&PS, 36, 501
Drażkowska, J., & Alibert, Y. 2017, A&A, 608, A92
Drażkowska, J., & Dullemond, C. P. 2014, A&A, 572, A78
Dufresne, E. R., & Anders, E. 1962, GeoCoA, 26, 1085
Dullemond, C. P., Stammler, S. M., & Johansen, A. 2014, ApJ, 794, 91
Duncan, M. J., Levison, H. F., & Budd, S. M. 1995, AJ, 110, 3073
Duncan, M., Quinn, T., & Tremaine, S. 1988, ApJL, 328, L69
Ebel, D. S., & Alexander, C. M. O. 2011, P&SS, 59, 1888
Edgeworth, K. E. 1949, MNRAS, 109, 600

Eisenhour, D. D., Daulton, T. L., & Buseck, P. R. 1994, Sci, 265, 1067
Elford, W. G., & Robertson, D. S. 1953, JATP, 4, 271
Elkins-Tanton, L. T. 2012, AREPS, 40, 113
Engrand, C., & Maurette, M. 1998, M&PS, 33, 565
Feierberg, M. A., Lebofsky, L. A., & Tholen, D. J. 1985, Icar, 63, 183
Feigelson, E. D., Garmire, G. P., & Pravdo, S. H. 2002, ApJ, 572, 335
Fernandez, J. A. 1980, MNRAS, 192, 481
Fischer, D. A., & Valenti, J. 2005, ApJ, 622, 1102
Fraser, H. J., Collings, M. P., McCoustra, M. R. S., et al. 2001, MNRAS, 327, 1165
Fraser, W. C., Bannister, M. T., Pike, R. E., et al. 2017, NatAs, 1, 0088
Fraser, W. C., Brown, M. E., Morbidelli, A., et al. 2014, ApJ, 782, 100
Fujiwara, A., Kawaguchi, J., Yeomans, D. K., et al. 2006, Sci, 312, 1330
Fujiya, W., Sugiura, N., Hotta, H., Ichimura, K., & Sano, Y. 2012, NatCo, 3, 627
Fu, R. R., Ermakov, A. I., Marchi, S., et al. 2017, E&PSL, 476, 153
Füri, E., & Marty, B. 2015, NatGe, 8, 515
Gaffey, M. J., & Gilbert, S. L. 1998, M&PS, 33, 1281
Gagné, J., Faherty, J. K., Mamajek, E. E., et al. 2017, ApJS, 228, 18
Gail, H.-P., & Trieloff, M. 2019, A&A, 628, A77
Garaud, P., & Lin, D. N. C. 2004, ApJ, 608, 1050
Gardiner, T. A., & Stone, J. M. 2005, JCoPh, 205, 509
Gardiner, T. A., & Stone, J. M. 2008, JCoPh, 227, 4123
Gehrels, G. E., Valencia, V. A., & Ruiz, J. 2008, GGG, 9, Q03017
Gerbig, K., Murray-Clay, R. A., Klahr, H., et al. 2020, ApJ, 895, 91
Gibbons, P. G., Mamatsashvili, G. R., & Rice, W. K. M. 2015, MNRAS, 453, 4232
Gladman, B., Marsden, B. G., & Vanlaerhoven, C. 2008, in The Solar System Beyond Neptune, ed. M. A. Barucci, et al. (Tucson, AZ: Univ. Arizona Press), 43
Goldreich, P., & Ward, W. R. 1973, ApJ, 183, 1051
Gole, D. A., Simon, J. B., Li, R., et al. 2020, ApJ, 904, 132
Gomes, R. S. 2003, Icar, 161, 404
Goodwin, S. P., Kroupa, P., Goodman, A., et al. 2007, in Protostars and Planets V, ed. B. Reipurth, et al. (Tucson, AZ: Univ. Arizona Press), 133
Gonzalez, G. 1997, MNRAS, 285, 403
Gounelle, M. 2015, A&A, 582, A26
Gounelle, M., & Meynet, G. 2012, A&A, 545, A4
Gradie, J., & Tedesco, E. 1982, Sci, 216, 1405
Greenwood, R. C., Barrat, J.-A., Scott, E. R. D., et al. 2015, GeoCoA, 169, 115
Greenwood, R. C., Franchi, I. A., Jambon, A., et al. 2005, Natur, 435, 916
Grimm, R. E., & McSween, H. Y. 1989, Icar, 82, 244
Grün, E., Krüger, H., & Srama, R. 2019, SSRv, 215, 46
Guilera, O. M., de Elía, G. C., Brunini, A., et al. 2014, A&A, 565, A96
Gundlach, B., & Blum, J. 2015, ApJ, 798, 34
Hall, C., Rice, K., Dipierro, G., et al. 2018, MNRAS, 477, 1004
Haba, M. K., Wotzlaw, J.-F., Lai, Y.-J., et al. 2019, NatGe, 12, 510
Hahn, J. M., & Malhotra, R. 2005, AJ, 130, 2392
Hansen, B. M. S. 2009, ApJ, 703, 1131
Hartlep, T., & Cuzzi, J. N. 2020, ApJ, 892, 120

Hartmann, W. K., Tholen, D. J., & Cruikshank, D. P. 1987, Icar, 69, 33
Hasegawa, Y., Wakita, S., Matsumoto, Y., et al. 2016, ApJ, 816, 8
Hawkins, G. S., Hemenway, C. L., & Whipple, F. L. 1956, AJ, 61, 179
Heck, P. R., Greer, J., Kööp, L., et al. 2020, PNAS, 117, 1884
Henke, S., Gail, H.-P., Trieloff, M., et al. 2012, A&A, 545, A135
Herique, A., Kofman, W., Zine, S., et al. 2019, A&A, 630, A6
Hevey, P. J., & Sanders, I. S. 2006, M&PS, 41, 95
Hewins, R. H. 1983, in Chondrules and Their Origins, ed. E. A. King (Houston, TX: Lunar and Planetary Inst.), 122
Hey, J. S., & Stewart, G. S. 1947, PPS, 59, 858
Hilton, C. D., Bermingham, K. R., Walker, R. J., et al. 2019, GeoCoA, 251, 217
Hood, L. L., Ciesla, F. J., Artemieva, N. A., et al. 2009, M&PS, 44, 327
Hopkins, P. F. 2016, MNRAS, 455, 89
Hopkins, P. F., & Squire, J. 2018, MNRAS, 479, 4681
Horan, M. F., Smoliar, M. I., & Walker, R. J. 1998, GeoCoA, 62, 545
Hsieh, H. H., & Jewitt, D. 2006, Sci, 312, 561
Huang, J., Andrews, S. M., Pérez, L. M., et al. 2018, ApJL, 869, L43
Hublet, G., Debaille, V., Wimpenny, J., et al. 2017, GeoCoA, 218, 73
Hughes, D. W. 1978, in Cosmic Dust, ed. J. A. M. McDonnell (Chichester, NY: Wiley), 123
Huss, G. R., Meyer, B. S., Srinivasan, G., et al. 2009, GeoCoA, 73, 4922
Ida, S., & Guillot, T. 2016, A&A, 596, L3
Ida, S., & Lin, D. N. C. 2004, ApJ, 604, 388
Inaba, S., Wetherill, G. W., & Ikoma, M. 2003, Icar, 166, 46
Irwin, M., Tremaine, S., & Zytkow, A. N. 1995, AJ, 110, 3082
Ishii, H. A., Bradley, J. P., Dai, Z. R., et al. 2008, Sci, 319, 447
Ishitsu, N., Inutsuka, S., & Sekiya, M. 2009, arXiv:0905.4404
Jacobsen, S. B. 2005, AREPS, 33, 531
Jacobsen, B., Yin, Q.-Z., Moynier, F., et al. 2008, E&PSL, 272, 353
Jackson, S. E., Pearson, N. J., Griffin, W. L., & Belousova, E. A. 2004, ChGeo, 211, 47
Jacquet, E. 2014, Icar, 232, 176
Jacquet, E., Balbus, S., & Latter, H. 2011, MNRAS, 415, 3591
Jarosewich, E. 1990, Metic, 25, 323
Jedicke, R., Larsen, J., & Spahr, T. 2002, Asteroids III (Tucson, AZ: Univ. Arizona Press), 71
Jewitt, D., & Luu, J. 1993, Natur, 362, 730
Jewitt, D., Luu, J., & Marsden, B. G. 1993, IAUC, 5865, 1
Johansen, A., Blum, J., Tanaka, H., et al. 2014, in Protostars and Planets VI, ed. H. Beuther, et al. (Tucson, AZ: Univ. Arizona Press), 547
Johansen, A., Klahr, H., & Henning, T. 2006, ApJ, 636, 1121
Johansen, A., Mac Low, M.-M., Lacerda, P., et al. 2015, SciA, 1, 1500109
Johansen, A., Oishi, J. S., Mac Low, M.-M., et al. 2007, Natur, 448, 1022
Johansen, A., & Youdin, A. 2007, ApJ, 662, 627
Johansen, A., Youdin, A. N., & Lithwick, Y. 2012, A&A, 537, A125
Johansen, A., Youdin, A., & Mac Low, M.-M. 2009, ApJL, 704, L75
Johnson, T. V., & Lunine, J. I. 2005, Natur, 435, 69
Johnson, B. C., Minton, D. A., Melosh, H. J., et al. 2015, Natur, 517, 339
Johnston, K. G., Hoare, M. G., Beuther, H., et al. 2020, A&A, 634, L11

Jourdan, F., Kennedy, T., Benedix, G. K., et al. 2020, GeoCoA, 273, 205
Kataoka, A., Tanaka, H., Okuzumi, S., et al. 2013, A&A, 557, L4
Kawasaki, N., Park, C., Sakamoto, N., et al. 2019, E&PSL, 511, 25
Kawasaki, N., Wada, S., Park, C., et al. 2020, GeoCoA, 279, 1
Kehm, K., Hauri, E. H., Alexander, C. M. O., et al. 2003, GeoCoA, 67, 2879
Kenyon, S. J., & Bromley, B. C. 2020, PSJ, 1, 40
Kenyon, S. J., & Luu, J. X. 1999, AJ, 118, 1101
Kieffer, S. W. 1975, Sci, 189, 333
King, T. V. V., & King, E. A. 1978, Metic, 13, 47
King, T. V. V., & King, E. A. 1979, Metic, 14, 91
Klahr, H., & Schreiber, A. 2020, ApJ, 901, 54
Kleine, T., Mezger, K., Palme, H., et al. 2005, GeoCoA, 69, 5805
Kleine, T., Mezger, K., Münker, C., et al. 2004, GeoCoA, 68, 2935
Kleine, T., Touboul, M., Bourdon, B., et al. 2009, GeoCoA, 73, 5150
Kobayashi, H., Tanaka, H., & Krivov, A. V. 2011, ApJ, 738, 35
Kokubo, E., & Ida, S. 1998, Icar, 131, 171
Kokubo, E., Kominami, J., & Ida, S. 2006, ApJ, 642, 1131
Konopliv, A. S., Asmar, S. W., Park, R. S., et al. 2014, Icar, 240, 103
Krapp, L., Youdin, A. N., Kratter, K. M., et al. 2020, MNRAS, 497, 2715
Krijt, S., Ormel, C. W., Dominik, C., et al. 2016, A&A, 586, A20
Krivov, A. V., & Wyatt, M. C. 2021, MNRAS, 500, 718
Krot, A. N., Amelin, Y., Cassen, P., et al. 2005, Natur, 436, 989
Krot, A. N., Amelin, Y., Russell, S.S., & Twelker, E. 2004, in Meteoritics & Planetary Science, Vol. 39, Supplement, Proc. 67th Annual Meeting of the Meteoritical Society, 5044
Krot, A. N., Nagashima, K., Yoshitake, M., et al. 2010, GeoCoA, 74, 2190
Kruijer, T. S., Burkhardt, C., Budde, G., et al. 2017, PNAS, 114, 6712
Kruijer, T. S., Fischer-Gödde, M., Kleine, T., et al. 2013, E&PSL, 361, 162
Kruijer, T. S., Touboul, M., Fischer-Gödde, M., et al. 2014, Sci, 344, 1150
Kuiper, G. P. 1951, PNAS, 37, 1
Kuiper, G. P. 1952, Liege International Astrophysical Colloquia, 4, 361
Küppers, M., O'Rourke, L., Bockelée-Morvan, D., et al. 2014, Natur, 505, 525
Laibe, G., & Price, D. J. 2014, MNRAS, 440, 2136
Lambrechts, M., Johansen, A., Capelo, H. L., et al. 2016, A&A, 591, A133
Larson, H. P., Feierberg, M. A., Fink, U., et al. 1979, Icar, 39, 257
Larson, H. P., & Fink, U. 1975, Icar, 26, 420
Lawler, S. M., Kavelaars, J. J., Alexandersen, M., et al. 2018, FrASS, 5, 14
Lawler, S. M., Pike, R. E., Kaib, N., et al. 2019, AJ, 157, 253
Lawler, S. M., Shankman, C., Kavelaars, J. J., et al. 2018, AJ, 155, 197
Lee, A. T., Chiang, E., Asay-Davis, X., et al. 2010, ApJ, 718, 1367
Lee, C.-F., Li, Z.-Y., & Turner, N. J. 2020, NatAs, 4, 142
Lee, T., Papanastassiou, D. A., & Wasserburg, G. J. 1977, ApJL, 211, L107
Leinhardt, Z. M., Dobinson, J., Carter, P. J., et al. 2015, ApJ, 806, 23
Leinhardt, Z. M., Marcus, R. A., & Stewart, S. T. 2010, ApJ, 714, 1789
Leinhardt, Z. M., & Stewart, S. T. 2012, ApJ, 745, 79
Lenz, C. T., Klahr, H., Birnstiel, T., et al. 2020, A&A, 640, A61
Levison, H. F., Kretke, K. A., Walsh, K. J., et al. 2015, PNAS, 112, 14180

Levison, H. F., Morbidelli, A., Van Laerhoven, C., et al. 2008, Icar, 196, 258
Levy, E. H., & Araki, S. 1989, Icar, 81, 74
Li, R., Youdin, A. N., & Simon, J. B. 2019, ApJ, 885, 69
Libourel, G., & Krot, A. N. 2007, E&PSL, 254, 1
Lichtenberg, T., Golabek, G. J., Dullemond, C. P., et al. 2018, Icar, 302, 27
Lin, M.-K. 2021, ApJ, 907, 64
Lin, M.-K., & Youdin, A. N. 2015, ApJ, 811, 17
Liou, J.-C., Zook, H. A., & Dermott, S. F. 1996, Icar, 124, 429
Loesche, C., Teiser, J., Wurm, G., et al. 2014, ApJ, 792, 73
Loesche, C., Wurm, G., Teiser, J., et al. 2013, ApJ, 778, 101
Lorek, S., Lacerda, P., & Blum, J. 2018, A&A, 611, A18
Love, S. G., & Brownlee, D. E. 1993, Sci, 262, 550
Lu, C. X., Schlaufman, K. C., & Cheng, S. 2020, AJ, 160, 253
Lugaro, M., Ott, U., & Kereszturi, Á. 2018, PrPNP, 102, 1
Luhman, K. L. 2018, AJ, 156, 271
Lykawka, P. S., & Ito, T. 2019, ApJ, 883, 130
Lyra, W., & Umurhan, O. M. 2019, PASP, 131, 072001
Luu, J., Jewitt, D., & Marsden, B. G. 1993, IAUC, 5730, 1
MacPherson, G. J., Bullock, E. S., Janney, P. E., et al. 2010, ApJL, 711, L117
Magna, T., Wiechert, U., & Halliday, A. N. 2006, E&PSL, 243, 336
Maldonado, J., Eiroa, C., Villaver, E., et al. 2012, A&A, 541, A40
Maldonado, J., Eiroa, C., Villaver, E., et al. 2015, A&A, 579, A20
Malhotra, R. 1995, AJ, 110, 420
Marsset, M., Fraser, W. C., Pike, R. E., et al. 2019, AJ, 157, 94
Maurette, M., Duprat, J., Engrand, C., et al. 2000, P&SS, 48, 1117
McCord, T. B., Adams, J. B., & Johnson, T. V. 1970, Sci, 168, 1445
McKeegan, K. D., Aléon, J., Bradley, J., et al. 2006, Sci, 314, 1724
McKinnon, W. B. 1989, ApJL, 344, L41
McKinnon, W. B., Richardson, D. C., Marohnic, J. C., et al. 2020, Sci, 367, aay6620
McNally, C. P., Hubbard, A., Mac Low, M.-M., et al. 2013, ApJL, 767, L2
McNally, C. P., Lovascio, F., & Paardekooper, S.-J. 2021, MNRAS, 502, 1469
McSween, H. Y., Emery, J. P., Rivkin, A. S., et al. 2018, M&PS, 53, 1793
McSween, H. Y., Mittlefehldt, D. W., Beck, A. W., et al. 2011, SSRv, 163, 141
Megeath, S. T., Gutermuth, R., Muzerolle, J., et al. 2012, AJ, 144, 192
Menten, K. M., Reid, M. J., Forbrich, J., et al. 2007, A&A, 474, 515
Messenger, S., Keller, L. P., Stadermann, F. J., et al. 2003, Sci, 300, 105
Michel, P., Benz, W., & Richardson, D. C. 2004, Icar, 168, 420
Monnereau, M., Toplis, M. J., Baratoux, D., et al. 2013, GeoCoA, 119, 302
Morgan, J. W., & Anders, E. 1980, PNAS, 77, 6973
Morbidelli, A., Bottke, W. F., Nesvorný, D., et al. 2009, Icar, 204, 558
Morbidelli, A. 1997, Icar, 127, 1
Morbidelli, A., Levison, H. F., Tsiganis, K., et al. 2005, Natur, 435, 462
Morbidelli, A., & Rickman, H. 2015, A&A, 583, A43
Morbidelli, A., Thomas, F., & Moons, M. 1995, Icar, 118, 322
Morris, M. A., Boley, A. C., Desch, S. J., et al. 2012, ApJ, 752, 27
Mosqueira, I., & Estrada, P. R. 2003, Icar, 163, 198

Movshovitz, N., Nimmo, F., Korycansky, D. G., et al. 2016, Icar, 275, 85
Mullane, E., Russell, S. S., & Gounelle, M. 2005, E&PSL, 239, 203
Najita, J. R., & Kenyon, S. J. 2014, MNRAS, 445, 3315
Nakagawa, Y., Sekiya, M., & Hayashi, C. 1986, Icar, 67, 375
Nanne, J. A. M., Nimmo, F., Cuzzi, J. N., et al. 2019, E&PSL, 511, 44
Napier, K. J., Gerdes, D. W., Lin, H. W., et al. 2021, PSJ, 2, 59
Nelson, R. P., Gressel, O., & Umurhan, O. M. 2013, MNRAS, 435, 2610
Nesvorný, D., Bottke, W. F., Levison, H. F., et al. 2003, ApJ, 591, 486
Nesvorný, D., Bottke, W. F., Vokrouhlický, D., et al. 2006, in IAU Symp. 229, Asteroids, Comets, Meteors, ed. D. Lazzaro, S. Ferraz-Mello, & J. A. Fernández (Cambridge: Cambridge Univ. Press), 289
Nesvorný, D. 2015, in Asteroids IV (Tucson, AZ: Univ. Arizona Press), 297
Nesvorný, D., Li, R., Youdin, A. N., et al. 2019, NatAs, 3, 808
Nesvorný, D., & Vokrouhlický, D. 2016, ApJ, 825, 94
Nesvorný, D., Vokrouhlický, D., Deienno, R., et al. 2014, AJ, 148, 52
Neumann, W., Breuer, D., & Spohn, T. 2012, A&A, 543, A141
Neumann, W., Jaumann, R., Castillo-Rogez, J., et al. 2020, A&A, 633, A117
Neumann, W., Henke, S., Breuer, D., et al. 2018a, Icar, 311, 146
Neumann, W., Kruijer, T. S., Breuer, D., et al. 2018b, JGRE, 123, 421
Neveu, M., & Vernazza, P. 2019, ApJ, 875, 30
Nimmo, F., & Kleine, T. 2007, Icar, 191, 497
Nixon, C. J., King, A. R., & Pringle, J. E. 2018, MNRAS, 477, 3273
O'Brien, D. P., Morbidelli, A., & Levison, H. F. 2006, Icar, 184, 39
Okuzumi, S., Tanaka, H., Kobayashi, H., et al. 2012, ApJ, 752, 106
Ormel, C. W., & Kobayashi, H. 2012, ApJ, 747, 115
Oshino, S., Hasegawa, Y., Wakita, S., et al. 2019, ApJ, 884, 37
Paardekooper, S.-J., McNally, C. P., & Lovascio, F. 2020, MNRAS, 499, 4223
Pan, M., & Schlichting, H. E. 2012, ApJ, 747, 113
Parker, A. H., & Kavelaars, J. J. 2010, ApJL, 722, L204
Parker, A. H., & Kavelaars, J. J. 2012, ApJ, 744, 139
Paton, C., Woodhead, J. D., Hellstrom, J. C., et al. 2010, GGG, 11, Q0AA06
Pearson, V. K., Sephton, M. A., Franchi, I. A., et al. 2006, M&PS, 41, 1899
Peixinho, N., Delsanti, A., Guilbert-Lepoutre, A., et al. 2012, A&A, 546, A86
Pérez, L. M., Carpenter, J. M., Andrews, S. M., et al. 2016, Sci, 353, 1519
Petit, J.-M., Kavelaars, J. J., Gladman, B. J., et al. 2011, AJ, 142, 131
Phuong, N. T., Dutrey, A., Di Folco, E., et al. 2020, A&A, 635, L9
Pilipp, W., Hartquist, T. W., Morfill, G. E., et al. 1998, A&A, 331, 121
Pinilla, P., & Youdin, A. 2017, in Formation, Evolution, and Dynamics of Young Solar Systems, ed. M. Pessah, et al. (New York: Springer), 91
Plane, J. M. C. 2012, ChSRv, 41, 6507
Pollack, H. N., Hurter, S. J., & Johnson, J. R. 1993, RvGeo, 31, 267
Poole, G. M., Rehkämper, M., Coles, B. J., et al. 2017, E&PSL, 473, 215
Poulet, F., Lucchetti, A., Bibring, J.-P., et al. 2016, MNRAS, 462, S23
Prombo, C. A., & Clayton, R. N. 1993, GeoCoA, 57, 3749
Qin, L., Alexander, C. M. O., Carlson, D., et al. 2010, GeoCoA, 74, 1122
Qin, L., Dauphas, N., Wadhwa, M., et al. 2008, E&PSL, 273, 94

Raettig, N., Klahr, H., & Lyra, W. 2015, ApJ, 804, 35
Rafikov, R. R. 2004, AJ, 128, 1348
Raymo, M. E., Ruddiman, W. F., & Froelich, P. N. 1988, Geo, 16, 649
Raymond, S. N., Quinn, T., & Lunine, J. I. 2004, Icar, 168, 1
Raymond, S. N., Quinn, T., & Lunine, J. I. 2005, ApJ, 632, 670
Rice, W. K. M., Lodato, G., Pringle, J. E., et al. 2004, MNRAS, 355, 543
Robertson, D. S., Liddy, D. T., & Elford, W. G. 1953, JATP, 4, 255
Robinson, J. E., Fraser, W. C., Fitzsimmons, A., et al. 2020, A&A, 643, A55
Roche, É. 1847, Académie des sciences de Montpellier: Mémoires de la section des sciences, 1, 243
Ros, K., & Johansen, A. 2013, A&A, 552, A137
Roszjar, J., Whitehouse, M. J., Srinivasan, G., et al. 2016, E&PSL, 452, 216
Rubin, A. E., Sailer, A. L., & Wasson, J. T. 1999, GeoCoA, 63, 2281
Ruesch, O., Platz, T., Schenk, P., et al. 2016, Sci, 353, aaf4286
Russell, S. S., Krot, A. N., Huss, G. R., et al. 2005, in Chondrites and the Protoplanetary Disk, ed. A. N. Krot, et al. (San Francisco, CA: ASP), 341
Sanders, I. S., & Scott, E. R. D. 2012, M&PS, 47, 2170
Sanders, I. S., & Taylor, G. J. 2005, in Chondrites and the Protoplanetary Disk, ed. A. N. Krot, et al. (San Francisco, CA: ASP), 915
Sandford, S. A., & Bradley, J. P. 1989, Icar, 82, 146
Sarafian, A. R., Nielsen, S. G., Marschall, H. R., et al. 2014, Sci, 346, 623
Sarafian, A. R., Nielsen, S. G., Marschall, H. R., et al. 2017, GeoCoA, 212, 156
Schäfer, M., Schäfer, T., Izawa, M. R. M., et al. 2018, M&PS, 53, 1925
Schäfer, U., Yang, C.-C., & Johansen, A. 2017, A&A, 597, A69
Scherstén, A., Elliott, T., Hawkesworth, C., et al. 2006, E&PSL, 241, 530
Schlichting, H. E., Fuentes, C. I., & Trilling, D. E. 2013, AJ, 146, 36
Schoonenberg, D., & Ormel, C. W. 2017, A&A, 602, A21
Schrader, D. L., Nagashima, K., Krot, A. N., et al. 2017, GeoCoA, 201, 275
Scott, E. R. D. 2007, AREPS, 35, 577
Scott, E. R. D., Keil, K., Goldstein, J. I., et al. 2015, in Asteroids IV (Tucson, AZ: Univ. Arizona Press), 573
Seeds, M., & Backman, D. 2019, The Solar System (10th ed.; Boston: Cengage Learning)
Sengupta, D., Dodson-Robinson, S. E., Hasegawa, Y., et al. 2019, ApJ, 874, 26
Shankman, C., Kavelaars, J. J., Bannister, M. T., et al. 2017, AJ, 154, 50
Shankman, C., Kavelaars, J., Gladman, B. J., et al. 2016, AJ, 151, 31
Shannon, A., & Wu, Y. 2011, ApJ, 739, 36
Shi, J.-M., & Chiang, E. 2013, ApJ, 764, 20
Shu, F. H. 1977, ApJ, 214, 488
Shu, F. H., Shang, H., Glassgold, A. E., et al. 1997, Sci, 277, 1475
Simon, J. B., Armitage, P. J., Li, R., et al. 2016, ApJ, 822, 55
Simon, J. B., Armitage, P. J., Youdin, A. N., et al. 2017, ApJL, 847, L12
Singer, K. N., McKinnon, W. B., Gladman, B., et al. 2019, Sci, 363, 955
Slater-Reynolds, V., & McSween, H. Y. 2005, M&PS, 40, 745
Spitzer, L. 1965, Interscience Tracts on Physics and Astronomy (2nd rev. ed.; New York: Interscience Publication)
Squire, J., & Hopkins, P. F. 2018, MNRAS, 477, 5011
Squire, J., & Hopkins, P. F. 2018, ApJL, 856, L15

Stern, S. A. 1991, Icar, 90, 271
Stewart, S. T., & Leinhardt, Z. M. 2009, ApJL, 691, L133
Stone, J. M., Gardiner, T. A., Teuben, P., et al. 2008, ApJS, 178, 137
Stoyanovskaya, O. P., Okladnikov, F. A., Vorobyov, E. I., et al. 2020, ARep, 64, 107
Sugiura, N., & Fujiya, W. 2014, M&PS, 49, 772
Tang, H., & Dauphas, N. 2012, E&PSL, 359, 248
Tegler, S. C., & Romanishin, W. 1998, Natur, 392, 49
Teiser, J., & Wurm, G. 2009, MNRAS, 393, 1584
Thomas, P. C., Burns, J. A., Helfenstein, P., et al. 2007, Icar, 190, 573
Thommes, E. W., Matsumura, S., & Rasio, F. A. 2008, Sci, 321, 814
Tiscareno, M. S., & Malhotra, R. 2009, AJ, 138, 827
Tomeoka, K., & Buseck, P. R. 1985, GeoCoA, 49, 2149
Touboul, M., Kleine, T., Bourdon, B., et al. 2009, E&PSL, 284, 168
Travis, B. J., & Schubert, G. 2005, E&PSL, 240, 234
Trinquier, A., Elliott, T., Ulfbeck, D., et al. 2009, Sci, 324, 374
Trujillo, C. A., & Brown, M. E. 2002, ApJL, 566, L125
Urey, H. C., & Craig, H. 1953, GeoCoA, 4, 36
Umurhan, O. M., Estrada, P. R., & Cuzzi, J. N. 2020, ApJ, 895, 4
Usui, F., Hasegawa, S., Ootsubo, T., et al. 2019, PASJ, 71, 1
Uyama, T., Muto, T., Mawet, D., et al. 2020, AJ, 159, 118
van Kooten, E., Cavalcante, L., Wielandt, D., et al. 2020, M&PS, 55, 575
van Kooten, E. M. M. E., Wielandt, D., Schiller, M., et al. 2011, PNAS, 113, 2011
Verbiscer, A. J., Skrutskie, M. F., & Hamilton, D. P. 2009, Natur, 461, 1098
Vernazza, P., Marsset, M., Beck, P., et al. 2015, ApJ, 806, 204
Vilas, F., & Gaffey, M. J. 1989, Sci, 246, 790
Vockenhuber, C., Oberli, F., Bichler, M., et al. 2004, PhRvL, 93, 172501
Vorobyov, E. I., Akimkin, V., Stoyanovskaya, O., et al. 2018, A&A, 614, A98
Wadhwa, M., Amelin, Y., Davis, A. M., et al. 2007, in Protostars and Planets V, ed. B. Reipurth, et al. (Tucson, AZ: Univ. Arizona Press), 835
Wahlberg Jansson, K., & Johansen, A. 2014, A&A, 570, A47
Wakita, S., Matsumoto, Y., Oshino, S., et al. 2017, ApJ, 834, 125
Walsh, K. J., & Levison, H. F. 2016, AJ, 152, 68
Walsh, K. J., Morbidelli, A., Raymond, S. N., et al. 2011, Natur, 475, 206
Winn, J. N., & Fabrycky, D. C. 2015, ARA&A, 53, 409
Wang, H., Weiss, B. P., Bai, X.-N., et al. 2017, Sci, 355, 623
Ward, W. R. 1981, Icar, 46, 97
Warren, P. H. 2011, E&PSL, 311, 93
Weidenschilling, S. J. 1980, Icar, 44, 172
Weidenschilling, S. J. 1997, Icar, 127, 290
Weiss, B. P., Berdahl, J. S., Elkins-Tanton, L., et al. 2008, Sci, 322, 713
Wetherill, G. W., & Chapman, C. R. 1988, in Meteorites & the Early Solar System, ed. D. S. Lauretta, & H. Y. McSween Jr. (Tucson, AZ: Univ. Arizona Press), 35
Wetherill, G. W., & Stewart, G. R. 1993, Icar, 106, 190
Whipple, F. L. 1966, Sci, 153, 54
Wiechert, U., & Halliday, A. N. 2007, E&PSL, 256, 360
Williams, J. G. 1979, Asteroids, 1040

Wise, A. W., & Dodson-Robinson, S. E. 2018, ApJ, 855, 145
Wolff, S., Dawson, R. I., & Murray-Clay, R. A. 2012, ApJ, 746, 171
Wyatt, M. C. 2008, ARA&A, 46, 339
Yang, C.-C., & Johansen, A. 2014, ApJ, 792, 86
Yang, C.-C., & Johansen, A. 2016, ApJS, 224, 39
Yang, C.-C., Johansen, A., & Carrera, D. 2017, A&A, 606, A80
Yang, J., Goldstein, J. I., & Scott, E. R. D. 2007, Natur, 446, 888
Yang, J., Goldstein, J. I., & Scott, E. R. D. 2008, GeoCoA, 72, 3043
Yang, J., Goldstein, J. I., Scott, E. R. D., et al. 2014, GeoCoA, 124, 34
Yoder, C. F. 1995, Global Earth Physics: A Handbook of Physical Constants, ed. T. J. Ahrens (Washington, DC: AGU), 1
Yoshida, M. 2010, PolSc, 3, 272
Yoshida, M., Ando, H., Omoto, K., Naruse, R., & Ageta, Y. 1971, Antarctic Record, 39, 62
Youdin, A. N., & Chiang, E. I. 2004, ApJ, 601, 1109
Youdin, A. N., & Goodman, J. 2005, ApJ, 620, 459
Youdin, A., & Johansen, A. 2007, ApJ, 662, 613
Youdin, A. N., & Shu, F. H. 2002, ApJ, 580, 494
Zappalà, V., Bendjoya, P., Cellino, A., et al. 1995, Icar, 116, 291
Zuber, M. T., McSween, H. Y., Binzel, R. P., et al. 2011, SSRv, 163, 77
Zhu, K., Moynier, F., Wielandt, D., et al. 2019, ApJL, 877, L13

www.ingramcontent.com/pod-product-compliance
Ingram Content Group UK Ltd.
Pitfield, Milton Keynes, MK11 3LW, UK
UKHW051856100926
8556IPUK00005B/653

9 780750 321341